IEE POWER SERIES 13

Series Editors: Professor A. T. Johns
J. R. Platts
Dr. D. Aubrey

STATISTICAL TECHNIQUES FOR HIGH-VOLTAGE ENGINEERING

Other volumes in this series:

Volume 1 **Power circuits breaker theory and design** C. H. Flurscheim (Editor)

Volume 2 **Electric fuses** A. Wright and P. G. Newbery

Volume 3 **Z-transform electromagnetic transient analysis in high-voltage networks** W. Derek Humpage

Volume 4 **Industrial microwave heating** A. C. Metaxas and R. J. Meredith

Volume 5 **Power system economics** T. W. Berrie

Volume 6 **High voltage direct current transmission** J. Arrillaga

Volume 7 **Insulators for high voltages** J. S. T. Looms

Volume 8 **Variable frequency AC motor drive systems** D. Finney

Volume 9 **Electricity distribution network design** E. Lakervi and E. J. Holmes

Volume 10 **SF_6 Switchgear** H. M. Ryan and G. R. Jones

Volume 11 **Conduction and induction heating** E. J. Davies

Volume 12 **Overvoltage protection of low voltage systems** P. Hasse

STATISTICAL TECHNIQUES FOR HIGH-VOLTAGE ENGINEERING

W. HAUSCHILD
AND
W. MOSCH

Translated from the German by P. Perkins

Peter Peregrinus Ltd. on behalf of the Institution of Electrical Engineers

Originally published in German by VEB Verlag Technik, Berlin, 1984

English edition published by Peter Peregrinus Ltd., London, United Kingdom

Peter Peregrinus Ltd.,
Michael Faraday House,
Six Hills Way, Stevenage,
Herts. SG1 2AY, United Kingdom

British Library Cataloguing in Publication Data
Hauschild, Wolfgang
Statistics for electrical engineers.
1. Statistical mathematics
I. Title II. Mosch, Wolfgang
519.5

ISBN 0 86341 205 X

Printed in England by J. W. Arrowsmith Ltd., Bristol

Contents

Preface

The far-ranging tasks of increasing productivity, careful use of raw materials and economical energy consumption are forcing engineers engaged in the laboratory, in construction and in engineering to give far greater consideration to influencing factors that have a random action. Only in this way is it possible to ensure high reliability and safety of products in service, with the least possible cost to society.

This principle naturally applies to products in the electricity industry, and especially in the high-voltage apparatus industry. They govern an extremely important economic process, from production, through transportation, right up to the consumption of electrical energy.

A need has consequently developed, within the framework of the training of graduate engineers in electrical engineering and that of further training of practising electrical engineers, for the provision of an aid in the field of mathematical statistics, relevant to the discipline and ready for use, so that comprehensive and informative results can be obtained with the minimum of training, from the often expensive experiments involved.

This book was compiled from experience of experimental high-voltage research, and after many years trying out the methods of mathematical statistics that are of importance to the electrical engineer. It should inspire both the student of electrical engineering (especially students of high-voltage engineering) and practising engineers (particularly in laboratories and test plants) to use statistical methods to open up possibilities of rationalisation and increased efficiency of engineering work. Most of the contents relate to the field of evaluatory statistics. The book will therefore be particularly useful when experimental material has to be evaluated as comprehensively as possible. Since high-voltage engineering is a fundamental field of electrical engineering which has a decidedly experimental orientation, the book has been written by high-voltage engineers primarily for high-voltage engineers, but electrical engineers in other fields will find useful advice in it for their work.

When we were trying out the methods provided by mathematical statistics, we were helped by many students, graduates and PhDs in the 'High-voltage engineering' area of science in the Department of Electrical Engineering at Dresden Technical University. To all of those who assisted, the authors are sincerely grateful. We received a great deal of advice, help and hints, in particular from Dr.-Ing. J. Speck, by virtue of his thorough mathematical knowledge as a trained mathematician and of his active participation in the 'High-voltage engineering' area of science.

This book would not have seen the light of day, however, without close co-operation for many years with the 'Probability theory and mathematical statistics' area of science in the Mathematics Department of Dresden TU.

First, we would like to thank Prof. Dr. rer. nat. habil. P. H. Müller who, through his continuous and generous support, has always been an inspirational, critical and therefore extremely welcome colleague in discussions. He has carefully read through the manuscript of the book.

Our colleagues in the office of the Electrical Engineering Section, particularly Eva-Maria Biesslich, deserve our heartfelt thanks for the trouble and care they have taken in preparing the manuscript, as of course — in long-established collaboration — do our colleagues at the GmbH Verlag Technik Institute of Berlin, especially our proofreader, Inge Epp.

Wolfgang Hauschild
Wolfgang Mosch

List of major symbols

In specialist literature on mathematical statistics a strict logic can usually be observed within a work with regard to the symbols used, but differences from book to book can be considerable. We shall adhere closely to the notation used in many standard collections of statistical tables, but the simultaneous use of symbols common in electrical engineering has resulted in a few compromises. All the symbols used, as well as their indices, are explained when they are introduced in the text, so we have only listed important symbols below.

Basically, one must remember that random events and the variates describing them are represented by capital letters (e.g. X), and the realisations of variates (individual values) by lower-case letters (e.g. x). Parameter estimates are indicated by an asterisk after the symbol for the (theoretical) parameter (e.g. γ^*). A symbol with a bar over it (e.g. $\bar{u}_d$) represents the estimate for the mean value of the variate concerned.

A	area
a_{xy}, a_{yx}	position coefficient of regression lines
b_{xy}, b_{yx}	regression coefficient (rise of regression lines)
$C = 0.5772...$	Euler's constant
D^2X	dispersion of variate X
d	electrode gap
EX	expectation of variate X
E_d	electric strength
E_{dh}	maximum breakdown field strength
e_h	curvature factor
$F(x)$	distribution function of variate X at x
$F_{m1;m2;q}$	quantile of F distribution (order q; degrees of freedom m_1 and m_2)
$f(x)$	density function of variate X at x
g_u ; g_0	limit of a confidence region
H_0	hypothesis for a test
h	relative frequency
h_Σ	relative cumulative frequency
i	serial number
j	serial number
k	number of events observed
k_α	length
l_v	relative service-life consumption
m	number of values observed
m	stage number in constant-voltage experiment
m_k	empirical kth moment

n	size of sample
n	enlargement (scaling) factor
P	probability (for event)
p	probability (concrete value)
p_{20}	insulating-gas pressure at 20°C
q	order of a quantile
q_i	pulse charge (apparent charge)
R	spread
R	error probability (operand for statistical insulation co-ordination)
R_x ; R_y	rank
r	life coefficient
r; r_i ; r_a	radius
$S(u_d) = F(u_d)$	cumulative frequency function of breakdown voltage
s	mean square deviation
s_{xy}	covariance of two samples
t	test quantity
t_d	breakdown time
$t_{m;q}$	quantile of t distribution (order q; degree of freedom m)
t_n	stress time; test duration
U_{nst}	rated withstand voltage
u	test quantity of U test
u_d	breakdown voltage
V	volume
$V(u_d)$	performance function of breakdown voltage
$v = s\bar{x}$	variation coefficient
v_u	rate of rise of voltage
X	general variate
x	realisation of a general variate
x_0	initial value; parameter of Weibull distribution
Y	general variate
y	realisation of a general variate
$z = (x - \mu)/\sigma$	realisation from the standard normal distribution
α	error probability; significance level
β	statistical certainty (confidence coefficient)
$\Gamma(p)$	gamma function
γ	measure of dispersion of double exponential distribution
Δu	voltage difference
Δt	time difference
δ	measure of dispersion of Weibull distribution (Weibull exponent)
ε	statistical certainty (confidence coefficient)
η	63% quantile; parameter of Weibull and double exponential distribution
η	Schwaiger's degree of uniformity
λ	parameter of Poisson and exponential distributions
λ_q	normal-distribution quantile
μ	expectation; normal-distribution parameter

μ_k	kth moment
ρ	correlation coefficient
σ	standard deviation; normal-distribution parameter
σ^2	variance
τ	tolerance limit
$\Phi(x)$	distribution function of normal distribution
$\phi(x)$	density function of normal distribution
$\chi^2_{m;q}$	quantile of chi-squared distribution (order q; degree of freedom m)

Chapter 1
Introduction

Most phenomena in nature, society and technology are subject to random variation. This randomness is often ignored when considering such phenomena; what tends to be concentrated on instead is the average trends, in order to interpret the nature of a relationship being investigated. Quite often, however, it is not the mean value, but an extreme value that determines the performance of a system. In the past this has generally been taken into account in technology by multiplying known mean values by 'safety factors'. This is the method used, for example, to match the mechanical strength of extremely varied structures to anticipated static and dynamic stresses. We need a rather better approach to the problems outlined, both for a more precise description of the phenomena themselves and also for the economical use of raw materials and energy. The deterministic description of average trends needs to be replaced by a thorough statistical treatment of stochastic/random phenomena — right down to the constructional design of technical structures. This task is far easier to set than to solve, however. It often encounters experimental problems in the specialist field concerned and also major difficulties of mathematical formulation. No complete solution to the problem can therefore be expected in any technical field: what one must always aim at is rather to arrive at technically effective partial solutions, through close collaboration between technical specialists and mathematicians.

This development can also be seen in electrical engineering. To explain the development further, and at the same time to outline the aim of the present book, which is a very limited part of the whole problem, we intend now to consider a few details more closely, by reference to electrical engineering examples.

Before one can reach the stage of statistical formulation of a random (stochastic) process, it needs to be 'phenomenologically known'. One therefore has to start with experiments whose results (measured quantities) vary within certain random limits (see Section 2.1.1). The causes of variations of experimental results can be inherent to the process, can arise from its boundary conditions or may reside in random measuring errors. This last category should be minimised and be determined as accurately as possible. They are not taken into consideration in the following deliberations.

The interplay between the course of a random process and its boundary conditions is often at the heart of experimental investigations. The aim is to adjust the boundary conditions in a defined manner and keep them constant (see Section 3.1.1.3). This is often only possible with a great deal of trouble and is sometimes not possible at all. For example, the breakdown of an insulating clearance in atmospheric air is affected by climatic conditions, ultra-violet and cosmic rays, air flows and dust, in a rather random manner. The effects of these boundary conditions can, admittedly, be studied separately in time-consuming experiments, but one needs to bear in mind that such air insulation has to operate under the complex action of these and other boundary conditions. While it is useful, in order to understand relationships, to know the effects of special influencing quantities, it can be technically expedient not to differentiate between the randomness inherent in the processes (stochastic) and the action of random boundary conditions. The boundary conditions should only be adjusted to the extent required in practice. A random experiment then supplies measured values varying in a random manner (realisations of variates — see Section 2.1.4), which have to be evaluated and technically interpreted (Fig. 1.1). This book is limited to such evaluatory statistics, although of course statistical evaluations also, to a certain extent, involve suitable planning and execution of the experiments.

After a comprehensive and generally applicable review of the mathematical/statistical fundamentals needed (Section 2), special explanations, hints for use, parameters for random processes, relationships to standard test methods and dimensioning methods are given, directly applicable to high-voltage insulation engineering (Sections 3 to 6). There are analogies throughout to other experimental areas of power engineering — from low-voltage insulation engineering, plasma technology, electrical energy (overvoltages and overcurrents), through to the problems of electrical machines (commutation processes; mechanical performance) — so that this chapter will also be of general interest to electrical engineers.

If statistically evaluated data are available for models, components or assemblies, it is also possible to assess the reliability of electrical engineering apparatus constructed from them. Reliability considerations are extremely important in all technical fields: their theory and application are extensive and have been highly developed, particularly for electrical engineering.

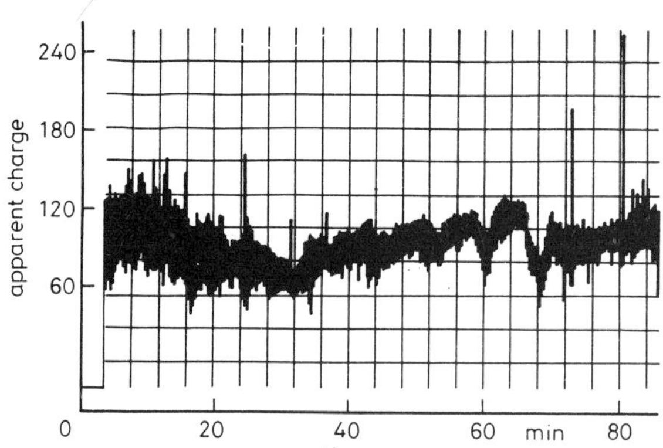

Fig. 1.1 Partial-discharge signal varying in a random manner with respect to time (original recording of apparent charge)

The problem of reliability considerations has not been included in the book: in this connection reference should be made to more specialist literature.

We cannot conclude this short introductory chapter without indicating how the application of mathematical statistics is likely to develop in electrical engineering in the future.

As the number of parameters increases, so the work involved in experimental investigations becomes greater and consequently more expensive. If the phenomena concerned are modelled on the basis of a few experiments and are extrapolated to other parameter combinations using these models, the work and expense can be reduced. The stochastic nature of the processes naturally has to be preserved when modelling. In many cases, e.g. when dealing with overvoltage problems, a Monte Carlo simulation has proved suitable. The physical process often has to be dealt with as a stochastic process: the theory of stochastic processes provides the requisite mathematical tools. As an example, let us consider the development of electron avalanches in the insulating gas SF_6.

Every electron avalanche starts with an initiatory electron produced by cosmic radiation, by the release of negative ions or by emission from the cathode in a supply process that is generally dependent on time and field strength. Through the Poisson process, the theory of stochastic processes provides data for the probability of a certain number of initiatory electrons being released per unit time.

Subsequent avalanche build-up is, in its turn, a random process: when an initiatory electron collides with a gas molecule, either a further electron is freed (ionisation; electron growth), or the electron is attached (attachment;

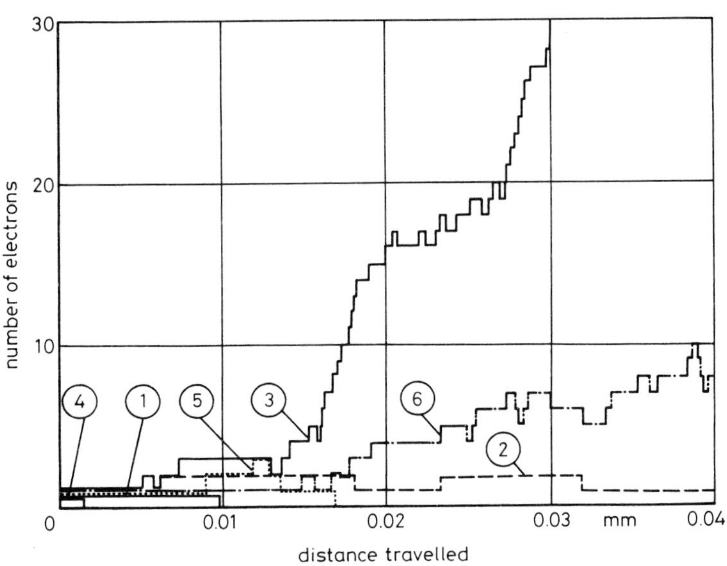

Fig. 1.2 Simulation of six electron avalanches in a uniform field in SF_6

$p_{20} = 0.1$ MPa; $E = 90$ kV/cm

electron loss), or it is deflected (elastic collision; electron constancy). The number of electrons consequently develops in discrete random steps (Fig. 1.2), with only a relatively small number of initiatory electrons managing to form an avalanche of a sufficiently high intensity for breakdown to occur ($n_k \approx 10^8$ charge carriers). In the mathematical formulation of the problem, the coefficients for ionisation and attachment are regarded as the intensities of a birth-and-death process. By this means, one can calculate the probability of the critical electron number n_k being exceeded, as a function of field strength and the number of initiatory electrons. Since this probability can

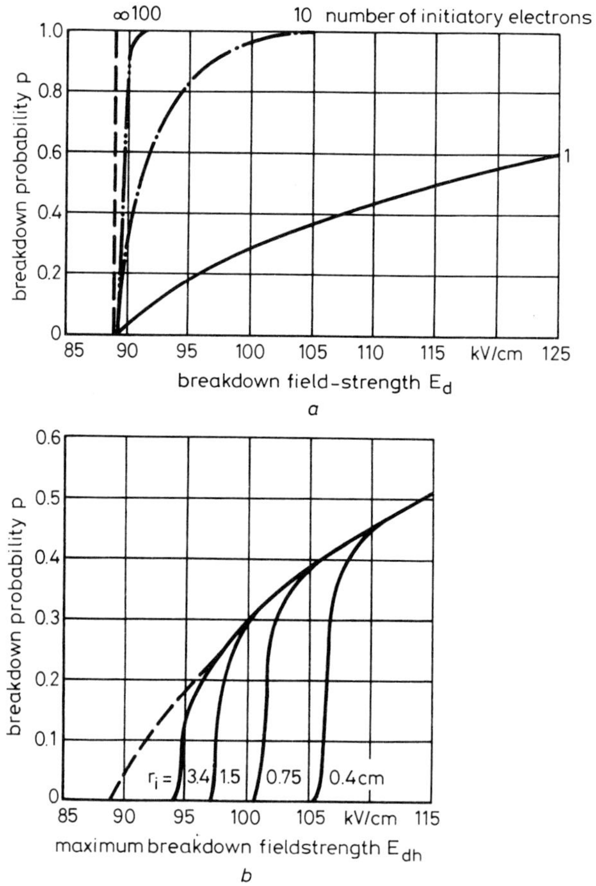

Fig. 1.3 Distribution function of maximum breakdown field-strength in SF_6 calculated by a stochastic process

$p_{20} = 0.1$ MPa

a Uniform field

b Coaxial-cylinder field for an initiatory electron (inside cylinder negative; $r_a = 5$ cm; dashed line relates to uniform field)

be regarded as a breakdown probability, its relationship to the field strength producing breakdown provides the technically interesting distribution function of the breakdown field-strength (e.g. of the uniform field in Fig. 1.3*a*). It is readily apparent that randomness is lost as the number of initiatory electrons increases. If, as in the case of impulse voltages, there are very few initiatory electrons present, the inherent randomness of the process is very pronounced. On the other hand a very large number of initiatory electrons — as with continuous voltages — results in a deterministic behaviour of the process. (If fluctuating breakdown voltages are nevertheless measured, these arise from random boundary conditions, perhaps dust in the system.) The birth-and-death process does not provide complete solutions for non-uniform fields — elaborate mathematical calculations are required. The distribution functions of the maximum breakdown field strength also depend on the electrode geometry (Fig. 1.3*b*, as an example for an initiatory electron), and approximate to that in a uniform field in the sense of a limiting value. There is also a dependence on electrode polarity, since this governs the direction of the electron avalanche. Stochastic modelling enables the breakdown probability to be investigated as a function of the starting point of an initiatory electron (Fig. 1.4). It demonstrates that there must be a polarity effect of the breakdown voltage.

The two random processes mentioned — the presentation of one or more initiatory electrons and avalanche build-up — naturally still need to be

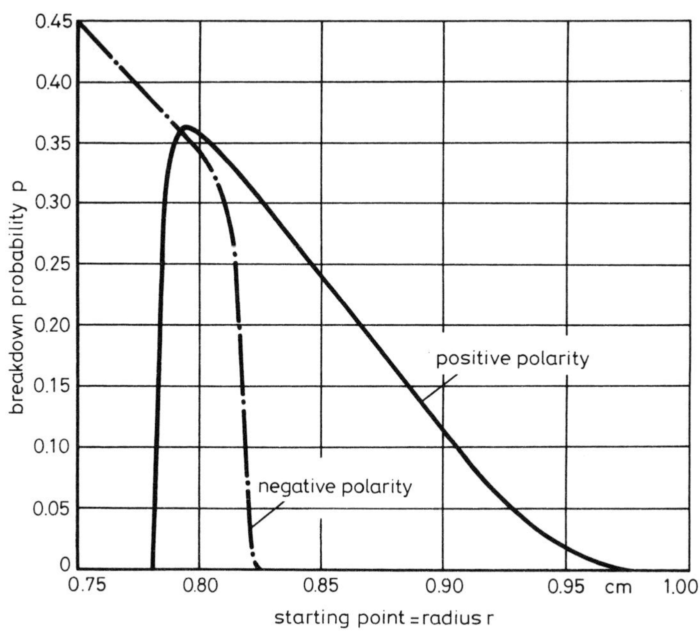

Fig. 1.4 Dependence of breakdown probability in SF_6 on the starting point of an initiatory electron and on the polarity of the inner electrode

superimposed. The model produced is mathematically extremely complicated; with present possibilities one can draw valuable qualitative conclusions but no quantitative conclusions. As physical random processes can generally only be modelled stochastically, the development of certain methods will have to advance further for results of direct technical use to be achieved. This presupposes that the engineer knows how to think in statistical categories. The statistics, which are still relatively simple to evaluate mathematically, and which are treated in the present book from an engineering standpoint, should therefore go further than their narrower application and help to pave the way for greater penetration of the many valuable statistical methods into electrical engineering.

Chapter 2
Review of fundamentals

From the proliferation of mathematical/statistical concepts, trends and methods, we have in the following, and without bothering to supply evidence or proof, considered only those important in understanding the practical chapters or of particular significance in high-voltage engineering practice. We have given the mere outline, supplementing it on the one hand by bibliographical references relating to the fundamentals, and on the other by examples from, and advice on, applications in high-voltage engineering.

2.1 Basic concepts of probability theory

2.1.1 Trial and random event

A trial (random experiment) is an experiment in a very general sense, the result of which is uncertain within certain possibilities dependent on the physical circumstances and which, under the same external conditions, can notionally be repeated as often as one wishes.

During the trial, random (stochastic) processes occur. Under the action of random influences, the result of the trial is a random event. In probability calculations, the uncertainties that exist in the occurrence of a certain event are described by dimensional figures (see Section 2.1.1). Random events are represented by capital letters A, B, C, and so on.

Example 2.1.1.1 An insulation arrangement is repeatedly stressed with a voltage of a prescribed waveform and amplitude (constant-voltage test). During the stress, the random processes of preparation of an initiatory electron, charge-carrier increase up to the intensity required for independent partial discharges, and the development of partial discharges into breakdown can occur. Various events can be considered as the result of the trial: either the alternative events of breakdown (D) and non-breakdown (N), or detailed consideration of the events breakdown (D), stable partial discharges (T) and freedom from discharge (F). In the case of repeated constant-voltage tests, the number of events N will be equal to the sum of events T and F. It is possible to provide a measure of the probability of occurrence of events D and N, and of D, T and F.

Example 2.1.1.2 An insulation arrangement is stressed as described in Example 2.1.1.1, but the voltage is sufficiently high for breakdown to occur with certainty. In this case the *time* to breakdown, which fluctuates within certain limits because of the random breakdown process, can be regarded as a random event. The result of the trial is a non-negative, real, dimensional number.

Example 2.1.1.3 An insulation arrangement is repeatedly stressed with a steadily increasing voltage until breakdown occurs (rising-voltage test). With a physically similar sequence to that in Example 2.1.1.1, the event of breakdown is in this instance certain, but the numerical value of the breakdown voltage is random.

The relationships between random events are usually illustrated by expressions from 'set theory' (theory of quantities) ([2.1–2.6]). Perhaps a few terms need to be explained:

- An event $V(I)$ is called a certain event if it occurs in each trial (e.g. breakdown in Example 2.1.1.3).
- An event $V(\Phi)$ is called an impossible event if it does not occur in any trial (e.g. non-breakdown in Example 2.1.1.3).
- An event $\bar{A}$ is called a complementary event to an event A if $\bar{A}$ occurs precisely when A does not occur (e.g. D and N in Example 2.1.1.1 are complementary to each other).
- Two events are referred to as uncombinable or disjoint when their simultaneous occurrence is impossible (e.g. events T and F in Example 2.1.1.1 are precluded in this sense).
- The logical sum $(A \cup B)$ of two events is called a union. It means that at least one of events A or B has occurred (e.g. in Example 2.1.1.1 the relationship $N = T \cup F$ applies).
- The logical product $(A \cap B)$ of two events is referred to as an intersection or meet. If, for example, it is required that two parallel insulators 1 and 2 should not break down, then neither of the two should break down, i.e. $N = N_1 \cap N_2$.

If, with the occurrence of an event A, a second event B occurs with certainty, we say that A draws B after it: $A \subset B$ (e.g. an electrical breakdown always presupposes the inception of discharges).

2.1.2 Relative frequency and probability

A trial is repeated n times; the random event A occurs m times. The reference number

$$h_n(A) = \frac{m}{n} \tag{2.1}$$

is referred to as the relative frequency. Because $0 \leqq m \leqq n$, the relative frequency is

$$0 \leqq h_n \leqq 1. \tag{2.2}$$

It follows that $h_n(I) = 1$ for the certain event and $h_n(\Phi) = 0$ for the impossible event. The relative frequency for two disjointed events B and C is their logic sum $h_n(B \cup C) = h_n(B) + h_n(C)$. In relation to important high-voltage engineering problems, the relative frequency for two complementary events A and $\bar{A}$ is of particular significance:

$$h_n(\bar{A}) = 1 - h_n(A). \tag{2.3}$$

Example 2.1.2.1 An insulation arrangement in the constant-voltage test in Example 2.1.1.1 is stressed $n = 20$ times, the event 'breakdown' (D) occurring $m = 7$ times. According to eqn. 2.1, the relative frequency $h_n(D)$ is consequently 0.35. The event 'non-breakdown' (N) is complementary to it, so that its relative frequency $h_n(N) = 1 - h_n(D) = 0.65$ is easily derived using eqn. 2.3.

If the test series consisting of n trials is repeated several times, the same value will always be obtained for $h_n(A)$. It transpires that the relative frequency $h_n(A)$ will vary about a fixed value, to which it approaches if n increases (Fig. 2.1) and reaches it as the limiting value.

$$\lim_{n \to \infty} h_n(A) = p(A). \tag{2.4}$$

This limiting value is referred to as the (statistical) probability of the random event. The probability $p(A)$ can be assigned to any random event A as a dimensional figure for the degree of certainty of its occurrence. As $p(A)$ itself cannot be determined experimentally, the probability is estimated through the relative frequency $h_n(A)$. Section 3.2 contains a detailed discussion of how such estimates can be expediently conducted in physics and high-voltage engineering. The properties established for the relative frequency are then axiomatically also assumed for the probability (for theory,

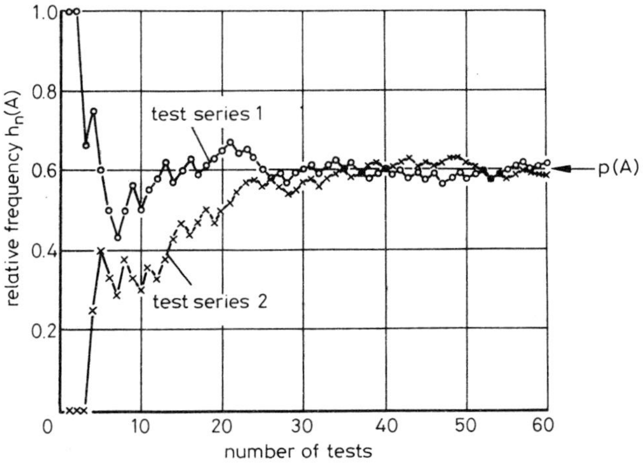

Fig. 2.1 Relative frequency as a function of the number of tests performed. Discrete measured values have only been joined up for the sake of clarity

see more detailed works in the Bibliography), i.e.:

$$0 \leqq P(A) \leqq 1,^{1}$$

$$P(I) = 1, \tag{2.5}$$

$$P(\Phi) = 0,$$

$$P(\bar{A}) = 1 - P(A). \tag{2.6}$$

To a large extent, the general addition expression

$$P(A \cup B) = P(A) + P(B) - P(A \cap B), \tag{2.7}$$

applies to probabilities, from which it follows directly that

$$P(A \cup B) \leqq P(A) + P(B)$$

wherein the equals sign applies to disjointed events, i.e. A and B are mutually exclusive. Besides the (statistical) definition of probability used here, an important part is played by a classical definition, derived from Laplace and starting with equally probable random events (e.g. when playing dice).

2.1.3 Conditional probabilities and independent events

An event A is considered in a trial. It is known at the same time that an event B has also occurred. The fact that the occurrence of B is known gives additional information for the investigations of A. It is therefore appropriate to investigate the probability of A under the condition that B has occurred. For this conditional probability we write $P(A/B)$.

Example 2.1.3.1 In the constant-voltage tests in Example 2.1.1.1, the non-breakdowns (N) may be conditioned by further events. There are non-breakdowns with which stable partial discharges (T) are simultaneously observed and to which the conditional probability $P(N/T)$ has to be assigned.

The conditional probability $P(N/F)$ is representative of non-breakdowns with which freedom from discharge (F) prevails. Naturally $P(N) = P(N/T) + P(N/F)$.

Example 2.1.3.2 When investigating the probability with which electron avalanches can produce breakdown, e.g. in a uniform field in gases, the initial state must be presupposed as a condition. For example, the probability of a certain number of electrons n_k being reached or exceeded at a point x in the field can only be indicated if the number n_0 of electrons initiated at the cathode ($x = 0$) is known as an initial condition. We can write $P(n(x) \geqq n_k / n(0) = n_0)$ where $n_0 = 1$.

The conditional probability of A under condition B is defined as the quotient

[1] In general probability considerations, this is represented by capital letters; lower-case letters are used when indicating concrete values.

of the probability of the logical product $A \cap B$ and of event B

$$P(A/B) = \frac{P(A \cap B)}{P(B)} \tag{2.8}$$

with $P(B) > 0$. By analogy,

$$P(B/A) = \frac{P(A \cap B)}{P(A)} \tag{2.9}$$

with $P(A) > 0$. It follows directly from eqns. 2.8 and 2.9 that

$$P(A \cap B) = P(B)P(A/B) = P(A)P(B/A) \tag{2.10}$$

which is the multiplication expression for probabilities.

From this it is possible to derive a property that is fundamental to almost all further considerations: the definition of the independence of random events.

Two events are independent of each other precisely when they are not dependent on the conditions:

$$P(A/B) = P(A) \quad \text{and} \quad P(B/A) = P(B). \tag{2.11}$$

For two independent events, it follows from eqns. 2.10 and 2.11 that

$$P(A \cap B) = P(A)P(B) \tag{2.12}$$

which is the multiplication expression for independent probabilities.

The multiplication expression, eqn. 2.12, enables the so-called 'enlargement (scaling) effects' of breakdown probability (number or area or volume of insulation arrangements; duration of stress) to be handled statistically (see Section 6), under the assumption of independence of the process occurring.

Example 2.1.3.3 For two insulators in constant-voltage tests, the probability of flashover is estimated as $p_1 = 0.45$ and $p_2 = 0.60$. How great is the probability $p_{ü}$ of flashover if the two insulators are connected in parallel and the discharges on the insulators are independent of each other? The complementary events of non-flashover $(1 - p_1) = 0.55$ and $(1 - p_2) = 0.40$ are considered for the solution. Non-flashover of the arrangement as a whole $(1 - p_{ü})$ presupposes the non-flashover of each of the independent insulators, i.e. using eqn. 2.12 it follows that

$$(1 - p_{ü}) = (1 - p_1)(1 - p_2) = 0.22$$

$$p_{ü} = 1 - 0.22 = 0.78.$$

The probability of flashover is increased by parallel connection.

The independence of random events is particularly important when determining important statistical characteristic quantities, especially when many tests are made on a test piece. In this case, a previous event (e.g. a breakdown) can change the test piece in such a way that the result of the subsequent test is affected by the different conditions. This can be avoided by appropriate experimental precautions (Sections 3.2.2 and 3.3.2), and the effectiveness of these measures should be checked by independence tests (Section 2.5.3).

2.1.4 Variate and sample

The investigation of a random event frequently involves obtaining a dimensional figure. If the random event is represented by a real, usually dimensional figure, a variate (random variable) is produced. Variates are represented by capital letters, e.g. $X, Y, Z, \ldots$. The quantity chosen as the variate for a random process is of subordinate importance for mathematical treatment and should be decided according to physical or technical criteria.

Example 2.1.4.1 If an insulation arrangement is stressed in a rising-voltage test (see Example 2.1.1.3), the numerical value of the breakdown voltage U_d kV can be used directly as the variate. If the results are to be randomised in a physical/technical sense, it is expedient to calculate the variates maximum breakdown field-strength E_{dh} kV · cm^{-1} or electric strength E_d kV · cm^{-1} from U_d, taking the field geometry into account (e.g. [2.9]), and to consider these quantities further. In certain problems, such as service-life problems, it is advisable to use as the variate, not the breakdown voltage U_d kV, but the time elapsing from voltage application up to breakdown, i.e. the breakdown time T_d s.

A realisation x of the variate X is obtained in the result of a random test. Realisations are indicated by lower-case letters.[1] All the possible realisations constitute the population, which can embrace a finite or infinite number of values. If the realisations assume only as many numerical values as can be counted, we talk of a discrete variate or discontinuous variate. If, on the other hand, the values move about in a continuous range, the variate is a continuous variate.

Example 2.1.4.2 In a constant-voltage test (Example 2.1.1.1), when considering the events breakdown and non-breakdown, there is a discrete variate with a finite population: the numbers 0 and 1 can be assigned to the events, for example.

One would similarly be faced with a discrete variate if one were investigating how many insulators on an overhead line flashover in a certain period of time. On the other hand, the breakdown time in the constant-voltage test in Example 2.1.1.2 is a continuous variate from an infinite population $[0 < t_d \leqq \infty]$. In this sense, all rising-voltage tests (Example 2.1.1.3) also provide continuous variates.

To analyse a random process, it is in practice only possible to conduct a finite number of trials and to obtain only a finite number of values from

[1] In high voltage engineering practice it is often difficult to be consistent in the allocation of capitals and lower-case letters, since the symbols have been traditionally assigned. This notation is used in the present presentation, however.

them. If these trials are not dependent on each other, such a series of realisations x_i ($i = 1, \ldots, n$) is called a (concrete) sample of sample size n.

Example 2.1.4.3 In constant-voltage tests (Examples 2.1.1.1 and 2.1.4.2) a sample of size n is obtained. m breakdowns ($m \leqq n$) are recorded, for example, to which the number 1 would be assigned, and ($n - m$) non-breakdowns, to be marked 0. For evaluation purposes, the relative frequency is formed as in Example 2.1.2.1. In rising-voltage tests (Example 2.1.1.3) the size n sample consists of n values of breakdown voltage U_{di} ($i = 1, \ldots, n$). In both cases one is dealing with concrete samples of size n.

The example itself indicates that, in the course of evaluation, mathematical operations are performed using the sample values, in order to draw conclusions about probability statements concerning the population.

These operations are assisted by sample functions (e.g. the known formulae for the arithmetic mean, variance, etc. — see Sections 2.2.3 and 2.5), which are derived using accepted 'mathematical samples'. A mathematical sample of size n consists of n independent, identically distributed variates (distribution function: see Section 2.2.1). The concrete samples considered here are then taken as realisations of the mathematical sample.

The main aim of this book is to explain how concrete samples can be advantageously planned and evaluated in high-voltage engineering. The requisite sample functions are given in Sections 2.2 to 2.5.

2.2 Distribution functions

2.2.1 Concept and properties

The distribution law of a variate is best described by a so-called distribution function. For this, events of the form $A_x = X < x$ and the associated probabilities $P(A_x) = P(X < x)$ are considered for real numerical values of x.

The distribution function is defined by

$$F(x) = P(X < x). \tag{2.13}$$

The distribution function at point x indicates the probability with which the variate X will assume a value below the boundary x. The distribution function possesses the following properties:

(i) The definition range for the distribution function is derived from the definition range for probabilities (eqn. 2.5):

$$0 \leqq F(x) \leqq 1. \tag{2.14}$$

(ii) The distribution function increases monotonically (not decreasing):

$$F(x_1) \leqq F(x_2) \quad \text{for } x_1 < x_2. \tag{2.15}$$

(iii) The boundary conditions

$$\lim_{x \to \infty} F(x) = F(-\infty) = 0 \tag{2.16}$$

and

$$\lim_{x\to\infty} F(x) = F(+\infty) = 1. \tag{2.17}$$

apply.

(iv) Every distribution function defined by eqn. 2.13 is continuous on the left-hand side.

Conversely, every function possessing these four properties can be regarded as a distribution function of a variate and be characterised by eqn. 2.13. This explains the large number of possible distribution functions, but from

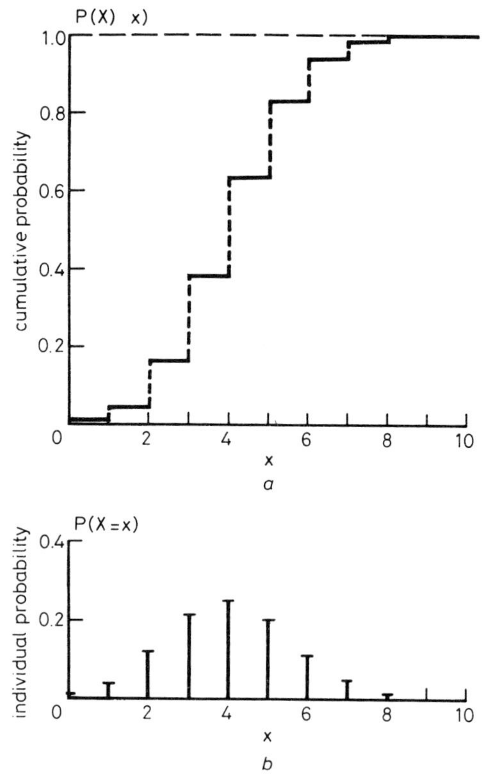

X:	0	1	2	3	4	5	6	7	8	9	10
	0.006	0.040	0.121	0.215	0.251	0.201	0.112	0.043	0.011	0.002	0.000

c

Fig. 2.2 Distribution function of a discrete variate

a discrete distribution function (cumulative probability)
b discrete density function (individual probability)
c probability table

a theoretical and practical standpoint preference is only given to certain distribution functions (see Section 2.3). It is appropriate to differentiate between the distribution functions of discrete and continuous variates.

Distribution functions of discrete variates (Fig. 2.2*a*) permit the representation

$$F(x) = P(X < x) = \sum_{x_i < x} P(X = x_i) = \sum_{x_i < x} p_i, \tag{2.18}$$

in which x_i are the discrete values which the variate can assume, and p_i are the associated individual probabilities. If the population is finite, then

$$\sum_{i=1}^{n} p_i = 1,$$

if it is infinite, it follows that:

$$\sum_{i=1}^{\infty} p_i = 1.$$

Instead of the discrete distribution function, the individual probabilities can also be adopted as a discrete density function

$$f(x) = P(X = x_i) = p_i \tag{2.19}$$

to characterise the discrete variate (Fig. 2.2*b*). Discrete variates can be described by the probability table

$$X: \begin{pmatrix} x_1 & x_2 & x_3 & \ldots \\ p_1 & p_2 & p_3 & \ldots \end{pmatrix}$$

in which the individual or cumulative probabilities are correlated with the discrete values x_i (Fig. 2.2*c*).

With discrete variates, it is often advantageous in calculations not to use the distribution function (eqn. 2.18), but to use the often more convenient individual probabilities p_i directly from the probability table.

Two important information quantities relating to a distribution function are the expectation and the variance. The expectation (position parameters)[1] of a discrete variate X is given by

$$\mu = EX = \sum_i x_i p_i \tag{2.20}$$

and the variance (extent parameter)[2] by

$$\sigma^2 = D^2 X = \sum_i (x_i - \mu)^2 p_i. \tag{2.21}$$

The expectation of the expression $(X-a)^k$, i.e.

$$\mu_k = E[(X-a)^k] = \sum_i (x_i - a)^k p_i, \tag{2.22}$$

[1] If total mass = 1 is allocated to the distribution function, the expectation is the centre of gravity of the mass distribution.

[2] According to the same model as in [1], the variance is the moment of inertia about the centre of gravity.

is generally referred to as the kth moment of variate X with respect to a. The expectation is consequently the first moment (or first-order moment) for $a=0$, while the variance is the second moment with respect to $a=\mu=EX$ (second central moment). The square root of the variance is referred to as the standard deviation σ.

The following rules apply to calculation using expectations and variances:

$$E(\alpha X+\beta)=\alpha EX+\beta, \tag{2.23}$$

$$E(X_1+X_2)=EX_1+EX_2, \tag{2.24}$$

$$D^2X=E(X^2)-(EX)^2, \tag{2.25}$$

$$D^2(\alpha X+\beta)=\alpha^2 D^2 X \tag{2.26}$$

and for independent variates:

$$E(X_1X_2)=E(X_1)E(X_2), \tag{2.27a}$$

$$D^2(X_1+X_2)=D^2(X_1)+D^2(X_2). \tag{2.27b}$$

Distribution functions of continuous variates (Fig. 2.3) can be represented in the form

$$F(x)=P(x<x)=\int_{-\infty}^{x} f(t)\,dt \tag{2.28}$$

where the expression

$$f(x)=\frac{d}{dx}F(x) \tag{2.29}$$

is referred to as a density function (probability density). The density function possesses the properties

$$f(x)\geqq 0 \tag{2.30}$$

for all x and

$$\int_{-\infty}^{+\infty} f(x)\,dx=1. \tag{2.31}$$

Every function with both these properties can be considered as a density function of a continuous variate. Whether a variate ought to be described by its distribution function or its density function should be decided less according to mathematical criteria than with regard to physical or technical data. Generally speaking, and especially in high-voltage engineering, it is an advantage to work with the distribution function.

The probability of the continuous variate X falling in an interval $[a, b]$ is given by

$$P(a\leqq X\leqq b)=\int_{a}^{b} f(x)\,dx=F(b)-F(a). \tag{2.32}$$

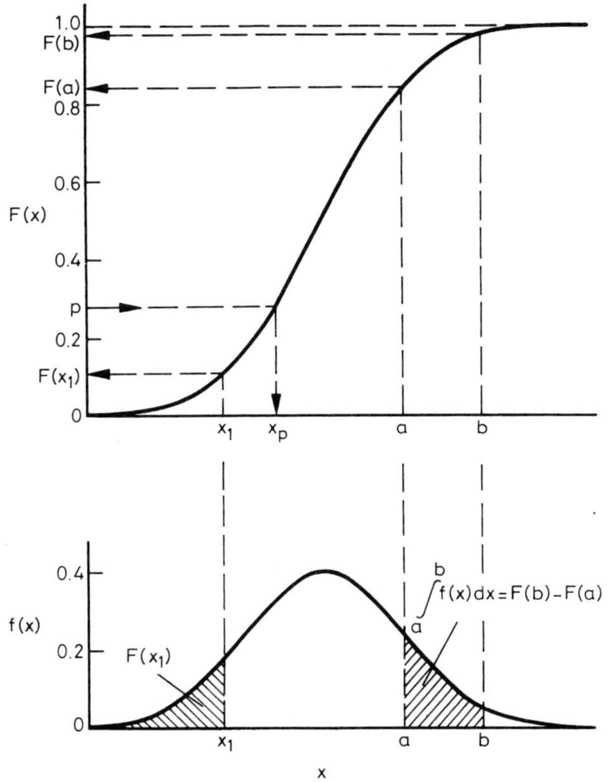

Fig. 2.3 Distribution function of a continuous variate

a Distribution function
b Density function

If it is a small interval Δx, then, in approximate terms:

$$P(x \leqq X \leqq x + \Delta x) \approx f(x)\Delta x. \tag{2.33}$$

A realisation $x = q_p$ associated with a certain value of the distribution function $F(q_p) = p$ is referred to as a p quantile (quantile of order p).

Example 2.2.1.1 The 10% quantile of breakdown voltage U_{d10} associated with a breakdown probability $p = 0.10$ ($\hat{=}10\%$), is, according to current national and international standards, defined as a statistical withstand voltage and is adopted for insulation co-ordination.

Moments can also be indicated for continuous variates, where, by analogy with 2.22, the following applies for the kth moment[1] with respect to a:

$$\mu_k = E[(X-a)^k] = \int_{-\infty}^{+\infty} (x-a)^k f(x)\, dx \qquad (2.34)$$

and the special cases have the following significance:

$k-1, a=0$ is the expectation of μ as the mean value of the distribution function;

$k=2, a=\mu$ is the variance σ^2 as a measure of the width of the distribution function;

$k=3, a=\mu$ is the third central moment μ_3, which is used to calculate the skewness $\gamma=\mu_3/\sigma^3$ as a measure of the asymmetry of the distribution function;

$k=4, a=\mu$ is the fourth central moment μ_4, which is used to calculate the excess $\sigma=\mu_4/\sigma^4-3$ as a measure of the deviation from a normal distribution (see Section 2.3.2.2).

The derived rules (eqns. 2.23 to 2.27) apply to calculation using the expectation and variance. By functional combinations, it is possible, from one or more variates with a known distribution function or density function, to generate new variates, the distributions of which are listed in Table 2.1. The explicit representation of these functions often presents difficulties, however.

There are a large number of mathematically well prepared distribution functions of discrete variates (eqns. 2.14 to 2.17 and 2.29 to 2.31). Such distributions, such as the binomial distribution, Poisson distribution, normal distribution, Weibull distribution and others (dealt with in Section 2.3), characterise a particular mathematical model and in our situation ought to be described as 'theoretical' distribution functions. These theoretical distribution functions are fully characterised by their parameters. The formulae for calculating the moments from the parameters are known: with a normal distribution, the mean value and variance moments are even identical to the parameters. Using the moments, it is possible to compare variates distributed in different ways (e.g. normal-distribution and Weibull-distribution breakdown times). Tables are available of theoretical distribution functions in a generalised (standardised) form. There are therefore many advantages in describing experimentally investigated relationships by such theoretical distribution functions.

The samples obtained from experiments, however, initially provide empirical distribution functions, i.e. a relationship between the individual realisations and the probability of their occurrence in the sense of eqn. 2.13. The empirical distribution function exists in tabular or graphical representation, e.g. a polygon graph (see Section 2.2.2). It is more convenient, however, to approximate the empirical distribution function by a theoretical distribution function with established parameters. For this purpose, the parameters

[1] It should be pointed out that, for the moments, their existence has to be presupposed (eqns. 2.22 and 2.34).

Table 2.1 Functions of independent variates and their density and distribution functions

Mathematical operation	Initial quantities	Function	Density function or distribution function of function $f(y)$ or $F(y)$
Linear transformation	X; $f(x)$; $F(x)$ a; b constants $a>0$	$Y=aX+b$	$F(y)=F\left(\frac{y-b}{a}\right)$; $f(y)=f\left(\frac{y-b}{a}\right)\frac{1}{\lvert a\rvert}$
Sum	X_1; $f_1(x_1)$ X_2; $f_2(x_2)$	$Y=X_1+X_2$	$f(y)=\int_{-\infty}^{+\infty} f_1(x_1)f_2(y-x_1)\,dx_1$ (Convolution integral)
Product	X_1; $f_1(x_1)$ X_2; $f_2(x_2)$	$Y=X_1X_2$	$f(y)=\int_{-\infty}^{+\infty} f_1(x_1)f_2\left(\frac{y}{x_1}\right)\frac{1}{\lvert x_1\rvert}\,dx_1$
Quotient	X_1; $f_1(x_1)$ X_2; $f_2(x_2)$	$Y=\frac{X_1}{X_2}$	$f(y)=\int_{-\infty}^{+\infty} f_1(yx_1)f_2(x_1)\lvert x_1\rvert\,dx_1$
Difference	X_1; $f_1(x_1)$ X_2; $f_2(x_2)$	$Y=X_1-X_2$	$f(y)=\int_{-\infty}^{+\infty} f_1(x_1)f_2(x_1-y)\,dx_1$ *(Convolution integral)*
Square	X; $f_x(x)$	$Y=X^2$	$f(y)=\frac{1}{2\sqrt{y}}[f_x(\sqrt{y})+f_x(-\sqrt{y})]$
Arbitrary function	X; $f_x(x)$	$Y=\varphi(X)$	$f(y)=f_x[\varphi^{-1}(y)]\frac{d\varphi^{-1}(y)}{dy}$

are estimated from the existing realisations (see Section 2.2.3). By means of the approximation, the theoretical distribution used is randomly accepted as being valid for the unknown distribution law of the variate under investigation. Since, however, the parameters are only more-or-less reliable estimates, the empirical nature of the distribution function is retained even after the approximation.

2.2.2 Empirical distribution functions and their representation

In trials, a sample of size n is produced. The realisations x_i are recorded in the order they occur and in this form constitute the list of individual test values, which is the basis for any further statistical treatment.

Example 2.2.2.1 Alternating breakdown voltages are measured in rising-voltage tests (Example 2.1.1.3) on a coaxial cylinder arrangement in SF_6. The sample of size $n = 24$ to be evaluated has the following individual test values:

u_{di}/kV = 210; 208; 208; 175; 182; 206; 190; 194; 198; 205; 212; 200; 205; 202; 207; 210; 202; 201; 188; 205; 209; 201; 216; 196.

The primary distribution table can now be derived from the list of test values: all values of x_k between the lowest (x_{min}) and highest (x_{max}) realisations are recorded, graded according to the smallest measuring unit present or simply in order of magnitude. The absolute frequencies $h_{mk}(x_i = x_k)$, the relative frequencies

$$h_k = \frac{h_{mk}}{n+1} \tag{2.35}$$

and the relative cumulative frequencies

$$h_{\Sigma k} = \sum_{j=1}^{k} h_j \tag{2.36}$$

are correlated with the values using a 'tally'.

In contrast to eqn. 2.1, eqn. 2.35 contains not the sample size n, but $(n+1)$. This procedure is particularly advantageous with a small sample size n, as all the realisations can be included in the evaluation when using probability papers (see Section 2.5.1). As n increases, the relative frequency according to eqn. 2.35 converges with that from eqn. 2.1: when $n > 50$, one need only refer to n.

Example 2.2.2.2 According to the list of individual test values in Example 2.2.2.1, $u_{dmin} = 175$ kV and $u_{dmax} = 216$ kV. The smallest unit of measurement is $\Delta_u = 1$ kV (e.g. $\Delta_u = (210-209)$ kV). In the primary distribution table the absolute and relative frequencies, as well as the relative cumulative frequencies, are then correlated with values from 175 kV to 216 kV (Table 2.2).

To reduce the amount of writing involved, it is also possible just to plot the realisations x_i in order of magnitude in the primary distribution table, but the representation of all the values of $x_k = u_{dk}$ used in Table 2.2 facilitates sorting of the realisations. Working by hand with the primary distribution table is very tedious with a large sample. The amount of calculation can be reduced by arranging classes between x_{min} and x_{max}, instead of the individual values of x_k, and assigning frequencies to them. In this way, a frequency table is produced. Before going into the details of its compilation, it should be emphasised that the transition from primary distribution table to frequency table always involves a loss of information. Whenever outlay on calculation is of subordinate importance (e.g. with a small sample or when using digital computers), one should dispense with the compilation of frequency tables and attach further evaluations to the primary distribution table.

Table 2.2 Primary distribution table for alternating breakdown voltages measured in a rising-voltage test (Example 2.2.2.2)

Values between $u_{d\min}$ and $u_{d\max}$ u_{dk}/kV	Tally	Absolute frequency h_{mk}	Relative frequency h_k	Relative cumulative frequency $h_{\Sigma k}$
175	|	1	0.04	0.04
176		0	0	0.04
177		0	0	0.04
178		0	0	0.04
179		0	0	0.04
180		0	0	0.04
181		0	0	0.04
182	|	1	0.04	0.08
183		0	0	0.08
184		0	0	0.08
185		0	0	0.08
186		0	0	0.08
187		0	0	0.08
188	|	1	0.04	0.12
189		0	0	0.12
190	|	1	0.04	0.16
191		0	0	0.16
192		0	0	0.16
193		0	0	0.16
194	|	1	0.04	0.20
195		0	0	0.20
196	|	1	0.04	0.24
197		0	0	0.24
198	|	1	0.04	0.28
199		0	0	0.28
200	|	1	0.04	0.32
201	||	2	0.08	0.40
202	||	2	0.08	0.48
203		0	0	0.48
204		0	0	0.48
205	|||	3	0.12	0.60
206	|	1	0.04	0.64
207	|	1	0.04	0.68
208	||	2	0.08	0.76
209	|	1	0.04	0.80
210	||	2	0.08	0.88
211		0	0	0.88
212	|	1	0.04	0.92
213		0	0	0.92
214		0	0	0.92
215		0	0	0.92
216	|	1	0.04	0.96

The classification needed for the frequency table can be undertaken according to the following guide-values: the class number k should be chosen between 7 and 20, assuming roughly that

$$k \approx 5 \log n. \tag{2.37}$$

Having established the class number k, the class width d is chosen from the spread:

$$R = x_{max} - x_{min} \tag{2.38}$$

according to

$$d \approx R/k. \tag{2.39}$$

The initial position (lower limit of first class) x_{1u} should be below the lowest measured value ($x_{1u} < x_{min}$), and d should be established in such a way that none of the further class limits coincides with a measured value of x_i. Both the choice of class number k and that of the initial position (x_{1u}) determine the empirical distribution function. How problematical a classification is is demonstrated by the empirical distribution functions reproduced in Fig. 2.4, all of which are constructed on the basis of the same list of individual test values from Example 2.2.2.1 and which nevertheless differ considerably.

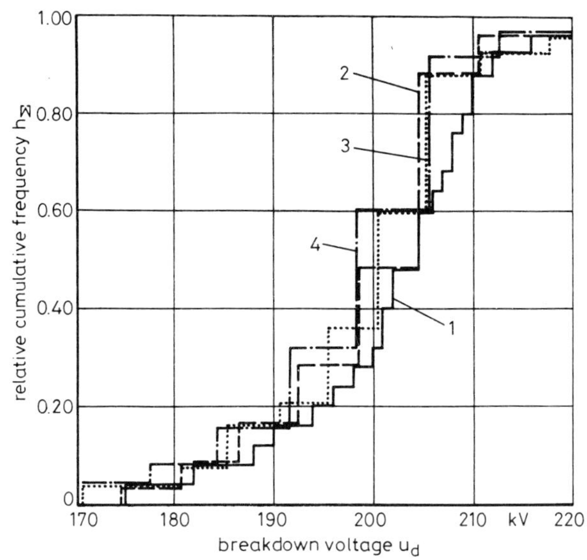

Fig. 2.4 Empirical distribution functions from primary distribution tables and different frequency tables with the same individual test values

1 from primary distribution table
2 from frequency table with $k = 7$; $d = 6$ kV; $u_{d1u} = 174.5$ kV
3 from frequency table with $k = 10$; $d = 5$ kV; $u_{d1u} = 170.5$ kV
4 from frequency table with $k = 7$; $d = 7$ kV; $u_{d1u} = 170.5$ kV

Example 2.2.2.3 For the individual test values in Example 2.2.2.1 it follows that:

$$k = 5 \times \log 24 = 6.9011; \qquad k = 7 \text{ selected;}$$

$$R = (216 - 175)\,\text{kV} = 41\,\text{kV};$$

$$d = R/k = 41\,\text{kV}/7 = 5.8571; \qquad d = 6\,\text{kV selected;}$$

initial position selected as $u_{d1u} = 174.5$ kV.

The classes, 'tally' and frequencies (eqns. 2.35 and 2.36) are listed in the frequency table (Table 2.3).

For an engineering-related judgment of empirical distribution functions, they need to be graphically represented. Representation of the empirical distribution function can be based both on the primary distribution table (Table 2.2) and on the frequency table (Table 2.3), and may take the form of a step function, a cumulative frequency curve or a cumulative frequency polygon. The procedure is explained in more detail in Table 2.4. The graphical representation with a linearly graduated ordinate (h_Σ) explained here is comparatively rarely used.

More commonly used are so-called probability papers, on which the ordinate is graduated according to the inverse function of a theoretical distribution function. If the empirical distribution function being considered, for which the cumulative frequency polygon is expediently used, corresponds to the type of the theoretical one, it will appear on the probability paper as a straight line. Probability paper is dealt with in Section 2.5.1.1.

When representing empirical density functions one has to start with the frequency table, since the primary distribution table is in most cases not sufficiently densely populated. The empirical density function can be represented as a histogram, frequency polygon or a frequency curve, as shown in Table 2.5. It should be pointed out that the class number k and class width d, as well as the starting point x_{1u}, have a far greater influence on

Table 2.3 Frequency table for alternating breakdown voltages measured in a rising-voltage test (Example 2.2.2.3)

Class number	Class limits lower	Class limits upper	Middle of class	Tally	Absolute frequency	Relative frequency	Relative cumulative frequency
l	u_{dlu}	u_{dlo}	u_{dlm}		h_{ml}	h_l	$h_{\Sigma l}$
1	>174.5	180.5	177.5	|	1	0.04	0.04
2	>180.5	186.5	183.5	|	1	0.04	0.08
3	>186.5	192.5	189.5	||	2	0.08	0.16
4	>192.5	198.5	195.5	|||	3	0.12	0.28
5	>198.5	204.5	201.5	~~||||~~	5	0.20	0.48
6	>204.5	210.5	207.5	~~||||~~ ~~||||~~	10	0.40	0.88
$k=7$	>210.5	216.5	213.5	||	2	0.08	0.96

Table 2.4 Representation of empirical distribution functions[1]

Form of representation	Representation based on primary distribution table (Table 2.3)	Frequency table (Table 2.3)	Example of representation
Stepped curve	The cumulative frequencies $h_{\Sigma k}$ (eqn. 2.36) are plotted as rectangles with the width of the smallest unit of measurement present [$\Delta x = \min(x_{i+1} - x_i)$] against all the values $x = x_k$ occurring in the primary distribution table (Table 2.2).	The cumulative frequencies $h_{\Sigma 1}$ (Table 2.3) are represented as rectangles over each class.	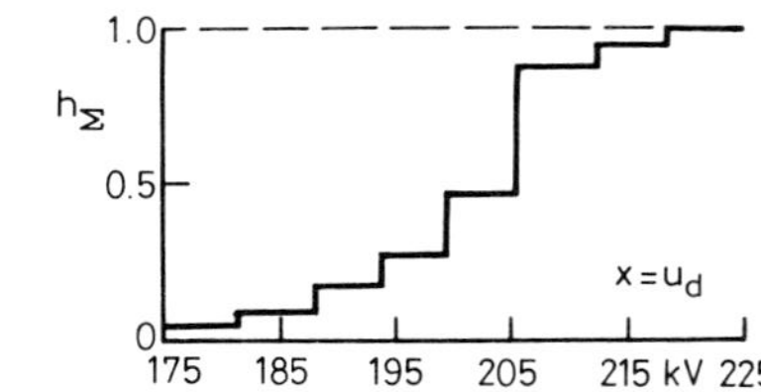
Cumulative frequency polygon (recommended by preference)	The cumulative frequencies $h_{\Sigma k}$ are plotted against the respective realisations of x_i, and are joined up by a polygon graph. A measured point must appear for each realisation (see Fig. 2.38, for example).	The cumulative frequencies $h_{\Sigma 1}$ are plotted over the upper class limit and are joined up by a polygon graph.	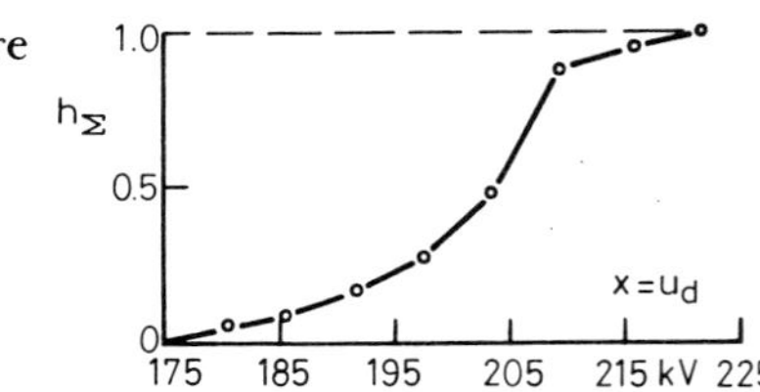
Cumulative frequency curve	The cumulative frequencies $k_{\Sigma k}$ are plotted as they were for the cumulative frequency polygon, and a line of best fit drawn.	The cumulative frequencies $h_{\Sigma 1}$ are plotted as they were for the cumulative frequency polygon, and a line of best fit is drawn through the points.	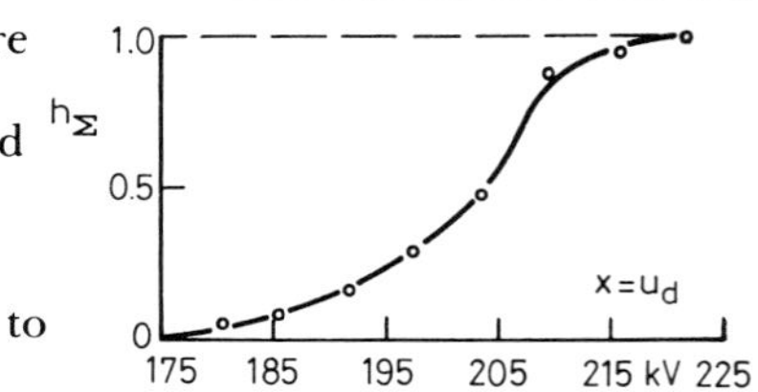
	N.B. When using probability paper, the curve becomes **a straight** line, to which linear regression can be applied (see Section 2.4).		

[1] The stepped curve is usually the direct empirical conversion of the definition of a distribution function (eqn. 2.13) on the basis of eqns. 2.1 and 2.36. A simplified representation, which is often more suitable for technical problems, is offered by the cumulative-frequency polygon and curve, which, in view of the anticipated fidelity of the graphical representation, must take place on the basis of eqns. 2.35 and 2.36.

Table 2.5 Representation of empirical density functions

Form of representation	Procedure, on basis of frequency table (Table 2.3)	Example of representation
Histogram	The related relative frequencies h_l/d are reproduced as a rectangle over each class. d = class width	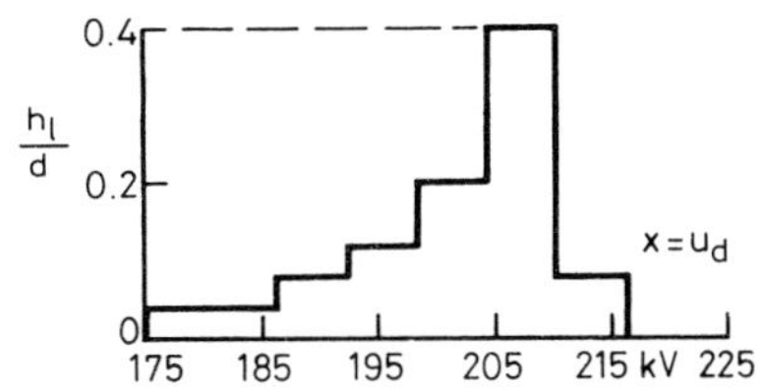
Frequency polygon	The related relative frequencies h_l/d are plotted as a point over the class middle x_{lm} and are joined up by a polygon graph.	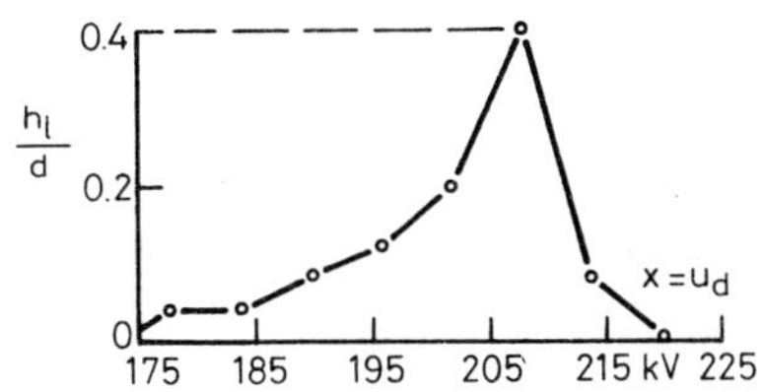
Frequency curve	The related relative frequencies h_l/d are plotted as they were for the frequency polygon, but they are joined up by a fitted curve. When using probability paper the path of the frequency curve can be simplified (e.g. by omitting inflection points).	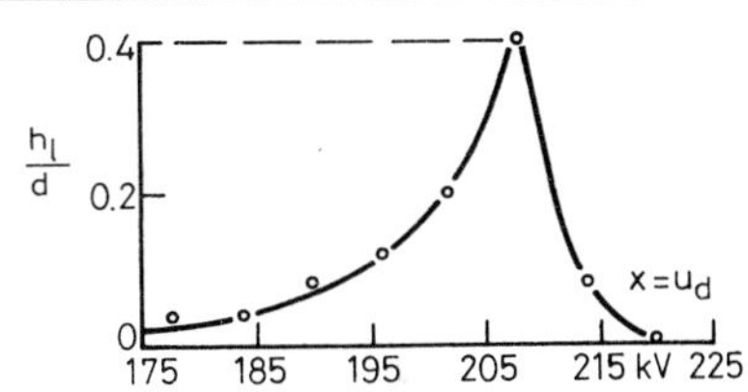

the density function than they do on the distribution function (Fig. 2.4). The representations are consequently not without a certain arbitrariness. The use of probability paper can also be advantageous here (see Section 2.5.1.1).

2.2.3 Parameter estimates

Parameters are often used to describe distribution functions. It is appropriate to differentiate between parameters of general interest in relation to all distribution functions — such as expectation, variance, quantile, mean value, etc. (functional parameters) — and parameters that occur explicitly in the formula for a theoretical distribution function (distribution parameters). There are, of course, close relationships between these two types of parameter, since the functional parameters can often be represented as a function of the distribution parameters and are sometimes even identical to them.

Among the functional parameters, the expectation, variance and quantile have already been dealt with as theoretical quantities (see Section 2.2.1). If they are to be determined empirically as samples, the individual realisations of the sample need to be mathematically processed in a sample function. Such a sample function is constituted, for example, by the empirical kth order moment with respect to a:

$$m_k = \frac{1}{n} \sum_{i=1}^{n} (x_i - a)^k, \qquad k = 1, 2, 3, \ldots \tag{2.40}$$

from which important functional parameters can be derived.

The following functional parameters can be derived as mean values:

(i) The arithmetic mean (also referrred to as the 'sample mean') is the empirical moment for $k = 1$ and $a = 0$

$$\bar{x} = \frac{1}{n} \sum_{i=1}^{n} x_i \tag{2.41}$$

and an estimated value for the expectation. With a class division, we can also write:

$$\bar{x} = \frac{1}{n} \sum_{l=1}^{k} h_{l\mathrm{m}} x_{l\mathrm{m}} = \sum_{l=1}^{k} h_l x_{l\mathrm{m}}, \tag{2.42}$$

where $x_{l\mathrm{m}}$ is the middle of the class, $h_{l\mathrm{m}}$ the absolute class-frequency and h_l the relative class-frequency. As most pocket calculators are programmed for $\bar{x}$, eqn. 2.42 and other simplified methods of calculation (multiplication processes, addition processes) will only be used in special cases. By virtue of its property of being an estimate of the expectation, the arithmetic mean is of fundamental importance.

(ii) The central value or median $\tilde{x} = x_{50}$ of a sample is the empirical quantile for a cumulative frequency $h = 0.5$ and can be easily determined from the primary distribution table. If n is an odd number, $\tilde{x}$ is the measured value in the middle. If n is even, $\tilde{x}$ is determined by the mean value of the two values in the middle. The central value is often indicated by simple determination, and also used as an estimated value for the arithmetic mean.

(iii) The mode or density mean $\hat{x}$ is the quantile associated with the maximum value of the density function. It can be estimated from the sample, using the frequency table, according to

$$\hat{x} = x_{uu} + \left(\frac{h_{mu} - h_{m(u-1)}}{2h_{mu} - h_{m(u-1)} - h_{m(u+1)}} \right) d \tag{2.43}$$

where x_{uu} is the lower limit of the most heavily populated class, h_{mu} the associated absolute frequency, $h_{m(u-1)}$ and $h_{m(u+1)}$ the absolute frequencies of the two adjoining classes and d the class width. In some distribution functions, the mode is also a distribution parameter (e.g. double exponential distribution, see Section 2.3.2.5), and is therefore of practical interest.

(iv) The geometric mean

$$\mathring{x} = \sqrt[n]{\prod_{i=1}^{n} x_i}, \tag{2.44}$$

$$\log \mathring{x} = \frac{1}{n} \sum_{i=1}^{n} \log x_i \tag{2.45}$$

is used in economics statistics. There are no known applications in high-voltage engineering.

(v) The harmonic mean

$$\breve{x} = \frac{n}{\sum_{i=1}^{n} \frac{1}{x_i}} \tag{2.46}$$

is used when one wishes to indicate mean values of related quantities, e.g. average speeds or mean particle densities. (If a distance A to B is covered at $v_1 = 30$ km/h and back from B to A at $v_2 = 60$ km/h, the average speed according to eqn. 2.46 is $\mathring{v} = 40$ km/h.)

The following sample functions are used as measures of dispersion:

(i) The spread (eqn. 2.38)

$$R = x_{\max} - x_{\min}$$

embraces all the measured values; with a very small sample (roughly <5) it is a preferred measure of dispersion. With a medium sample size ($10 \leqq n \leqq 100$), R is linked to the standard deviation s (see below) by $R \approx 4s$.

(ii) So-called interdecile ranges

$$I = x_p - x_q \tag{2.47}$$

for example $I_{80} = x_{90} - x_{10}$ are, by analogy to spread, formed as the difference between two selected empirical quantiles. The empirical standard deviation s can be estimated by the interdecile range $s = I_{50-16} = x_{50} - x_{16}$. Other distribution parameters can also be expressed by interdecile ranges (i.e. certain quantiles).

(iii) The mean square deviation (empirical standard deviation) is the square root of the empirical moment with $k = 2$ and $a = \bar{x}$ (empirical dispersion):

$$s = \sqrt{\frac{1}{n-1} \sum_{i=1}^{n} (x_i - \bar{x})^2} = \sqrt{\left(\frac{1}{n-1} \sum_{i=1}^{n} x_1^2 - n\bar{x}^2\right)}. \tag{2.48}$$

Because of the accuracy of the expectation, the reduced value $(n-1)$ is used instead of the sample size n. In a classification, using the symbols explained for eqn. 2.42:

$$s = \sqrt{\frac{1}{n-1} \sum_{l=1}^{k} h_{l\mathrm{m}}(x_{l\mathrm{m}} - \bar{x})^2}. \tag{2.49}$$

One can nowadays generally dispense with the use of eqn. 2.49 and of further simplified methods, since s can be easily determined using a pocket calculator. Like the arithmetic mean $\bar{x}$, the mean square deviation s is of fundamental importance in assessing samples.

One is often interested in the ratio of the variation of a variate to the mean value, rather than the functional parameter itself. The quotient

$$v = s/\bar{x} \tag{2.50}$$

is referred to as the variation coefficient and provides a clear measure of the relative fluctuation of the variate. Comparisons can be made to advantage using the variation coefficient.

Example 2.2.3.1 The following functional parameters can be determined from the sample in Example 2.2.2.1:

- arithmetic mean $\bar{x} = 201.25$ kV (eqn. 2.41)
- central value (median) $\tilde{x} = 203.5$ kV
- mode (density mean) $\hat{x} = \left[204.5 + \left(\frac{10-5}{20-5-2}\right)6\right]$ kV

$$= 206.8 \text{ kV (eqn. 2.43)}$$

- geometric mean (of no practical interest) $\mathring{x} = 201.02$ kV (eqn. 2.44)
- harmonic mean (of no practical interest) $\breve{x} = 200.77$ kV (eqn. 2.46)
- spread $R = 216$ kV $- 175$ kV $= 41$ kV
- interdecile range $I_{80} = 212$ kV $- 188$ kV $= 24$ kV (eqn. 2.47)
- mean square deviation $s = 9.76$ kV (eqn. 2.48)
- variation coefficient $v = 9.76/201.25 = 0.0485$ (eqn. 2.50)

The individual functional parameters are illustrated in Fig. 2.5.

We do not intend to go deeply into the distribution parameters in this section: the sample functions used to estimate them will be given when dealing with the individual theoretical distribution functions (see Section 2.3). The estimate can be first made as a point estimate, i.e. the parameter being sought is estimated by a discrete numerical value. Point estimates are obtained by expressing the parameters to be estimated, using theoretical moments or theoretical quantiles. The theoretical moments or quantiles are then replaced by the corresponding empirical moments (eqn. 2.40) or empirical quantiles (see Section 2.2.1), thereby obtaining the desired point estimates (moment or quantile method). Another way of obtaining point estimates is by the maximum likelihood method. Out of all the possible parameter expressions, the expression is sought for which the given sample has the maximum likelihood (probability).

The advantage of point estimates is that they supply concrete numerical values; the snag with them is that one never knows how accurate they are. The unknown 'true' value of the parameter is merely described by an estimated value that varies to a lesser or greater degree. The sample size,

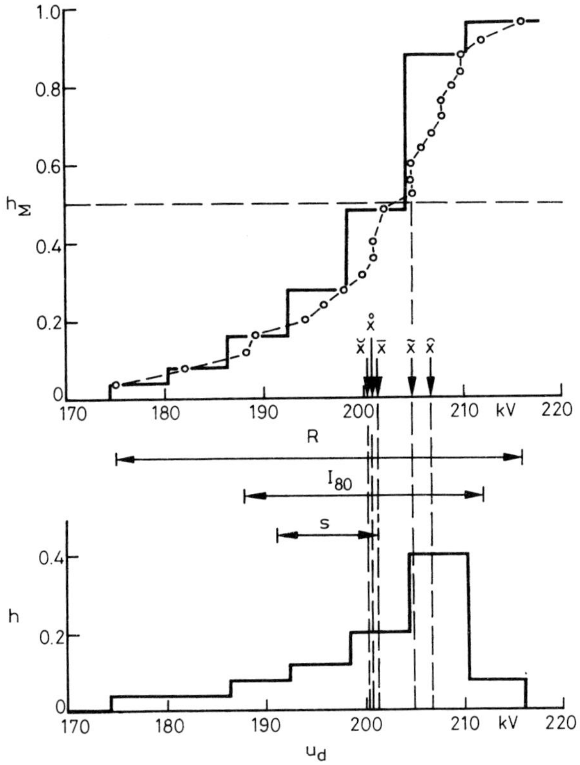

Fig. 2.5 Illustration of functional parameters for the sample in Example 2.2.2.1. For explanation of symbols see Example 2.2.3.1; circles represent realisations from primary distribution table (Table 2.2) (cumulative frequency polygon)

too, which governs the accuracy of the estimation, is not expressed in the estimated value.

In a confidence estimation an interval (confidence region) with a lower and upper limit $I[g_u; g_o]$ is calculated, covering the unknown parameter (discrete numerical value) with a given confidence coefficient (statistical certainty) ε. Confidence estimations presuppose point estimates. For example, for a parameter γ, we can write

$$P(g_u \leqq \gamma \leqq g_o) = \varepsilon = 1 - \alpha, \tag{2.51}$$

where α is complementary to ε and is referred to as the error probability.

The width of the confidence region is a function of the confidence coefficient ε chosen, the sample size n and the distribution — especially the dispersion — of the variate. The position and width of the confidence region are additionally determined by the (random) sample. Fig. 2.6 illustrates this with ten confidence regions for the arithmetic mean (as parameters of a normal distribution: see Section 2.3.2), calculated from ten samples, each of size $n = 10$ and all taken from the same population. Since $\varepsilon < 1$, not all

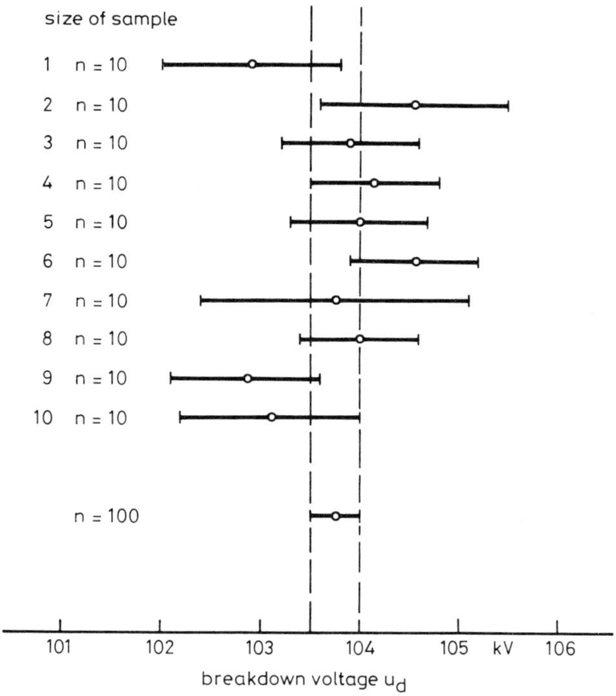

Fig. 2.6 Confidence regions for the arithmetic mean

10 samples of size $n = 10$ and 1 sample of size $n = 100$ from the same population

the regions need necessarily cover the true value. From a sample with $n = 100$, with the same confidence coefficient, a far narrower confidence region will be obtained.

The limits g_u ; g_o do not have to be finite. If one selects $g_u = -\infty$ or $g_o = +\infty$, we talk of a one-sided confidence region $I = [-\infty; g_o]$ or $I = [g_u ; +\infty]$. A two-sided confidence region, on the other hand, has finite limits $I = [g_u ; g_o]$ (Fig. 2.7). One has to establish whether one-sided or two-sided confidence limits ought to be used on the basis of the physical or technical problem concerned.

The sample functions for point estimation and for confidence estimation will be given when dealing with the theoretical distribution functions (see Section 2.3).

2.3. Selected theoretical distribution functions

Mathematical models for random processes produce variates to which certain 'theoretical' distribution functions are assigned (see Section 2.2.1). Starting with a mathematical model, we have in the following compiled those theoretical distribution functions that have been introduced into, and

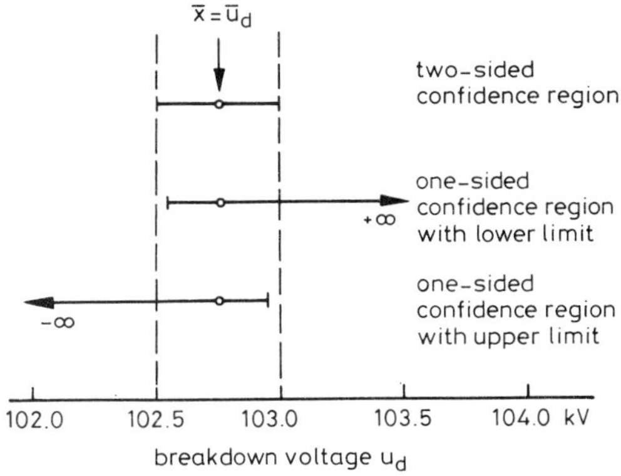

Fig. 2.7 Illustration of one-sided and two-sided confidence regions

proved their worth in, high-voltage engineering. In addition to these, we will be considering distribution functions that contribute to basic comprehension (Sections 2.3.1.1, 2.3.1.2, 2.3.2.1) or play a special role in estimation and test methods (Sections 2.3.2.6 to 2.3.2.9). The presentation of the individual distribution functions is deliberately sketchy, to make the section suitable as a source of reference. The table references relate in every case to the extremely clear and user-friendly *Mathematical Statistics Tables* of Müller, Neumann and Storm, which no high-voltage laboratory or high-voltage test plant should be without. The presentation is rounded-off with details of parameter estimation from samples and with advice concerning applications in high-voltage engineering.

2.3.1 Discrete variates

2.3.1.1. Single-point distribution

Model: The variate X can only assume a fixed value $x = a$, the entire unit mass of probability being concentrated in a (Fig. 2.8). Since this is not a genuine random (stochastic) event, it must be dealt with deterministically.

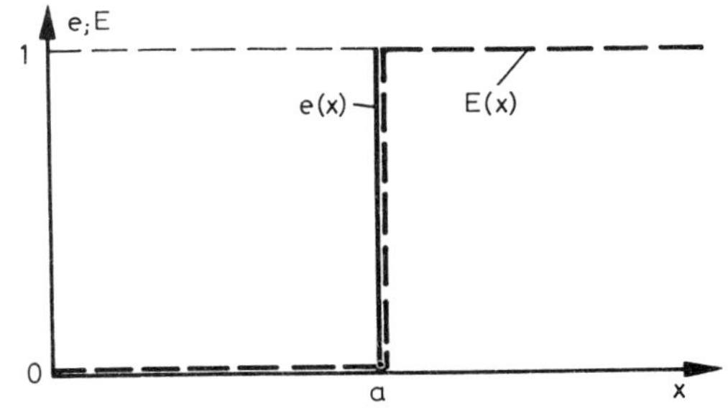

Fig. 2.8 Single-point distribution

Individual probability $e(x)$
Cumulative probability $E(x)$

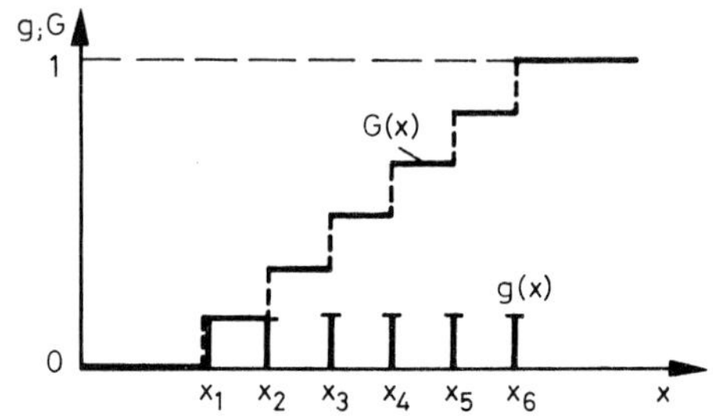

Fig. 2.9 Uniform distribution with parameter $n=6$
Individual probability $g(x)$
Cumulative probability $G(x)$

Individual probability (density function) $e(x)$:
Probability table

$$X:\begin{pmatrix} a \\ 1 \end{pmatrix}, \qquad (2.52a)$$

i.e., $p_1 = P(X=a) = 1$.

Cumulative probability (distribution function):

$$E(x) = \begin{cases} 0 & x \leqq a \\ 1 & x > a \end{cases}. \qquad (2.52b)$$

Parameter: a.

Expectation: $EX = a$.

Variance: $D^2X = 0$.

Application: When using the 'enlargement law' (see Section 6), distribution functions with an initial value x_0[1] converge towards a single-point distribution at x_0 for enlargement factors $n \to \infty$.

2.3.1.2. Discrete uniform distribution
Model: The variate X can assume n different values x_i $(i=1, \ldots, n)$ whose probabilities are all $p_i = 1/n$ (Fig. 2.9).

Individual probability (density function) $g(x)$: Probability table

$$X:\begin{pmatrix} x_1 & x_2 & \ldots & x_n \\ \dfrac{1}{n} & \dfrac{1}{n} & \ldots & \dfrac{1}{n} \end{pmatrix} \qquad (2.53)$$

[1] At the initial value x_0, a distribution function limited on the left-hand side assumes a function value greater than zero, i.e.

$$F(x) = 0 \text{ for } x \leqq x_0$$

$$F(x) > 0 \text{ for } x \geqq x_0 + \varepsilon \text{ with } \varepsilon > 0.$$

Cumulative probability (distribution function $G(x)$: Cumulative probability table

$$X: \begin{pmatrix} x_1 & x_2 & \dots & x_n \\ \frac{1}{n} & \frac{2}{n} & \dots & 1 \end{pmatrix}$$

with $$x_1 < x_2 < \dots x_n.$$

Parameters: $n, x_i \ (i = 1, \dots, n)$.

Expectation:

$$EX = \frac{1}{n} \sum_{i=1}^{n} x_i.$$

Variance:

$$D^2X = \frac{1}{n} \sum_{i=1}^{n} x_i^2 - (EX)^2.$$

Application: The number thrown with an 'ideal' dice is an example of a uniformly distributed discrete variate with $n = 6$. Uniform distributions are extremely important when generating random numbers (Table 21*a* of Müller *et al*). Random numbers are needed when sampling, for example. Generally speaking, one always has to reckon with uniform distributions when it is always one out of a finite number of equally probable random events that occurs. Hence the applications in the calculation of reliability.

2.3.1.3. Binomial distribution

Model: Using a so-called Bernouilli trial diagram, two complementary events A and $\bar{A}$ (e.g. breakdown and non-breakdown), occurring with probabilities $P(A) = p$ and $P(\bar{A}) = q$, $(p + q = 1)$, are investigated. The binomial distribution indicates the probability with which event A will occur precisely k times in n independent trials. Let us illustrate this by considering the following cases:

$$n = 1 \quad \left.\begin{aligned} P(k=1) &= p \\ P(k=0) &= 1 - p = q \end{aligned}\right\} p + q = 1$$

$$n = 2 \quad \left.\begin{aligned} P(k=2) &= P(A \cap A) = p^2 \\ P(k=1) &= P[(A \cap \tilde{A}) + (\tilde{A} \cap A)] = 2pq \\ P(k=0) &= P(\tilde{A} \cap \tilde{A}) = q^2 \end{aligned}\right\} (p+q)^2 = 1$$

$$n = 3 \quad (p+q)^3 = 1$$

$$\vdots$$

$$n \qquad (p+q)^n = 1.$$

In the generalisation of this derivation, the desired individual probabilities are obtained for any n and k $(k \leqq n)$, using the binomial coefficients

$$\binom{n}{k} = \frac{n!}{k!(n-k)!} = \frac{n(n-1) \cdot \ldots \cdot (n-k+1)}{1 \cdot 2 \cdot \ldots \cdot k}.$$

Individual probability (density function):

$$P(X=k)=\binom{n}{k}p^k(1-p)^{n-k}; \qquad k=0,1,2,\ldots,n. \tag{2.54}$$

Cumulative probability (distribution function):

$$P(X\leqq k)=\sum_{m=0}^{k}\binom{n}{m}p^m(1-p)^{n-m}. \tag{2.55}$$

Parameters: n (number of cases considered) p (probability of occurrence of A).

Expectation:

$$EX=np. \tag{2.56}$$

Variance:

$$D^2X=np(1-p). \tag{2.57}$$

Example: Fig. 2.10.

Table reference: Tables 1*a* and 1*b* of Müller *et al.*, indication of individual and cumulative probabilities from $n=1$ to $n=25$.

Point estimate for p: the relative frequency $h_n=k/n$ (eqn. 2.1) from n trials in which event A occurred k times is determined as a point estimate for parameter p (probability).

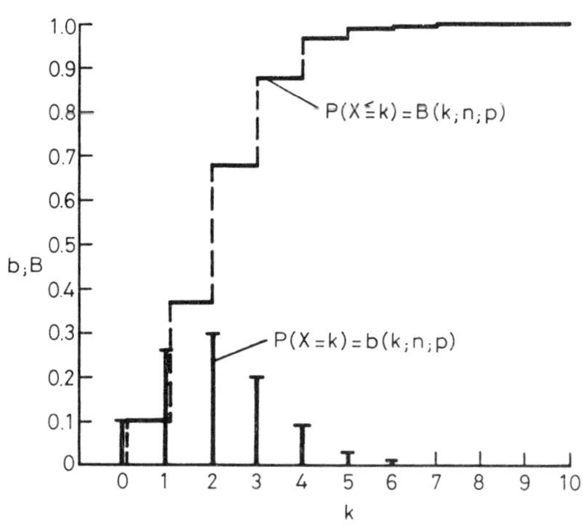

Fig. 2.10 Binomial distribution with parameters $p=0.2$ and $n=10$

Individual probability $P(X=k)=b(k;n;p)$
Cumulative probability $P(X\leqq k)=B(k;n;p)$

Confidence estimate for p:

(i) The F distribution (see Section 2.3.2.8) produces the one-sided confidence region with an upper limit

$$p_o = \frac{(k+1)F_{m_1;m_2;\varepsilon}}{n-k+(k+1)F_{m_1;m_2;\varepsilon}}, \tag{2.58}$$

where $F_{m1;m_2;\varepsilon}$ is the quantile of the F distribution of order $q=\varepsilon$ (ε = confidence coefficient) and with degrees of freedom $m_1 = 2(k+1)$ and $m_2 = 2(n-k)$($I = [0; p_o]$), as well as a one-sided confidence region with a lower limit

$$p_u = \frac{k}{k+(n-k+1)F_{m_3;m_4;\varepsilon}} \tag{2.59}$$

[$m_3 = 2(n-k+1)$; $m_4 = 2k$]. In this case $I = [p_u; 1]$.

(ii) A two-sided confidence region after Neumann is tabulated in [2.11, Table 1*c*] for $n = 1$ to $n = 30$. Larger values of n are covered by the asymptotic relationship:

$$\left.\begin{matrix} p_u \\ p_o \end{matrix}\right\} = \frac{1}{n+\lambda_q^2}\left(k+\frac{\lambda_q^2}{2} \mp \lambda_q\sqrt{\frac{k(n-k)}{n}+\frac{\lambda_q^2}{4}}\right), \tag{2.60}$$

where λ_q is the quantile of the standardised normal distribution (see Section 2.3.2.2) of order $q = (1+\varepsilon)/2$.

These confidence limits are represented graphically in Fig. 2.11. They can be taken directly from this nomogram.

Example 2.3.1.3.1 $k = 20$ breakdowns are recorded in constant-voltage tests with $n = 100$. We are to estimate breakdown probability p as a parameter of the binomial distribution:

(i) Point estimate (eqn. 2.1): $h_n = 20/100 = 0.2$.
(ii) Confidence estimate with F distribution ($\varepsilon = 0.95$) (eqns. 2.58 and 2.59):
Degrees of freedom: $m_1 = 42$; $m_2 = 160$; $m_3 = 162$; $m_4 = 40$
Quantiles: $F_{42;160;0.95} = 1.47$; $F_{162;40;95} = 1.56$
One-sided confidence region with upper limit: $I = [0; 0.278]$
One-sided confidence region with lower limit: $I = [0.137; 1]$.
(iii) Confidence region after Neumann ($\varepsilon = 0.95$): the two-sided confidence region $I = [0.13; 0.29]$ is derived directly from Fig. 2.11. The same result is supplied by eqn. 2.60.

Application: The binomial distribution is suitable for all testing problems in which yes/no decisions are involved and where the size of sample is small compared with the population.[1] (If sampling has a reactive effect on the

[1] The binomial distribution is precisely applicable when the population is completely unaffected by sampling, i.e. when the samples are returned to the population or the population is infinitely large.

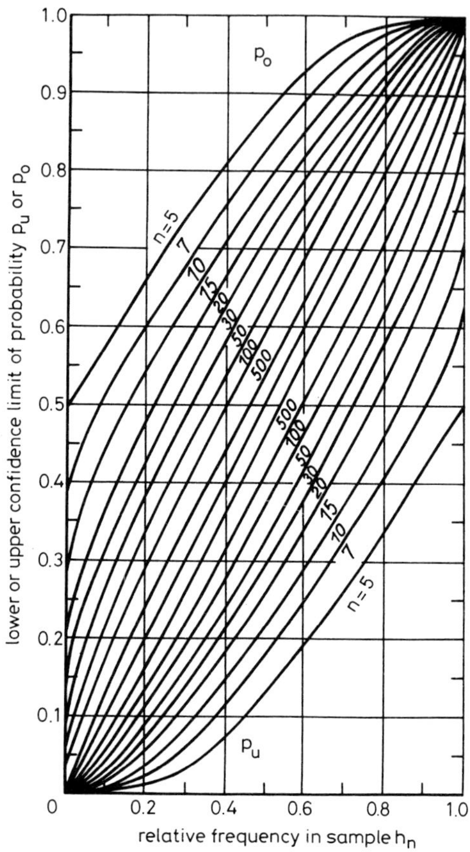

Fig. 2.11 Confidence limits for unknown probabilities after Neumann, with a confidence coefficient $\varepsilon = 0.95$ (2.11, Table 1*c*, p. 54)

population, the hypergeometric distribution, related to the binomial distribution, should be used.) Use of the binomial distribution is particularly important in constant-voltage tests with alternative questioning (Example 2.1.1.1 and Section 3.2), as well as for the evaluation of standardised test methods (see Section 4). In such breakdown measurements, which are repeated as often as is desired, the population is infinite.

Example 2.3.1.3.2 According to IEC, the test on self-restoring insulation is regarded as having been successful if two breakdowns at the most occur in 15 applications of the rated impulse withstand voltage (see Section 4.3). How great will the probability be of the test being successful if the breakdown probability at the rated impulse withstand voltage is $p = 0.05$? The test is successful if $k = 0$, $k = 1$ or $k = 2$ breakdowns occur in $n = 15$ voltage applications, i.e. according to the law of addition (eqn. 2.7), it follows that:

$$p_{\text{test}} = p_0(k=0;\ n=15) + p_1(k=1;\ n=15) + p_2(k=2;\ n=15).$$

Instead of the sum of the three individual probabilities, the cumulative frequency

$$p_{\text{test}} = p_{\Sigma}(k \leqq 2;\ n=15)$$

can also be used. It follows from Müller *et al.*, Table 1*a*, that $p_0 = 0.4633$, $p_1 = 0.3658$ and $p_2 = 0.1348$, so that $p_{\text{test}} = 0.9639$. The probability for the success of the test is then derived directly from Müller *et al.*, Table 1*b* as $p_{\Sigma} = 0.9638 = p_{\text{test}}$.

Besides the use of the binomial distribution itself, the confidence estimates for probabilities (Example 2.3.1.3.1) are very important in evaluating high-voltage tests.

2.3.1.4. Poisson distribution

Model: The Poisson distribution is derived from the binomial distribution when n tends towards infinity and at the same time $np = \lambda$ remains constant (Poisson's limiting-value statement):

$$\lim_{\substack{n\to\infty \\ np=\lambda}} \binom{n}{k} p^k (1-p)^{n-k} = \frac{\lambda^k}{k!} e^{-\lambda} \tag{2.61}$$

($k = 0, 1, 2 \ldots$). Since the probability p is very small for large n and $np = \lambda =$ constant, the Poisson distribution describes rare events.

Individual probability (density function):

$$P(X=k) = \frac{\lambda^k}{k!} e^{-\lambda}. \tag{2.62}$$

Cumulative probability (distribution function):

$$P(X \leqq k) = \sum_{m=0}^{k} \frac{\lambda^m}{m!} e^{-\lambda}. \tag{2.63}$$

Parameter: λ (distribution has only one parameter)

Expectation:

$$EX = \lambda. \tag{2.64}$$

Variance:

$$D^2X = \lambda. \tag{2.65}$$

Example: Fig. 2.12.

Table reference: (Müller *et al.*, Tables 2*a* and 2*b*).

Indication of individual and cumulative probabilities for $0.01 \leqq \lambda \leqq 50$ and $0 \leqq k \leqq 80$.

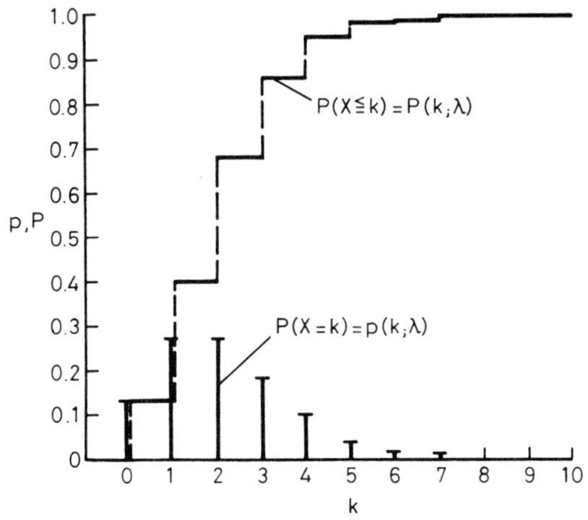

Fig. 2.12 Poisson distribution with parameter $\lambda = 2$

Individual probability $P(X = k) = p(k;\,\lambda)$
Cumulative probability $P(X \leqq k) = P(k;\,\lambda)$

Point estimate for λ: By virtue of eqn. 2.64, an estimated value can be derived directly:

$$\lambda = \frac{1}{n}\sum_{i=1}^{n} x_i. \tag{2.66}$$

Confidence estimate for λ: Using the sum $k = \sum_{i=1}^{n} x_i$ obtained from the sample, confidence factors can be obtained [2.11, Table 2*c*] for the lower ($\underline{\lambda}_{k;q}$) and upper ($\bar{\lambda}_{k;q}$) confidence limits. With $q = \varepsilon$ (confidence coefficient), one-sided confidence regions are obtained:

$$I = \left[\frac{\underline{\lambda}_{k;\varepsilon}}{n};\ \infty\right]; \qquad I = \left[0;\ \frac{\bar{\lambda}_{k;\varepsilon}}{n}\right] \tag{2.67a}$$

and, with $q = (1+\varepsilon)/2$, the two-sided confidence region:

$$I = \left[\frac{\underline{\lambda}_{k;(1+\varepsilon)/2}}{n};\ \frac{\bar{\lambda}_{k;(1+\varepsilon)/2}}{n}\right] \tag{2.67b}$$

The confidence factors are related to the quantiles of the χ^2 distribution (see Section 2.3.2.7), so that it is also possible to indicate confidence regions on the basis of this distribution function.

Application: According to Poisson's limiting-value statement, the binomial distribution can, for sufficiently large values of n and small values of p, be approximated by the Poisson distribution, which is simpler in formula terms. In practice, this is always an advantage when the binomial distribution is no longer available in tabular form ($n > 30$). It is also used in reliability theory and in the theory of elementary particles.

Example 2.3.1.4.1 The breakdown probability of an insulation arrangement is $p = 0.01$. What is the probability of one test piece at the most breaking down in a sample of size $n = 100$?

We need to calculate the cumulative frequency

$$P(k \leqq 1) = P(k = 0) + P(k = 1).$$

The binomial distribution, which is not available in tabular form for $n = 100$, gives

$$P(k \leqq 1, n = 100, p = 0.01) = \binom{100}{0} \cdot 0.01^0 \cdot (0.99)^{100} + \binom{100}{1} \cdot 0.01^1 \cdot (0.99)^{99}$$

$$= 0.3664 + 0.3701 = 0.7365.$$

The Poisson distribution supplies the desired cumulative probability from a table, with $\lambda = pn = 1$, without calculation:

$$P(k \leqq 1, \lambda = 1) = 0.7358.$$

It will be observed that the Poisson distribution gives a sufficiently precise value for the relationship being sought, which could actually be described by the binomial distribution.

2.3.2 Continuous variates

2.3.2.1. Continuous uniform distribution

Model: In an interval $a \leqq x \leqq b$, the variate X occurs with equal probability for all component intervals of equal length (Fig. 2.13).

Density function:

$$f(x) = \begin{cases} 0 & x < a \\ \dfrac{1}{b-a} & a \leqq x \leqq b. \\ 0 & x > b \end{cases} \tag{2.68}$$

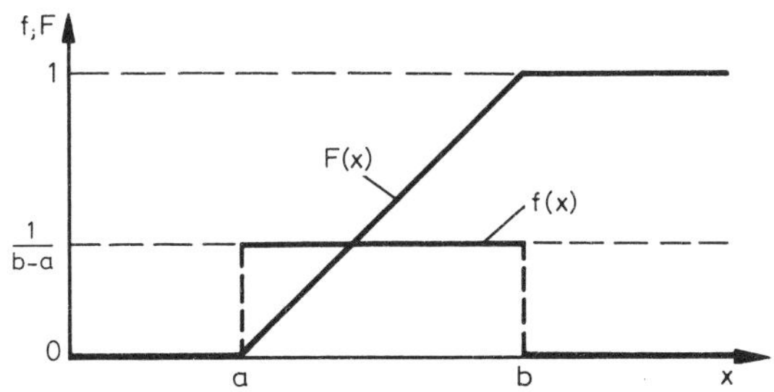

Fig. 2.13 Density function $f(x)$ and distribution function $F(x)$ of a continuous uniform distribution

Parameters a, b

Distribution function:

$$F(x)=\begin{cases} 0 & x<a \\ \dfrac{x-a}{b-a} & a\leqq x\leqq b \\ 1 & x>b. \end{cases} \tag{2.69}$$

Expectation: $EX=\frac{1}{2}(a+b)$.

Variance: $D^2X=\frac{1}{12}(b-a)^2$.

Application: Uniform distributions are always important when investigating physical processes having a more-or-less constant density (e.g. charge-carrier density). When considering discharge processes, the initiatory electrons are often assumed to be uniformly distributed in time and space.

2.3.2.2. Normal distribution

Model: A random process produces a normally distributed variate when the latter can be conceived as the sum of a large number of independent, randomly distributed variates, and when each of these variates makes only an insignificant contribution to the sum (central limiting-value statement). This model, which can be applied to many random phenomena (including discharge processes, measuring error and so on), brings us to the extraordinary significance of the normal distribution, derived from de Moivre (1667–1754), Laplace (1749–1827) and Gauss (1777–1855) (error and compensation calculations).

Density function:

$$\varphi(x;\,\mu;\,\sigma^2)=\frac{1}{\sqrt{2\pi}\sigma}\,e^{-(x-\mu)^2/2\sigma^2} \tag{2.70}$$

Distribution function:

$$\Phi(x;\,\mu;\,\sigma^2)=\frac{1}{\sqrt{2\pi}\sigma}\int_{-\infty}^{x} e^{-(t-\mu)^2/2\sigma^2}\,dt \tag{2.71}$$

Parameters: μ (actual, central value = median = mode), σ^2 (>0, variance; σ standard deviation).

Notation: $N(\mu;\,\sigma^2)=\Phi(x;\,\mu;\,\sigma^2)$.

Expectation: $EX=\mu$ (parameter as well).

Variance: $D^2X=\sigma^2$ (parameter as well).

Standardisation:

$$z=(x-\mu)/\sigma. \tag{2.72}$$

Normal distribution with $\mu=0$; $\sigma^2=1$; $N(0;\,1)$. Since $\Phi(x;\,\mu;\,\sigma^2)=\Phi(z)$ and $\varphi(x;\,\mu;\,\sigma^2)=\varphi(z)/\sigma$ only the $N(0;\,1)$ distribution is tabulated in each case.

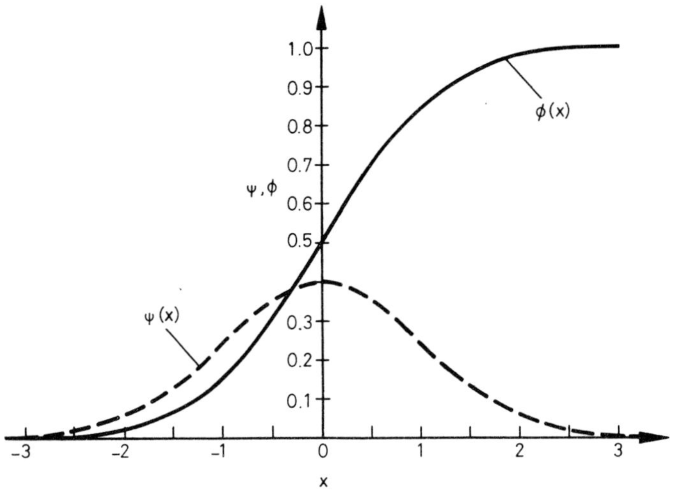

Fig. 2.14 Density function $\phi(x)$ and distribution function $\Phi(x)$ of standard normal distribution

$\mu = 0$; $\sigma^2 = 1$

Example: Fig. 2.14.

Symmetrical properties of the standard normal distribution

$$\varphi(-z) = \varphi(z)$$

$$\Phi(-z) = 1 - \Phi(z)$$

Table reference: In Tables 3*a* to 3*d* in Müller *et al.*, using the symmetrical properties, values are given of the density function (3*a*), distribution function (3*b*) and of quantiles which are particularly useful (3*c*) or which are needed for tests (3*d*).

A few selected quantiles λ_q of order q are given in Table 2.6.

Example 2.3.2.2.1 The breakdown voltage of an air spark-gap is normally distributed with $\mu = 200$ kV and $c = 10$ kV. Find the probability of realisations falling within the interval $170\text{ kV} \leqq u_d \leqq 185\text{ kV}$.

Using the standardisation (eqn. 2.72) and the symmetrical properties, it follows that

$$P(170\text{ kV} \leqq u_d \leqq 185\text{ kV}) = P\left(\frac{170-200}{10} \leqq z \leqq \frac{185-200}{10}\right)$$

$$= P(-3 \leqq z \leqq -1.5)$$

$$= \Phi(-1.5) - \Phi(-3)$$

$$= [1 - \Phi(1.5)] - [1 - \Phi(3)] = \Phi(3) - \Phi(1.5).$$

Table 2.6 Quantiles λ_q of standard normal distribution
Symmetrical property: $\lambda_q = -\lambda_{1-q}$

a)		b)	
Order q	Quantile q	Quantile q	Order q
0.9	1.281 552	0	0.500 000
0.95	1.644 854	0.5	0.691 462
0.975	1.959 964	1	0.841 345
0.99	2.326 348	1.5	0.933 193
0.995	2.575 829	2	0.977 250
0.999	3.090 232	2.5	0.993 790
		3	0.998 650

It then follows from Table 2.6 that $P(170\text{ kV} \leqq u_d \leqq 185\text{ kV} = 0.998650 - 0.933193 = 0.065457$.

Point estimate:

for

$$\mu : \mu^* = \bar{x} = \frac{1}{n} \sum_{i=1}^{n} x_i = x_{50}^* \tag{2.73}$$

for

$$\sigma^2 : \sigma^* = s^2 = \frac{1}{n-1} \sum_{i=1}^{n} (x_i - \bar{x})^2 = (x_{50}^* - x_{16}^*)^2 \tag{2.74}$$

(x_{50}^* and x_{16}^* are quantiles of the empirical distribution function).

Confidence estimate: It is assumed here that the two parameters are unknown: the confidence intervals for μ can then be derived from the t distribution (see Section 2.3.2.9), and for σ^2 from the χ^2 distribution (see Section 2.3.2.7). They are given in Table 2.7.

If the other parameter σ^2 is known at confidence estimation for parameter μ, more accurate confidence regions can be indicated.

Example 2.3.2.2.2 In a rising-voltage test with $n = 10$, a normally distributed breakdown voltage is determined; the point estimates (eqns. 2.73 and 2.74) supply the empirical parameters $\bar{x} = 21.10$ kV and $s = 0.55$ kV. The confidence limits for a confidence coefficient $\varepsilon = 0.95$ are calculated using the formulae given in Table 2.7, the confidence region for μ being calculated 'one-sidedly' with $t_{n-1;\varepsilon} = t_{9;0.95} = 1.83$ and 'two-sidedly' with $t_{n-1;(1+\varepsilon)/2} = t_{9;0.975} = 2.26$, while the confidence region for σ is calculated 'one-sidedly' with $\chi^2_{n-1;1-\varepsilon} = \chi^2_{9;0.05} = 3.33$ and $\chi^2_{n-1;\varepsilon} = \chi^2_{9;0.95} = 16.92$ and 'two-sidedly' with $\chi^2_{n-1;(1-\varepsilon)/2} = \chi^2_{9;0.025} = 2.70$ and $\chi^2_{n-1;(1+\varepsilon)/2} = \chi^2_{9;0.975} = 19.02$. The results are given in Table 2.8.

Table 2.7 Confidence limits for parameters μ and σ^2 of a normal distribution

Parameter	Confidence limits		
	One-sided with upper limit	lower limit	Two-sided
Mean value	$[-\infty; g_0]$	$[g_u; +\infty]$	$[g_u; g_0]$
μ		$g_u = \bar{x} - t_{n-1;\varepsilon}\frac{s}{\sqrt{n}}$	$g_u = \bar{x} - t_{n-1;(1+\varepsilon)/2}\frac{s}{\sqrt{n}}$
$t_{m,q}$ according to Table 2.22	$g_0 = \bar{x} + t_{n-1;\varepsilon}\frac{s}{\sqrt{n}}$		$g_0 = \bar{x} + t_{n-1;(1+\varepsilon)/2}\frac{s}{\sqrt{n}}$
Dispersion	$[0; g_0]$	$[g_u; +\infty]$	$[g_u; g_0]$
σ^2		$g_u = \frac{(n-1)s^2}{\chi^2_{n-1;\varepsilon}}$	$g_u = \frac{(n-1)s^2}{\chi^2_{n-1;(1+\varepsilon)/2}}$
$\chi^2_{m,q}$ according to Table 2.20	$g_0 = \frac{(n-1)s^2}{\chi^2_{n-1;1-\varepsilon}}$		$g_0 = \frac{(n-1)s^2}{\chi^2_{n-1;(1-\varepsilon)/2}}$

Table 2.8 Confidence estimates for parameters of a normal distribution (Example 2.3.2.2.2)

Parameter	Point estimate	Confidence estimates		
		One sided with upper limit	lower limit	Two-sided
Mean value μ	$\bar{u}_d = 21.10$ kV	$[-\infty; \bar{u}_{do1}]$	$[\bar{u}_{du1}; +\infty]$	$[u_{du2}; u_{do2}]$
			$\bar{u}_{du1} = 20.78$ kV	$\bar{u}_{du2} = 20.71$ kV
		$\bar{u}_{do1} = 21.42$ kV		$\bar{u}_{do2} = 21.49$ kV
Standard deviation σ	$s = 0.55$ kV	$[0; s_{o1}]$	$[s_{u1}; +\infty]$	$[s_{u2}; s_{o2}]$
			$s_{u1} = 0.40$ kV	$s_{u2} = 0.38$ kV
		$s_{o1} = 0.90$ kV		$s_{o2} = 1.00$ kV

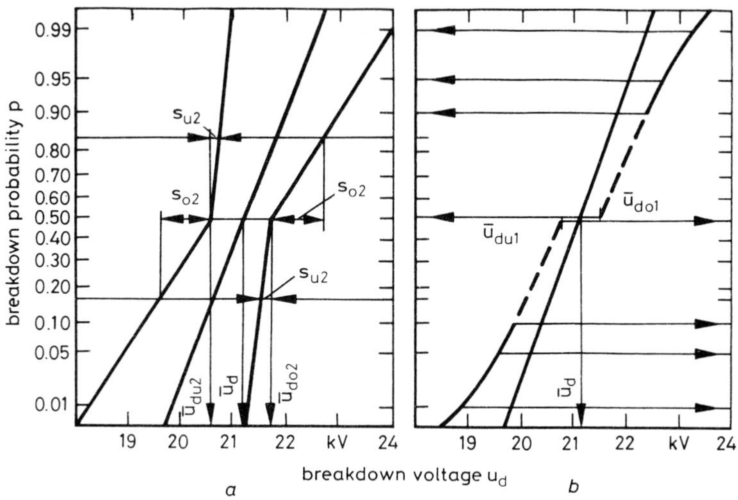

Fig. 2.15 Approximations for confidence limits of a normal distribution

a Superimposition of the least-favourable confidence limits of parameters μ and σ (two-sided)
b One sided tolerance limits according to [2.11]
Probability paper

Tolerance limits: Because of the superimposition of the least-favourable confidence limit, indication of confidence regions for the parameters μ and σ of a normal distribution only gives a rough impression of the confidence region for the distribution function as a whole and consequently an over-estimate (confidence region too wide: see Fig. 2.15*a*, numerical values from Example 2.3.2.2.2).

For statistically certain statements in relation to very small quantiles (e.g. statistical withstand voltages) or very large quantiles (e.g. statistical operating voltages), it would be desirable to have a usable approximation for confidence limits. Since precise confidence estimates of such quantiles are not available in a form prepared for technical application, it would appear to be advantageous at least to use an approximation using the tolerance ranges. The starting point can be the definition of the distribution function (eqn. 2.13) and the definition of a one-sided tolerance range that practically agrees with it: above the lower tolerance limit ($I = [\tau_u; +\infty]$) and below the upper tolerance limit ($I = [-\infty; \tau_0]$) there is at least a proportion γ of the population with a given confidence coefficient β.

For small (close to zero) quantiles of order $(1-\gamma)$, τ_u can therefore be interpreted as the lower one-sided confidence limit, and for large (close to unity) quantiles of order γ, τ_0 can be interpreted as the upper one-sided confidence limit. τ_u, τ_0 are calculated from the estimated parameters of the normal distribution ($\bar{x}$, s) and a tolerance factor $\tilde{k}_{n;\beta;\gamma}$ ([Müller *et al.*, Table

Table 2.9 Tolerance factor $\tilde{k}_{n;\beta;\gamma}$ for normal distribution from [2.11]. One-sided tolerance limits; confidence coefficient $\beta = 0.95$

Size of sample n	Proportion γ or $(1-\gamma)$ 0.90	0.95	0.99
5	3.413	4.209	5.746
10	2.355	2.911	3.981
15	2.069	2.566	3.519
20	1.926	2.396	3.294
30	1.778	2.220	3.063
50	1.646	2.065	2.862
100	1.527	1.927	2.684
500	1.386	1.763	2.475
∞	1.282	1.645	2.326

13*e*], an extract of which has been reproduced in Table 2.9):

$$\left.\begin{matrix}\tau_0\\ \tau_u\end{matrix}\right\} = \bar{x} \pm \tilde{k}_{n;\beta;\gamma} \cdot s. \tag{2.75}$$

For large values of n, the tolerance factor can also be calculated approximately using

$$\tilde{k}_{n;\beta;\gamma} \approx \frac{2(n-1)}{2(n-1)-\lambda_\beta^2}\left[\lambda_\gamma + \frac{\lambda_\beta}{\sqrt{2(n-1)}}\sqrt{2\frac{n-1}{n}+\lambda_\gamma^2-\frac{\lambda_\beta^2}{n}}\right] \tag{2.76}$$

where λ_β ; λ_γ are quantiles of order β and γ of normal distribution $N(0; 1)$.

Example 2.3.2.2.3 The tolerance limits are to be calculated for the empirical normal distribution ($\bar{u}_d = 21.10$ kV; $s = 0.55$ kV) used in Example 2.3.2.2.2 with $n = 10$. With a given confidence coefficient $\beta = 0.95$, the tolerance factors $\tilde{k}_{10;0.95\gamma}$ can be derived from Table 2.9 for $\gamma = 0.90$ as 2.355, for $\gamma = 0.95$ as 2.911 and for $\gamma = 0.99$ as 3.981. The tolerance limits calculated using eqn. 2.75 are reproduced graphically in Fig. 2.15*b*. The broken lines clearly demonstrate that the one-sided tolerance limits coincide with the one-sided confidence region for the mean value. (Details of the calculation and application of tolerance limits can be found in Müller *et al.* Section 13.) Sample sizes can also be planned using distribution-free tolerance limits (see Section 3.3.1).

Application: The normal distribution is of fundamental importance in many statistical estimates and tests. It is often suitable for describing random processes in high-voltage engineering, especially in the case of breakdown in air and flashover on insulators (see Section 5.2), but it can also be applied successfully to other insulation arrangements. Many statistical considerations in high-voltage engineering start with a normal distribution since, in many respects, it represents the distribution function that is easiest for an engineer to handle.

2.3.2.3. Log-normal distribution
Model: A variate $Y > 0$ is termed log-normally distributed when the variate produced by the transformation $X = \log Y$ (or $X = \ln Y$) is normally distributed. The model of the normal distribution consequently applies to the

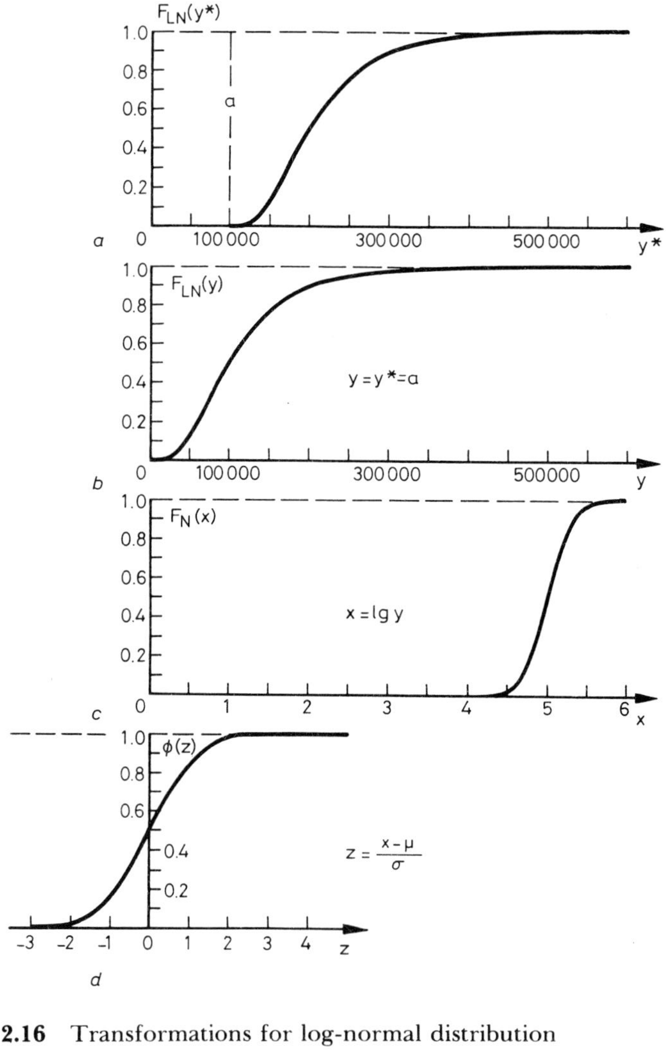

Fig. 2.16 Transformations for log-normal distribution

a Three-parameter log-normal distribution with initial value $a = 100\,000$
b Log-normal distribution ($a = 0$; $\mu = \log 100\,000 = 5$; $\sigma = \log 1.7783 = 0.25$)
c Normal distribution ($\mu = 5$; $\sigma = 0.25$)
d Standard normal distribution ($\mu = 0$; $\sigma = 1$)

transformed variate X. All the relationships available for $N(\mu; \sigma^2)$ (see Section 2.3.2.2) can be applied to all variates X.

If a variate $Y^* = Y + a$ is considered instead of variate Y, one has a three-parameter log-normal distribution with the initial parameter a, which can be converted into a two-parameter distribution by the transformation $Y = Y^* - a$.

Fig. 2.16 shows the transformations needed for the transition from three-parameter log-normal distribution to a standard normal distribution $N(0; 1)$. At the same time, it is evident that an unlimited distribution is produced from the limited log-normal distribution (initial value a for Y^* or 0 for Y).

Density function (two-parameter):

for $X = \ln Y$:

$$f(y) = \underbrace{\begin{cases} 0 & y \leqq 0 \\ \frac{1}{y}\varphi(\ln y;\ \theta;\ \sigma^2) & y > 0 \end{cases}}_{\text{eqn. 2.70}} \tag{2.77a}$$

for $X = \log Y$:

$$f(y) = \begin{cases} 0 & y \leqq 0 \\ \frac{1}{\ln y 10}\varphi(\log y;\ \mu;\ \sigma^2) & y > 0 \end{cases} . \tag{2.77b}$$

Distribution function (two-parameter):
for $X = \ln Y$:

$$F(y) = \underbrace{\begin{cases} 0 & y \leqq 0 \\ \Phi(\ln y;\ \mu;\ \sigma^2) & y > 0 \end{cases}}_{\text{eqn. 2.71}} \tag{2.78a}$$

for $X = \log Y$:

$$F(y) = \begin{cases} 0 & y \leqq 0 \\ \Phi(\log y;\ \mu\ \sigma^2) & y > 0 \end{cases} . \tag{2.78b}$$

Parameters: μ (actual; mean value of transformed variate), σ^2 (> 0; variance of transformed variate).

Expectation:

for $X = \ln Y$: $$EY = \exp\left(\mu + \frac{\sigma^2}{2}\right) \tag{2.79a}$$

for $X = \log Y$: $$EY = \exp(\mu \ln 10 + \tfrac{1}{2}\sigma^2 \ln^2 10) \tag{2.79b}$$

Variance:

for $X = \ln Y$: $$D^2 Y = [\exp(\sigma^2) - 1] \exp(2\mu + \sigma^2) \tag{2.80a}$$

for $X = \log Y$: $$D^2 Y = [\exp(\sigma^2 \ln^2 10) - 1] \exp(2\mu \ln 10 + \sigma^2 \ln^2 10) \tag{2.80b}$$

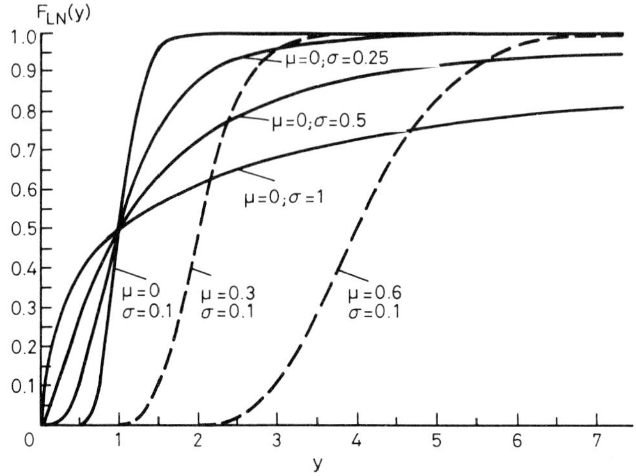

Fig. 2.17 Log-normal distributions of variate Y for different values of μ and σ

Example (Fig. 2.17): Unlike the normal distribution, the log-normal distribution is asymmetrical: variance and expectation are determined by two parameters (μ, σ^2).

Table reference: Since one is only working with the transformed variate, all the tables of the normal distribution $N(0; 1)$ can be used: (Müller *et al.*, Tables 3*a* to 3*d*).

Point estimates and confidence estimates: The parameters are estimated for the distributions of the transformed variates $X = \ln Y$ and $X = \log Y$, and the parameters μ and σ (see Section 2.3.2.2) are calculated for these, using the relationships for the normal distribution.

Application: In high-voltage engineering, preferably for breakdown-time and service-life problems, but nowadays often replaced by the Weibull distribution (see Section 2.3.2.4).

2.3.2.4. Weibull distribution and exponential distribution

Model: The Weibull distribution belongs to the group of extreme-value distributions, all derived from the same model. The realisation is in each case an extreme value (minimum or maximum) of all the possible realisations. For example, the breakdown of parallel insulating clearances always takes place at the (randomly) weakest point in dielectric terms.

To derive the extreme-value distributions, n elementary events with variates X_J are considered, all being identically distributed in accordance with $F_A(x)$. The variate X considered should then give, as the minimum of all those possible:

$$X = \min (X_J).$$

From the multiplication expression (see Section 2.1.3) or from the multiplication expression derived from it (see Section 6), the following is then obtained for the non-occurrence of event X:

$$P(X \geqq x) = [1 - F_A(x)]^n$$

and for its occurrence, and hence for the distribution function of variate X:

$$F_n(x) = P(X < x) = 1 - [1 - F_A(x)]^n. \qquad (2.81)$$

The extreme-value functions are obtained from the boundary transition $n \to \infty$ and simultaneous convergence of $F_A(x) \to 0$

$$\lim_{\substack{n \to \infty \\ F_A(x) \to 0}} \{1 - [1 - F_A(x)]^n\} = F(x). \qquad (2.82)$$

What extreme-value distribution $F(x)$ is produced will depend on $F_A(x)$, the so-called initial distribution. $F_A(x)$ does not have to meet all the requirements of a distribution function, since, because of the boundary transition, it is only of interest in the vicinity of $F_A(x) \to 0$. Here it is sufficient for the function to be growing on the right-hand side, i.e. $F(x_0) = 0$ and $F(x_0 + \varepsilon) > 0$ for $\varepsilon > 0$.

This requirement is satisfied by a great number of functions, but Gnedenko has shown that with any $F_A(x)$ only three types of extreme-value distribution are produced:

- Type I (Unlimited; when applied to minimum problems: double exponential distribution; see Section 2.3.2.5).
- Type II (Limited in an upward direction, unlimited in a downward direction; not used in high-voltage engineering).
- Type III (Limited in a downward direction, unlimited in an upward direction; when applied to minimum problems, Weibull distribution).

The initial distribution of the Weibull distribution can generally be written in the form

$$F_A(x) = \frac{1}{n}\left(\frac{x - x_0}{\eta}\right)^{\delta}$$

where $x > x_0$, $\eta > 0$, $\delta > 0$. For $x_0 = 0$ and $\delta = 1$, an exponential distribution is obtained as a special case of Weibull distribution.

Density function:

$$f(x) \begin{cases} \dfrac{\delta}{\eta}\left(\dfrac{x - x_0}{\eta}\right)^{\delta - 1} \exp\left[-\left(\dfrac{x - x_0}{\eta}\right)^{\delta}\right] & x > x_0 \\ 0 & x \leqq x_0 \end{cases} \qquad (2.83)$$

Special-case exponential distribution:

$$x_0 = 0, \qquad \delta = 1, \qquad \eta = \frac{1}{\lambda}.$$

Distribution function:

$$F(x)=\begin{cases} 1-\exp\left[-\left(\dfrac{x-x_0}{\eta}\right)^{\delta}\right] & x>x_0 \\ 0 & x\leqq x_0 \end{cases} \tag{2.84}$$

Special-case exponential distribution:

$$x_0=0; \qquad \delta=1; \qquad \eta=\frac{1}{\lambda}.$$

Parameters:

$\eta = x_{63}-x_0$ (63% quantile of associated two-parameter distribution)
δ (measure of dispersion, Weibull exponent)
x_0 (initial value).

For $x_0=0$, the two-parameter Weibull distribution is derived from the three-parameter distribution. Because $x_0=0$ and $\delta=1$, the exponential distribution has only one parameter, usually represented by $\lambda(\lambda=1/\eta)$.

Expectation:

$$EX=x_0+\eta\Gamma\left(\frac{1}{\delta}+1\right) \tag{2.85}$$

where $\Gamma(1/\delta+1)$ is the Gamma function available in tabular form.

Exponential distribution: $EX=\eta=1/\lambda$.

Variance:

$$D^2X=\eta^2\left[\Gamma\left(\frac{2}{\delta}+1\right)-\Gamma^2\left(\frac{1}{\delta}+1\right)\right]. \tag{2.86}$$

Exponential distribution: $D^2X=\eta^2=1/\lambda^2$.

Example (Fig. 2.18): The Weibull distribution is highly adaptable in structure, especially in its three-parameter form. Unlike the two-parameter version, the three-parameter distribution represents a linear transformation which, in Fig. 2.18 for example, is expressed by a displacement along the abscissa.

Table 2.10 gives selected quantiles $w_{q;\delta}$ for the reduced Weibull distribution ($\eta=1$; $x_0=0$). Quantiles of the general Weibull distribution (η; x_0) can then be obtained using

$$w_{q;\delta;\eta;x_0}=x_0+\eta w_{q;\delta}. \tag{2.87}$$

Table reference: Müller *et al.*, Tables 15*a* and *b*.

Point estimates: With a Weibull distribution, point estimates are rather complicated. If the initial value is known or is zero, the problem is reduced to a two-parameter distribution with parameters η and δ.

A simple estimate can be obtained from the empirical quantiles, which, in connection with the graphical representation of the sample (see Section

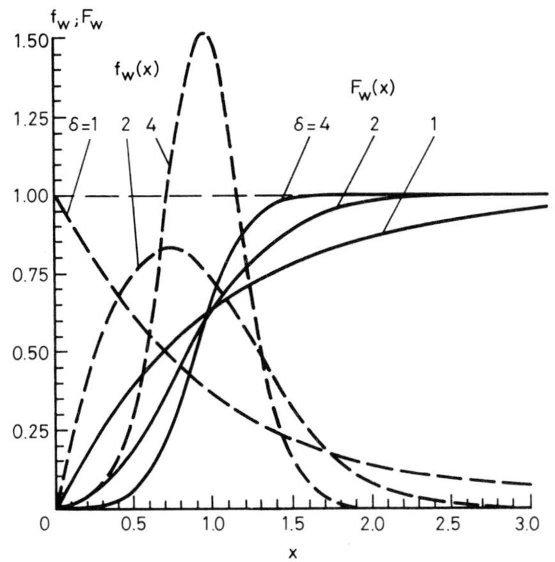

Fig. 2.18 Density function $f_w(x)$ and distribution function $F_w(x)$ of a two-parameter Weibull distribution with parameters $x_0 = 0$ and $\eta = x_{63} = 1$ ($\delta = 1$, exponential distribution)

Table 2.10 Quantiles $w_{q;\delta}$ of reduced Weibull distribution. $\eta = w_{63} = 1$; $x_0 = 0$; δ

Order q	Quantiles $w_{q;\delta}$ for $\delta = 0.5$	$\delta = 1.0$	$\delta = 1.5$	$\delta = 2.0$	$\delta = 3.0$	$\delta = 4.0$
0.01	0.0001	0.0101	0.0466	0.1003	0.2158	0.3166
0.05	0.0026	0.0513	0.1381	0.2265	0.3716	0.4759
0.10	0.0111	0.1054	0.2231	0.3246	0.4723	0.5697
0.20	0.0498	0.2231	0.3679	0.4724	0.6065	0.6873
0.30	0.1272	0.3567	0.5029	0.5972	0.7092	0.7728
0.40	0.2609	0.5108	0.6390	0.7147	0.7994	0.8454
0.50	0.4805	0.6931	0.7832	0.8326	0.8850	0.9124
0.60	0.8396	0.9163	0.9434	0.9572	0.9713	0.9784
0.70	1.4496	1.2040	1.1317	1.0973	1.0638	1.0475
0.80	2.5903	1.6094	1.3734	1.2686	1.1719	1.1263
0.90	5.3019	2.3026	1.7437	1.5174	1.3205	1.2318
0.95	8.9744	2.9957	2.0781	1.7308	1.4416	1.3156
0.99	21.2076	4.6052	2.7680	2.1460	1.6637	1.4649

Table 2.11 Point estimates for Weibull-distribution parameters [2.37]

Parameter	Point estimate	Auxiliary quantity, explanation	
a) Weibull exponent δ	$\delta^* = \frac{\pi}{\sqrt{6} s_y} M$	$\bar{y} = \frac{1}{n} \sum_{i=1}^{n} \ln (x_i - x_0)$ $s_y = \sqrt{\frac{\sum_{i=1}^{n} (\ln (x_i - x_0) - \bar{y})^2}{n-1}}$	Initial value x_0 known
63%-quantile η	$\eta^* = \exp\left(\bar{y} + \frac{C}{\delta^*}\right)$	M = correction factor dependent on size of sample ($\lim_{n\to\infty} M = 1$) (Table 2.12) $C = 0.577\ 226$ Euler's constant	

Parameter	Estimate	Notation	Case
b) 63%-quantile η	$\eta^* = \frac{s}{g_b}$	$\bar{x}$ = arithmetic mean (eqn. 2.41) s = mean square deviation (eqn. 2.48) g_b = correction factor dependent on Weibull exponent δ (Table 2.13) k_b = correction factor dependent on Weibull exponent δ (Table 2.13) $x_1 = \bar{x} - \eta^* k_b$ $x_{\min}$ = minimum realisation in sample	Weibull exponent δ known; preferably $\delta \geqq 1$
Initial value x_0	$x_0^* = \begin{cases} x_1 & \text{for } x_1 < x_{\min} \\ x_{\min} & \text{for } x_1 \geqq x_{\min} \end{cases}$		
Weibull exponent δ known, but $\delta < 1$	$x_0^* = x_{\min}$ is estimated and η is determined by the maximum-likelihood method [2.37]		
c) Weibull exponent δ	δ^*: Table 2.13 is used with $g = \gamma_b$ and the associated table-value of δ is determined as an estimated value	$\bar{x}$ = arithmetic mean (eqn. 2.41) s = mean square deviation (eqn. 2.48) $g = \dfrac{\frac{n}{(n-1)(n-2)} \sum_{i=1}^{n} (x_i - \bar{x})^3}{s^3}$ empirical skewness k_b, g_b, γ_b tabular values dependent on Weibull exponent δ (Table 2.13)	All three parameters unknown
63%-quantile η	$\eta^* = \frac{s}{g_b}$		
Initial value x_0	$x_0^* = \bar{x} - \eta^* k_b$		

2.5.1.1), arise in most high-voltage engineering evaluations:

$$\eta^* = x_{63}, \tag{2.88}$$

$$\delta^* = \frac{\log\left(\ln \dfrac{1}{1-F(x_1)}\right) - \log\left(\ln \dfrac{1}{1-F(x_2)}\right)}{\log\left(\dfrac{x_1}{x_2}\right)} \tag{2.89}$$

and with the special quantiles x_{63} and x_{05}

$$\delta^* = \frac{1.2898}{\log\left(\dfrac{x_{63}}{x_{05}}\right)}. \tag{2.90}$$

$F(x_1)$ and $F(x_2)$ are the orders (i.e. the values of the distribution function) for the empirical quantiles x_1 and x_2 (see Table 2.10 and eqn. 2.84).

Other point estimates can be derived by the moments method (Table 2.11*a*) or the maximum-likelihood method. If the Weibull exponent δ is known, the moments method (Table 2.11*b*) supplies usable estimates for the two remaining parameters (η and x_0) for $\delta \geqq 1$, whereas the maximum-likelihood method should be used for $\delta < 1$.

With the general three-parameter Weibull distribution, estimation of the initial value x_0 is extremely critical, and of great technical significance. Here again we ought first to refer to a graphical method. If realisations of a three-parameter Weibull distribution are available, these are reproduced in the probability grid of the two-parameter distribution (see Section 2.5.1.1), which produces a curved graph that extends into a straight line when the reduction $y = x - x_0^*$ (x_0^* = estimate for x_0) takes place. The estimation then takes place through appropriate assumptions for x_0^* and a graphical check.

Table 2.12 Correction factor for Weibull exponent [2.37]

n	M
5	0.738
7	0.808
10	0.863
15	0.906
20	0.928
30	0.950
40	0.961
50	0.969
60	0.974
80	0.980
100	0.984
120	0.986

If the realisations form a straight line, the assumption for x_0^* is a good estimate of the parameter. As the initial assumption for x_0^*, one should choose a value that, with a small sample size, is about 20% below the smallest realisation, and with a large sample is about 10% below the smallest realisation.

Veverka and Kvasnicka estimate x_0 from three arbitrary quantiles, using the relationship

$$\frac{\log\left(\ln\dfrac{1}{1-F(x_1)}\right)-\log\left(\ln\dfrac{1}{1-F(x_2)}\right)}{\log\left(\ln\dfrac{1}{1-F(x_1)}\right)-\log\left(\ln\dfrac{1}{1-F(x_3)}\right)}=\frac{\log\dfrac{x_1-x_0}{x_2-x_0}}{\log\dfrac{x_1-x_0}{x_3-x_0}}, \tag{2.91}$$

which is solved numerically with respect to x_0.

In a supplementary work, it is shown that the estimated value x_0^* determined in this way is subject to considerable scatter and can even assume values that are physically nonsensical. It is advisable to combine all the realisations by computer for the solution of eqn. 2.91 and to investigate the distribution of the estimated values $F(x_0)$. The mode (modal value) of this function should provide the best estimate.

The moments method offers a more accurate way of estimating three parameters (Table 2.11*c*) if one starts from the consideration that the theoretical skewness γ_b (see Section 2.2.1) is only connected with the Weibull exponent δ. If the empirical skewness g is equated with the theoretical skewness γ_0 (Table 2.11*c*), Table 2.13 provides the estimated value for δ directly. The further calculation process corresponds to that described above for known Weibull exponents δ.

Example 2.3.2.4.1 In a rising-voltage test (see Section 3.3), 39 realisations of the breakdown voltage of an oil spark-gap are obtained, falling on 15 voltage values (Table 2.14 shows the primary distribution table). Voltage steps 2, 6 and 11 are used to calculate the initial value, and the estimated value $u_{d0}^{(1)}=490$ kV is obtained after solving eqn. 2.91. Using transformation $y=u_{di}-u_{d0}$, the realisations in the probability grid can be joined by a straight line (Fig. 2.19, curve *a*). Evaluation of all the possible $N=\binom{15}{3}=455$ combinations shows that 127 combinations provide no solution and a further 80 combinations are physically meaningless ($u_{d0}<0$ or $u_{d0}>u_{d1}$ — smallest realisation). The distribution-function mode of the remaining 248 combinations provides the estimated value $u_{d0}^{(2)}=524$ kV. A straight line can therefore also be indicated on the probability grid (Fig. 2.19, curve *b*). There is an appreciable difference between the two functions, as the parameter estimate using eqns. 2.88 and 2.90 shows

$$F^{(1)}(u_d)\colon\ u_{d0}^{(1)}=490\ \text{kV},\qquad F^{(2)}(u_d)\colon\ u_{d0}^{(2)}=524\ \text{kV}$$

$$\eta^{(1)}=u_{d63}^{(1)}-u_{d0}^{(1)}=123.5\ \text{kV}\qquad \eta^{(2)}=u_{d63}^{(2)}-u_{d0}^{(2)}=88\ \text{kV}$$

$$u_{d63}^{(1)}=613.5\ \text{kV}\qquad u_{d63}^{(2)}=612\ \text{kV}$$

$$\delta^{(1)}=4.162\qquad \delta^{(2)}=2.593.$$

Table 2.13 Principal characteristics of Weibull distribution (correction factors for parameter determination) [2.37]

Weibull exponent δ	Correction factor k_b $k_b = \Gamma(1+1/\delta)$	Correction factor g_b $g_b = \sqrt{\Gamma((1+2/\delta)-k_b^2)}$	Correction factor γ_d $\gamma_b = (1/g_b^3)(\Gamma(1+3/\delta) -3k_b \times \Gamma(1+2/\delta) +2k_b^3)$
0.20	120.0	1901.0	190.1
0.22	56.33	665.1	112.3
0.25	24.00	199.4	60.1
0.30	9.261	50.08	28.33
0.35	5.029	19.98	16.74
0.40	3.323	10.44	11.35
0.45	2.479	6.46	8.413
0.50	2.000	4.472	6.619
0.60	1.505	2.645	4.593
0.70	1.266	1.851	3.498
0.80	1.133	1.428	2.815
0.90	1.052	1.171	2.345
1.00	1.000	1.000	2.000
1.10	0.9649	0.8783	1.734
1.20	0.9407	0.7872	1.521
1.30	0.9236	0.7164	1.346
1.40	0.9114	0.6596	1.198
1.50	0.9027	0.6129	1.072
1.70	0.8922	0.5402	0.865
2.00	0.8862	0.4633	0.6311
2.50	0.8873	0.3797	0.3586
3.00	0.8930	0.3246	0.1681
3.50	0.8997	0.2847	0.0251
4.00	0.9064	0.2543	−0.0872
4.50	0.9126	0.2301	−0.1784
5.00	0.9182	0.2103	−0.2541
6.00	0.9277	0.1798	−0.3733
7.00	0.9354	0.1572	−0.4632
8.00	0.9417	0.1397	−0.5336
10.00	0.9514	0.1145	−0.6378
12.00	0.9583	0.0970	−0.7107
15.00	0.9657	0.0790	−0.7871

Table 2.14 Primary distribution table ($n = 39$) of breakdown voltages of point/plane spark-gap in insulating oil [2.38] $u_{d0}^{(1)} = 490$ kV; $u_{d0}^{(2)} = 524$ kV

Number of voltage step i	Breakdown voltage u_{di} kV	Number of breakdowns k at $u_d \leqq u_{di}$	Relative cumulative frequency h_Σ	Reduced variate	
				$u_{di} - u_{d0}^{(1)}$ kV	$u_{di} - u_{d0}^{(2)}$ kV
1	542.00	1	0.025	52.00	18.00
2	550.46	2	0.050	60.46	26.46
3	558.93	4	0.100	68.93	34.93
4	567.39	7	0.175	77.39	43.39
5	575.86	9	0.225	85.86	51.86
6	584.32	11	0.275	94.32	60.32
7	592.78	14	0.350	102.78	68.78
8	601.25	17	0.425	111.25	77.25
9	609.71	21	0.525	119.71	85.71
10	618.18	26	0.650	128.18	94.18
11	626.64	31	0.775	136.64	102.64
12	635.10	36	0.900	145.10	111.10
13	643.57	37	0.925	153.57	119.57
14	652.03	38	0.950	162.03	128.03
15	660.50	39	0.975	170.50	136.50

The moments method provides the empirical skewness $g = -0.3735$ via $\bar{u}_d = 605.4$ kV, $s = 29.5$ kV (Table 2.11*c*). With $g = \gamma_b$, it follows from Table 2.13 that $\delta = 6.00$, $k_b = 0.9277$ and $g_b = 0.1798$. Using these values, better estimates can be obtained with the formulae in Table 2.11*c*:

$u_{d0}^{(3)} = 453.2$ kV; $\quad \eta^{(3)} = 164.0$ kV; $\quad u_{d63}^{(3)} = 617.2$ kV and $\delta^{(3)} = 6.00$.

Changing one parameter (u_{d0}) produces a change in all the other parameters. It is interesting to observe that the value $u_{d0} + \eta = u_{dm}$ is more or less independent of u_{d0} (deviations are due only to drawing inaccuracies in the graphical solution), whereas δ is greatly influenced by the initial value.

Point estimate for exponential distribution:

$$\lambda = \frac{1}{\eta} = \frac{1}{x_{63}} = \frac{n}{\sum_{i=1}^{n} x_i}. \tag{2.92}$$

Confidence estimates: Confidence estimates for the parameters of a Weibull distribution present extreme difficulties: there are as yet hardly any estimation methods available for the initial value x_0 capable of being generalised. Confidence estimates can be indicated for the other two parameters, for which the calculation formulae are given in Table 2.15 and the requisite quantiles in Table 2.16. Reichelt gives confidence estimates different from these, but their accuracy cannot be regarded as adequately proved.

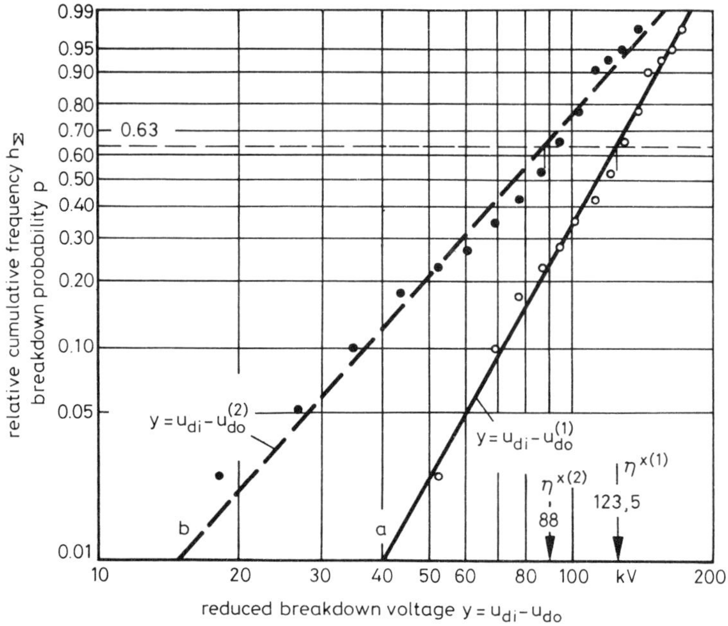

Fig. 2.19 Representation of sample in Table 2.14 as a three-parameter Weibull distribution with initial values (Weibull-distribution probability paper):

a $u_{d0}^{(1)} = 490\ \mathrm{kV}$

b $u_{d0}^{(2)} = 524\ \mathrm{kV}$

Example 2.3.2.4.2 The confidence limits for η and δ with a statistical certainty $q = 0.90$ have to be estimated for the sample in Example 2.3.2.4.1 with sample size $n = 39$ and estimated parameters $u_{d0}^{(3)} = 453.2\ \mathrm{kV}$, $\eta^{(3)} = 164\ \mathrm{kV}$ and $\delta^{(3)} = 6.00$. For a solution with $q = (1+\varepsilon)/2 = 0.95$ and $q = (1-\varepsilon)/2 = 0.05$, and with a sample size $n = 40$, the following can be read off from Table 2.16:

$$W_{40;0.95}^{(1)} = 0.285; \qquad W_{40;0.05}^{(1)} = -0.288;$$
$$W_{40;0.95}^{(2)} = 1.273; \qquad W_{40;0.05}^{(2)} = 0.839.$$

The confidence regions for η can be derived using the formulae in Table 2.15:

$$\left[164\ \mathrm{kV} \exp\left(\frac{-0.285}{6.00}\right);\ 164\ \mathrm{kV} \exp\left(\frac{+0.288}{6.00}\right)\right] = \left[156.4\ \mathrm{kV};\ 172.1\ \mathrm{kV}\right]$$

and for δ:

$$\left[\frac{6.00}{1.273};\ \frac{6.00}{0.839}\right] = [4.713;\ 7.151].$$

Table 2.15 Two-sided confidence limits for parameters η and δ of a Weibull distribution [2.37]
Initial value x_0^* known

Parameter	Confidence limits	Auxiliary quantities, explanation
63%-quantile η	$[g_u; g_o]$ $g_u = \eta^* \exp\left[\frac{-W^{(1)}_{n;(1+\varepsilon)/2}}{\delta^*}\right]$ $g_o = \eta^* \exp\left[\frac{-W^{(1)}_{n;(1-\varepsilon)/2}}{\delta^*}\right]$	$W^{(1)}_{n,q}$ is tabulated in [2.37] Extract in Table 2.16 $5 \leqq n \leqq 120$ $n > 120$; $W^{(1)}_{n,q} = \lambda_q \sqrt{\frac{1.108}{n}}$ λ_q quantile of normal distribution (Table 2.6)
Weibull exponent δ	$[g_u; g_o]$ $g_u = \frac{\delta^*}{W^{(2)}_{n;(1+\varepsilon)/2}}$ $g_o = \frac{\delta^*}{W^{(2)}_{n;(1-\varepsilon)/2}}$	$W^{(2)}_{n,q}$ is tabulated in [2.37] Extract in Table 2.16 $5 \leqq n \leqq 120$ $n > 120$: $W^{(2)}_{n,q} = 1 + \lambda_q \sqrt{\frac{0.608}{n}}$ λ_q quantile of normal distribution (Table 2.6)

Table 2.16 Factors for calculation of confidence limits for parameters of Weibull distribution [2.37]

Size of sample n	Factor $W^{(1)}_{n;q}$ (confidence limit for η) $q =$ 0.02	0.05	0.95	0.98	Factor $W^{(2)}_{n;q}$ (confidence limit for δ) $q =$ 0.02	0.05	0.95	0.98
5	−1.631	−1.247	1.107	1.582	0.604	0.683	2.779	3.518
7	−1.196	−0.874	0.829	1.120	0.639	0.709	2.183	2.640
10	−0.876	−0.665	0.644	0.851	0.676	0.738	1.807	2.070
15	−0.651	−0.509	0.499	0.653	0.716	0.770	1.564	1.732
20	−0.540	−0.428	0.421	0.549	0.743	0.791	1.449	1.579
30	−0.423	−0.338	0.334	0.435	0.778	0.820	1.334	1.429
40	−0.360	−0.288	0.285	0.371	0.801	0.839	1.273	1.351
50	−0.318	−0.254	0.253	0.328	0.817	0.852	1.235	1.301
60	−0.289	−0.230	0.229	0.297	0.830	0.863	1.208	1.267
80	−0.248	−0.197	0.197	0.255	0.848	0.878	1.173	1.222
100	−0.221	−0.174	0.175	0.226	0.861	0.888	1.150	1.192
120	−0.202	−0.158	0.159	0.205	0.871	0.897	1.133	1.171

Application: Together with the normal distribution, the Weibull distribution is the distribution function most widely used in high-voltage engineering, since it is eminently suitable for dealing with service-life problems, even in its two-parameter form. This applies primarily to the handling of breakdown-time problems in solid insulating materials. Application of the Weibull distribution to probems of breakdown voltage or electric strength can, on the one hand, in the case of high-polymer insulating materials, be derived from the service life expression (see Section 5.5.2), while on the other hand 'direct matching' is also possible. In such considerations, however, a usable match is usually only achieved by switching to a three-parameter Weibull distribution. The problems that arise were illustrated in Example 2.3.2.4.1, since a sample (Table 2.14) can be matched by greatly differing Weibull distributions (Fig. 2.19). The initial value, which can only be determined with uncertainty, still has the physical significance of an absolute withstand voltage (or field strength), so that because of this inadequate estimate of the initial value, erroneous technical conclusions are hard to avoid especially when using the 'enlargement law' (see Chapter 6).

If an exponent $\delta > 10$ occurs with the two-parameter Weibull distribution, the random process under investigation can also be conveniently represented by a double-exponential distribution (see Section 2.3.2.5).

2.3.2.5. Double exponential distribution

Model: Since the double exponential distribution belongs to the group of extreme-value distributions, the basic model is identical to that described for the Weibull distribution (see Section 2.3.2.4), the initial distribution now having the form

$$F_A(x) = \frac{1}{n}\exp\left(\frac{x-\eta}{\gamma}\right)$$

with η actual and $\gamma > 0$ (distribution function of minimal).

Density function:

$$f_d(x) = \frac{1}{\gamma}\exp\left(\frac{x-\eta}{\gamma} - \exp\left(\frac{x-\eta}{\gamma}\right)\right) \tag{2.93}$$

Distribution function:

$$F_d(x) = 1 - \exp\left(-\exp\left(\frac{x-\eta}{\gamma}\right)\right) \tag{2.94}$$

Parameters: η (63% quantile; mode), γ measure of dispersion.

Expectation:

$$EX = \eta - \gamma C \tag{2.95}$$

$C = 0.5772$ Euler's constant.

Variance:

$$D^2X = \tfrac{1}{6}\pi^2\gamma^2. \tag{2.96}$$

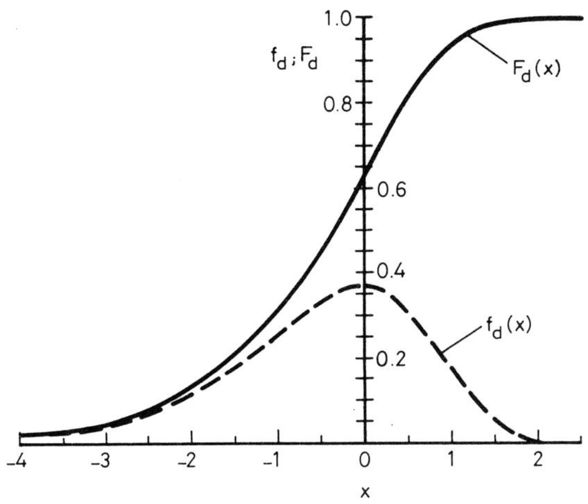

Fig. 2.20 Density function $f_x(x)$ and distribution function $F_d(x)$ of double exponential distribution ($\eta = 0$; $\gamma = 1$)

Example (Fig. 2.20): Using transformation $y = (x - \eta)/\gamma$, the double exponential distribution is converted to the standard form with $\eta = 0$ and $\gamma = 1$ reproduced in Table 2.17. Quantiles of order q of the general double

Table 2.17 Distribution function $F_d(x)$, density $f_d(x)$ and quantiles d_q of standard double exponential distribution. $\eta = x_{63} = 0$; $\gamma = 1$

a) Argument x	Distribution function $F_d(x)$	Density $f_d(x)$	b) Order q	Quantile d_q
−7	0.0009	0.0009	0.001	−6.9079
−6	0.0025	0.0025	0.01	−4.6001
−5	0.0067	0.0067	0.02	−3.9019
−4	0.0181	0.0180	0.05	−2.9702
−3.5	0.0297	0.0293	0.10	−2.2504
−3	0.0486	0.0474	0.20	−1.4999
−2.5	0.0788	0.0756	0.30	−1.0309
−2	0.1266	0.1178	0.40	−0.6717
−1.5	0.2000	0.1785	0.50	−0.3665
−1	0.3078	0.2546	0.60	−0.0874
−0.5	0.4548	0.3307	0.70	0.1856
0	0.6321	0.3679	0.80	0.4759
0.5	0.8077	0.3170	0.90	0.8340
1	0.9340	0.1794	0.95	1.0972
1.5	0.9887	0.0507	0.98	1.3641
2	0.9994	0.0046	0.99	1.5272
3	1.0000	0.0000		

exponential distribution can be calculated from the standard form d_q as

$$d_{q;\eta;\gamma} = d_q\gamma + \eta. \tag{2.97}$$

Point estimate: The parameters can first of all be estimated from quantiles of the empirical distribution function.

$$\eta^* = x_{63} \tag{2.98}$$

$$\gamma^* = \tfrac{1}{3}(x_{63} - x_{05}). \tag{2.99}$$

A point estimate is also possible using the empirical moments (eqns. 2.95 and 2.96), but the limited sample sizes need to be taken into account to obtain estimates in line with expectations. With the sample-size-dependent factors $k_n^{(1)}(\lim_{n\to\infty} k_n^{(1)} = \sqrt{\pi^2/6})$ and $k_n^{(2)}(\lim_{n\to\infty} k_n^{(2)} = C = 0.5772\ldots)$ from Table 2.18, the estimated parameters are:

$$\gamma^* = \frac{1}{k_n^{(1)}}\sqrt{\frac{1}{n-1}\sum_{i=1}^{n}(x_i - x)^2} = \frac{s}{k_n^{(1)}} \tag{2.100}$$

$$\eta^* = \left(\frac{1}{n}\sum_{i=1}^{n} x_i\right) + \gamma^* k_n^{(2)} = \bar{x} + \gamma^* k_n^{(2)}. \tag{2.101}$$

Table 2.18 Factors for point estimate of parameters of a double exponential distribution [2.33]

Size of sample n	Factor for γ^* estimate $k_n^{(1)}$	Factor for η^* estimate $k_n^{(2)}$
8	0.9043	0.4843
10	0.9497	0.4952
12	0.9833	0.5030
15	1.0206	0.5128
20	1.0628	0.5236
25	1.0915	0.5309
30	1.1124	0.5362
35	1.1285	0.5403
40	1.1413	0.5436
45	1.1519	0.5463
50	1.1607	0.5485
60	1.1747	0.5521
70	1.1854	0.5548
80	1.1938	0.5569
100	1.2065	0.5600
200	1.2360	0.5672
300	1.2479	0.5699
500	1.2588	0.5724
1000	1.2685	0.5745
∞	$\pi^2/6 = 1.282\ldots$	$C = 0.5772\ldots$

Confidence limits: Confidence estimates for the parameters of the double exponential distribution are not yet known. It is possible to calculate a 'control region', which in the central region of the double exponential distribution can be regarded as a confidence region.

Application: The double exponential distribution is successfully used to describe the variates breakdown voltage and electric strength, being particularly suitable for compressed-gas insulation. For large Weibull exponents ($\delta \to \infty$), the two-parameter Weibull distribution converges towards the double exponential distribution: as soon as $\delta > 20$, it is more expedient to work with a double-exponential distribution. When using the enlargement law (see Section 6.3), the normal distribution converges towards a double exponential distribution as the enlargement factor increases ($n \to \infty$). This convergent behaviour leads to further possible applications.

2.3.2.6. Double-limit distribution

Model: The unlimited nature of the normal distribution is a disadvantage in a few industrial applications. The normal distribution is therefore sometimes terminated at points $x_u = \mu - k\delta$ and $x_0 = \mu + k\delta$ (k being usually between 2 and 5), or other limits are sought [2.63]. The double-limit distribution presented by Wohlmuth [2.40, 2.64] is mathematically advantageous: it approximates well to the normal distribution over wide ranges, is limited at both ends and has only two parameters (Fig. 2.21).

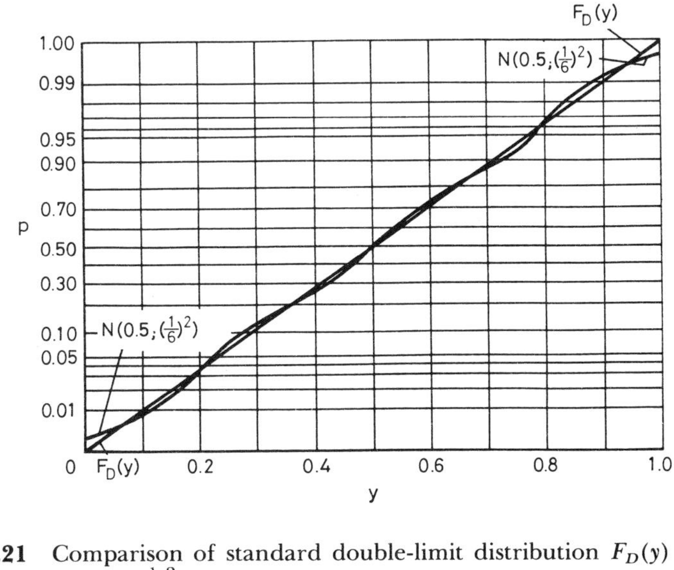

Fig. 2.21 Comparison of standard double-limit distribution $F_D(y)$ and a normal distribution $N(0.5; (\frac{1}{6})^2)$ [2.64]

Probability paper for double-limit distribution

Distribution function:

$$F_D(x)=\begin{cases}0 & \text{for } x\leqq x_0\\ \frac{1}{2}[2^{-((x_{100}-x)/(x-x_0))^{1.635}}-2^{((x-x_0)/(x_{100}-x))^{1.635}}+1] & \text{for } x_0<x<x_{100}.\\ 1 & \text{for } x\geqq x_{100}\end{cases} \quad (2.102)$$

Parameters: x_0 (initial value), x_{100} (final value).

Standardisation: $y_0=0$ and $y_{100}=1$

$$y=\frac{x-x_0}{x_{100}-x_0}; \qquad 1-y=\frac{x_{100}-x}{x_{100}-x_0}.$$

Since the density function is symmetrical with respect to $y=0.5$, it is sufficient to tabulate $F_D(y)$ only in the range $0\leqq y\leqq 0.50$ (Table 2.19).

Expectation:

$$EX=x_{50}=\frac{x_0+x_{100}}{2}.$$

Variance:

$$D^2X=\left(\frac{x_{100}-x_0}{6}\right)^2$$

Point estimate:

$$x_0=\bar{x}-3s \quad (2.103)$$

$$x_{100}=\bar{x}+3s, \quad (2.104)$$

where $\bar{x}$ is the arithmetic mean (eqn. 2.41) and s the mean square deviation (eqn. 2.48).

Application: It is possible to work with the double-limit distribution instead of the normal distribution. It is advisable, however, not to use the enlargement law with the double-limit distribution, since the distribution converges towards a single-point distribution at x_0.

2.3.2.7. Gamma distribution with χ^2 distribution

Model: The distribution function is based on the Gamma function $\Gamma(p)\int_0^\infty x^{p-1}\,e^{-x}\,dx$, available in tabular form.

As a special case of the Gamma distribution, the χ^2 distribution is the distribution of the sum of the squares of n independent $N(0;1)$-distributed variates.

Density function:
Gamma distribution:

$$f(x)=\begin{cases}0 & \text{for } x\leqq 0\\ \dfrac{b^p}{\Gamma(p)}x^{p-1}\,e^{-bx} & \text{for } x>0\end{cases} \quad (2.105)$$

Table 2.19 Standard double-limit distribution after Wohlmuth [2.40] [2.64]

$F_D(y)$ for $0 \leqq y \leqq 0.5$; $1 - F_D(y)$ for $0.5 \leqq y \leqq 1$

y	0	2	4	6	8	10	
0.00	0.00000	0.00001	0.00004	0.00008	0.00013	0.00019	0.99
0.01	0.00019	0.00026	0.00033	0.00041	0.00050	0.00060	0.98
0.02	0.00060	0.00070	0.00081	0.00093	0.00105	0.00118	0.97
0.03	0.00118	0.00131	0.00145	0.00160	0.00176	0.00192	0.96
0.04	0.00192	0.00208	0.00225	0.00243	0.00261	0.00280	0.95
0.05	0.00280	0.00300	0.00320	0.00341	0.00362	0.00384	0.94
0.06	0.00384	0.00406	0.00430	0.00453	0.00477	0.00502	0.93
0.07	0.00502	0.00528	0.00554	0.00580	0.00607	0.00535	0.92
0.08	0.00635	0.00663	0.00692	0.00722	0.00752	0.00783	0.91
0.09	0.00783	0.00814	0.00846	0.00878	0.00911	0.00945	0.90
0.10	0.00945	0.00979	0.01014	0.01050	0.01086	0.01123	0.89
0.11	0.01123	0.01160	0.01198	0.01237	0.01276	0.01316	0.88
0.12	0.01316	0.01357	0.01398	0.01439	0.01482	0.01525	0.87
0.13	0.01525	0.01569	0.01613	0.01658	0.01704	0.01750	0.86
0.14	0.01750	0.01797	0.01845	0.01894	0.01943	0.01992	0.85
0.15	0.01992	0.02043	0.02094	0.02146	0.02199	0.02252	0.84
0.16	0.02252	0.02307	0.02362	0.02418	0.02474	0.02532	0.83
0.17	0.02532	0.02591	0.02650	0.02711	0.02772	0.02835	0.82
0.18	0.02835	0.02898	0.02963	0.03029	0.03096	0.03165	0.81
0.19	0.03165	0.03235	0.03306	0.03379	0.03453	0.03529	0.80
0.20	0.03529	0.03607	0.03686	0.03768	0.03851	0.03936	0.79
0.21	0.03936	0.04024	0.04114	0.04205	0.04300	0.04396	0.78
0.22	0.04396	0.04496	0.04598	0.04702	0.04810	0.04920	0.77
0.23	0.04920	0.05033	0.05150	0.05269	0.05392	0.05518	0.76
0.24	0.05518	0.05647	0.05780	0.05916	0.06056	0.06199	0.75
0.25	0.06199	0.06343	0.06497	0.06652	0.06810	0.06973	0.74
0.26	0.06973	0.07139	0.07310	0.07484	0.07663	0.07846	0.73
0.27	0.07846	0.08032	0.08224	0.08419	0.08618	0.08822	0.72
0.28	0.08822	0.09030	0.09243	0.09459	0.09680	0.09905	0.71
0.29	0.09905	0.10135	0.10369	0.10607	0.10849	0.11096	0.70
0.30	0.11096	0.11346	0.11601	0.11861	0.12124	0.12392	0.69
0.31	0.12392	0.12663	0.12939	0.13219	0.13503	0.13791	0.68
0.32	0.13791	0.14082	0.14378	0.14678	0.14981	0.15289	0.67
0.33	0.15289	0.15600	0.15914	0.16233	0.16555	0.16880	0.66
0.34	0.16880	0.17209	0.17542	0.17878	0.18217	0.18560	0.65
0.35	0.18560	0.18906	0.19255	0.19607	0.19962	0.20321	0.64
0.36	0.20321	0.20682	0.21046	0.21414	0.21784	0.22157	0.63
0.37	0.22157	0.22532	0.22910	0.23291	0.23674	0.24060	0.62
0.38	0.24060	0.24449	0.24839	0.25232	0.25628	0.26025	0.61
0.39	0.26025	0.26425	0.26827	0.27231	0.27637	0.28045	0.60
0.40	0.28045	0.28455	0.28867	0.29280	0.29696	0.30113	0.59
0.41	0.30113	0.30532	0.30952	0.31374	0.31798	0.32223	0.58
0.42	0.32223	0.32650	0.33078	0.33507	0.33938	0.34370	0.57
0.43	0.34370	0.34803	0.35238	0.35673	0.36110	0.36548	0.56
0.44	0.36548	0.36987	0.37427	0.37867	0.38309	0.38752	0.55
0.45	0.38752	0.39195	0.39639	0.40084	0.40530	0.40977	0.54
0.46	0.40977	0.41424	0.41872	0.42320	0.42769	0.43218	0.53
0.47	0.43218	0.43668	0.44119	0.44569	0.45021	0.45472	0.52
0.48	0.45472	0.45924	0.43676	0.46829	0.47281	0.47734	0.51
0.49	0.47734	0.48187	0.48640	0.49093	0.49547	0.50000	0.50
	10	8	6	4	2	0	y

χ^2 distribution:

$$b = \tfrac{1}{2}; \qquad p = m/2$$

$$f_X(x) = \begin{cases} 0 & \text{for } x \leqq 0 \\ \dfrac{(1/2)^{m/2}}{\Gamma(m/2)} x^{(m/2)-1} e^{-x/2} & \text{for } x > 0 \end{cases}. \tag{2.106}$$

Parameters: Gamma distribution: b, p. χ^2 distribution: m (number of degrees of freedom).

The characteristics of the density function and distribution function of the χ^2 distribution are reproduced in Fig. 2.22.

Expectation and variance: Gamma distribution: $EX = p/b$; $D^2X = p/b^2$.
χ^2 distribution: $EX = m$; $D^2X = 2m$.

Table reference: Müller *et al.*, Table 4 contains important quantiles for degrees of freedom in the range $1 \leqq m \leqq 100$. An extract is given in Table 2.20.

Application: The gamma distribution is used in reliability theory. The χ^2 distribution is important in confidence estimates for dispersion (see Section 2.3.2.2) and in tests (see Section 2.5): χ^2 matching test, χ^2 independence test, χ^2 dispersion test, etc.

2.3.2.8. F distribution

Model: If Y and Z are independent χ^2-distributed variates with m_1 and m_2 degrees of freedom respectively, the quotient $X = (Y/m_1)/(Z/m_2)$ is F-distributed with degrees of freedom m_1 and m_2.

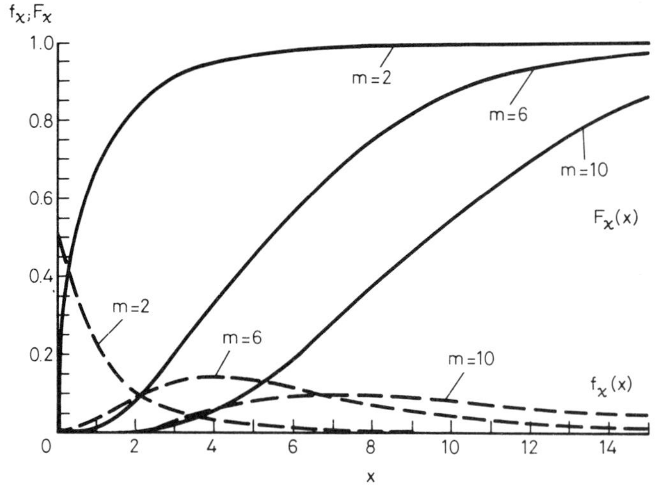

Fig. 2.22 Density function $f_X(x)$ and distribution function $F_X(x)$ of χ^2 distribution with m degrees of freedom

Table 2.20 Quantiles $\chi^2_{m;q}$ of chi-squared distribution

Degree of freedom m	Quantiles $\chi^2_{m;q}$ of order q: $q=0.025$	$q=0.05$	$q=0.95$	$q=0.975$
1	0.001	0.004	3.84	5.02
2	0.051	0.103	5.99	7.38
3	0.261	0.352	7.81	9.35
4	0.484	0.711	9.49	11.14
5	0.831	1.15	11.07	12.83
7	1.69	2.17	14.07	16.01
9	2.70	3.33	16.92	19.02
14	5.63	6.57	23.68	26.12
19	8.91	10.12	30.14	32.85
24	12.40	13.85	36.42	39.36
29	16.05	17.71	42.56	45.72
39	23.65	25.70	54.57	58.12
49	31.55	33.93	66.34	70.22
59	39.66	42.34	77.93	82.12
74	52.10	55.19	95.08	99.68
99	73.36	77.05	123.23	128.42

Density function:

$$f_F(x)=\begin{cases} 0 & \text{for } x\leqq 0 \\ \dfrac{\left(\dfrac{m_1}{m_2}\right)^{m_1/2} x^{(m_1/2)-1}}{B\left(\dfrac{m_1}{2};\dfrac{m_2}{2}\right)}\left(1+\dfrac{m_1}{m_2}x\right)^{-(m_1+m_2)/2} & \text{for } x>0 \end{cases} \tag{2.107}$$

with $B(r, s)=\int_0^1 x^{r-1}(1-x)^{s-1}\,dx$ (Beta function).

Parameters: Degrees of freedom m_1 and m_2.

The density-function characteristic is reproduced in Fig. 2.23.

Expectation:

$$EX=\frac{m_2}{m_2-2} \quad \text{for } m_2>2.$$

Variance:

$$D^2X=\frac{2m_2^2(m_i+m_2-2)}{m_1(m_2-2)^2(m_2-4)} \quad \text{for } m_2>4.$$

Table reference: Müller *et al.* Table 6 contains important quantiles for degrees of freedom in the range $1\leqq m_1\leqq\infty$ and $1\leqq m_2\leqq\infty$. An extract is given in Table 2.21.

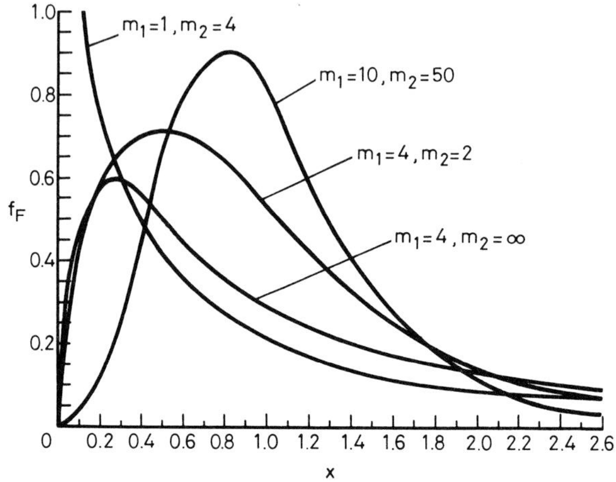

Fig. 2.23 Density function $f_F(x)$ of F distribution with m degrees of freedom

Table 2.21 Quantiles $F_{m1;m2:q}$ of F distribution

Degree of freedom m_2	Degree of freedom m_1: 6	8	10	13	16	20
	a) Order $q=0.95$					
6	4.28	4.15	4.06	3.98	3.92	3.87
8	3.58	3.44	3.35	3.26	3.20	3.15
10	3.22	3.07	2.98	2.89	2.83	2.77
13	2.92	2.77	2.67	2.58	2.51	2.46
16	2.74	2.59	2.49	2.40	2.33	2.28
20	2.60	2.45	2.35	2.25	2.18	2.12
	b) Order $q=0.975$					
6	5.82	5.60	5.46	5.33	5.24	5.17
8	4.65	4.43	4.30	4.16	4.08	4.00
10	4.07	3.85	3.72	3.58	3.50	3.42
13	3.60	3.39	3.25	3.11	3.03	2.95
16	3.34	3.12	2.99	2.85	2.76	2.68
20	3.13	2.91	2.77	2.64	2.55	2.46

Application: The F distribution is important in statistical tests (see Section 2.5.2.1): comparison of dispersion (F test), variance analysis, mean-value comparison, etc. [2.11].

2.3.2.9. t distribution

Model: If Y and Z are independent variates and if Y is $N(0;1)$-normally distributed and Z χ^2-distributed with m degrees of freedom, the quotient $X = Y/\sqrt{Z/m}$ has a t distribution with m degrees of freedom.

Moreover, the quotient of two independent, $N(0;1)$-normally distributed variates is t-distributed with $m = 1$ degree of freedom.

Density function:

$$f_t(x) = \frac{\Gamma\left(\frac{m+1}{2}\right)\left(1+\frac{x^2}{m}\right)^{-(m+1)/2}}{\Gamma\left(\frac{m}{2}\right)\sqrt{\pi \cdot m}}. \tag{2.108}$$

Parameter: m (degree of freedom).

The t distribution is symmetrical and unlimited: for $m \to \infty$ it converges towards a normal distribution $N(0;1)$ (Fig. 2.24).

Expectation:

$$EX = 0 \qquad (m > 1).$$

Variance:

$$D^2X = \frac{m}{m-2} \qquad (m > 2).$$

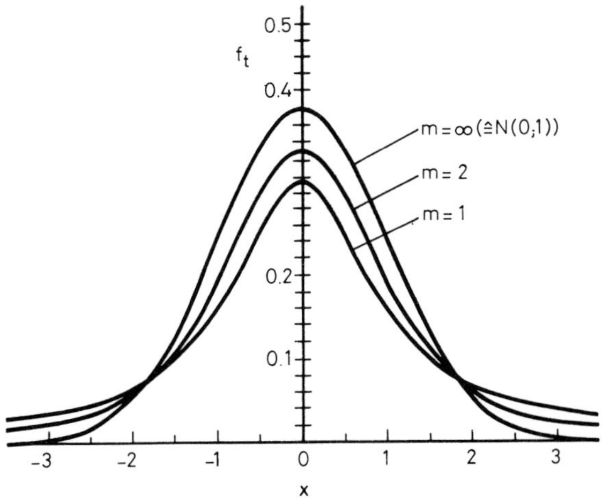

Fig. 2.24 Density function $f_t(x)$ of t distribution with m degrees of freedom

Table 2.22 Quantiles $t_{m;q}$ of t distribution

Degree of freedom m	Quantiles $t_{m;q}$ of order q $q=0.95$	$q=0.975$	$q=0.99$
1	6.31	12.7	31.8
2	2.92	4.30	6.96
3	2.35	3.18	4.54
4	2.13	2.78	3.75
5	2.02	2.57	3.37
7	1.89	2.36	3.00
9	1.83	2.26	2.82
14	1.76	2.14	2.62
19	1.73	2.09	2.54
24	1.71	2.06	2.49
40	1.68	2.02	2.42
120	1.66	1.98	2.36
∞	1.64	1.96	2.33

Table reference: Müller *et al.* Table 5 contains important quantiles $t_{m;q}$ of the t distribution for $1 \leqq m \leqq \infty$. Because of the symmetry of the distribution function, $t_{m;1-q} = -t_{m;q}$ applies. An extract is given in Table 2.22.

Application: The t distribution is widely used in confidence estimations and tests: confidence estimates for the mean value (see Section 2.3.2.2); testing the regression coefficient (see Section 2.4.3); comparison of mean values (see Section 2.5.2.2); independence tests, etc.

2.3.3 Mixed distributions

Many empirically obtained distribution functions represent a mixture of two or more theoretical distribution functions.

Example 2.3.3.1 The breakdown voltage of horizontally arranged coaxial SF_6 insulation is being considered, in which there is a disturbance in the form of a freely moving particle.

With direct voltage, the particle moves and initiates breakdown when it is located near to the inner electrode. The particle-ignited DC breakdown voltage is normally distributed (Fig. 2.25: $F(u_{dG})$). With switching voltage, the particle does not move: breakdown takes place as it does in an undisturbed field and at appreciably higher voltages than with direct voltage. The distribution function of the switching breakdown voltage is again approximated by a normal distribution (Fig. 2.25: $F(u_{dS})$). With a 'mixed voltage', consisting of DC and switching-voltage components, the breakdown process is determined by the instant of superimposition of the two voltages. If, as a result of its movement due to the DC component, the particle is situated by chance near to the inner electrode, breakdown will take place as it does with pure direct voltage; if it is near to the outer electrode, breakdown will take place as it does with pure switching voltage. The

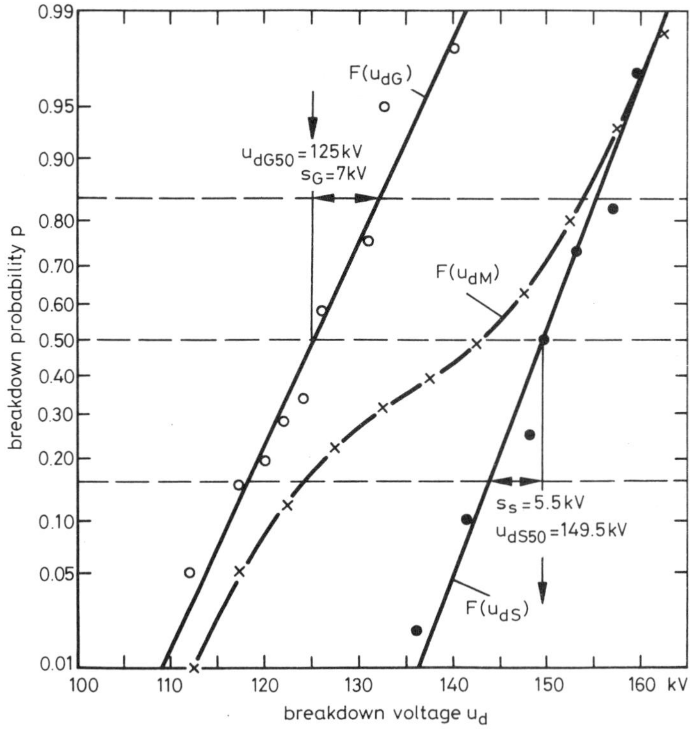

Fig. 2.25 Example of a mixed distribution

Breakdown of a coaxial cylinder arrangement in SF_6 with elliptical particles

$F(u_{dG})$ distribution function for direct voltage
$F(u_{dM})$ distribution function for mixed voltage
$F(u_{dS})$ distribution function for switching voltage

Normal-distribution probability grid

distribution function for mixed voltage is consequently a mixed distribution (Fig. 2.25: $F(u_{dM})$). Mixed distributions are likewise produced if, in particular insulation arrangements, breakdown can take place both with and without partial discharge or when breakdown is preceded by various forms of partial discharge.

The physical causes of such additively composed mixed distributions

$$F(x) = \sum_{i=1}^{j} \alpha_i F_i(x) \quad \text{with} \sum_{i=1}^{j} \alpha_i = 1 \tag{2.109}$$

are different mechanisms (described by the distribution functions $F_i(x)$), in accordance with which the entire process (described by the distribution functions $F(x)$) can take place.

Table 2.23 Evaluation of a mixed distribution (Fig. 2.26) with a sample size $n = 199$

Class number	Class limits		Absolute frequency	Relative cumulative frequency	Relative frequency	Relative frequency Universe I	Relative frequency Universe II	Relative cumulative frequency Universe I	Relative cumulative frequency Universe II
k	u_{dku}	u_{dko}	h_{mk}	$h_{\Sigma k}$	h_k	h_{kI}	h_{kII}	$h_{\Sigma kI}$	$h_{\Sigma II}$
1	>107.5	112.5	2	0.010	0.010	0.010	—	0.024	—
2	>112.5	117.5	4	0.050	0.040	0.040	—	0.119	—
3	>117.5	122.5	12	0.120	0.070	0.070	—	0.286	—
4	>122.5	127.5	20	0.220	0.100	0.100	—	0.525	—
5	>127.5	132.5	19	0.315	0.095	0.093	0.002	0.747	0.003
6	>132.5	137.5	15	0.390	0.075	0.062	0.013	0.895	0.026
7	>137.5	142.5	20	0.490	0.100	0.032	0.068	0.971	0.144
8	>142.5	147.5	28	0.630	0.140	0.010	0.130	0.995	0.370
9	>147.5	152.5	34	0.800	0.170	0.002	0.168	0.999	0.662
10	>152.5	157.5	26	0.930	0.130	—	0.130	—	0.888
11	>157.5	162.5	11	0.985	0.055	—	0.055	—	0.983
12	>162.5	167.5	2	0.995	0.010	—	0.010	—	0.999
					0.995	$\alpha_I = 0.419$ $\alpha \approx 0.42$	$\alpha_{II} = 0.576$ $1 - \alpha \approx 0.58$		

Besides these 'additive' mixed distributions, we also often speak of 'multiplicative' mixed distributions. This involves the application of the model described for the Weibull distribution (see Section 2.3.2.4) which is based on the multiplication expression of probability calculation (see Section 2.1.3). Multiplicative mixed distributions are therefore dealt with as an application of the enlargement law (see Sections 6.3 and 6.4).

The additive mixed distributions to be considered here are mathematically very difficult to process. Because of the five parameters involved (μ_I; σ_I; μ_{II}; σ_{II}; α)[1], mathematical breakdown of a mixed distribution consisting, say, of two normal distributions, is very complicated and subject to great uncertainty. There consequently remain empirical methods, which can be employed either using analogue computers or simply graphically using probability paper. These graphical methods are based on the empirical density function $f^*(x)$, which is divided by eye into density functions $f_i^*(x)$ of a particular type of distribution, in such a way that the initial density $f^*(x)$ is again obtained from the superimposition of $f_i^*(x)$. Matters are facilitated, for example, by the fact that, when using normal-distribution probability paper, the branches of the empirical density function $f^*(x)$ are almost normally distributed if the component densities $f_i^*(x)$ are also normally distributed. In a graphical approximation, these straight branches should be the starting point.

Example 2.3.3.2 The mixed distribution in Example 2.3.3.1 is broken down into its constituent parts (Fig. 2.25). The characteristic of the mixed distribution $F(u_{dM})$ and the model described suggest that $F(u_{dM})$ is composed of two normal distributions. Since the starting point has to be the empirical density function, the measured values are recorded in a frequency table (see Section 2.2.2) and processed (Table 2.23). The empirical density function is graphically represented (Fig. 2.26) and divided by eye into two 'universes' (populations). The relative frequency of each component universe is entered in the table. The sums of the frequencies supply the component parameters $\alpha_I = 0.42 = \alpha$ and $\alpha_{II} = 0.58 = (1-\alpha)$. The cumulative frequencies are then formed for each universe, according to

$$h_{\Sigma k} = \frac{1}{\alpha_{I;II}} \sum_{l=1}^{k} h_l \tag{2.110}$$

(Table 2.23), and represented in order to determine the parameters graphically as quantiles (u_{d50I} ; s_I ; u_{d50II} ; S_{II}). The result obtained is the following mixed distribution consisting of two normal distributions:

$$F(u_{dm}) = 0.42 \cdot \Phi(u_d;\, 127\text{ kV};\, (7.5\text{ kV})^2) + 0.58 \cdot \Phi(u_d;\, 149.5\text{ kV};\, (6.5\text{ kV})^2).$$

The parameters of the component distributions are in good agreement with the distribution functions for pure direct voltage or pure switching voltage (cf. Figs. 2.25 and 2.26).

[1] Parameter α is derived from eqn. 2.109 with $\alpha_I = \alpha$ and $\alpha_{II} = 1-\alpha$.

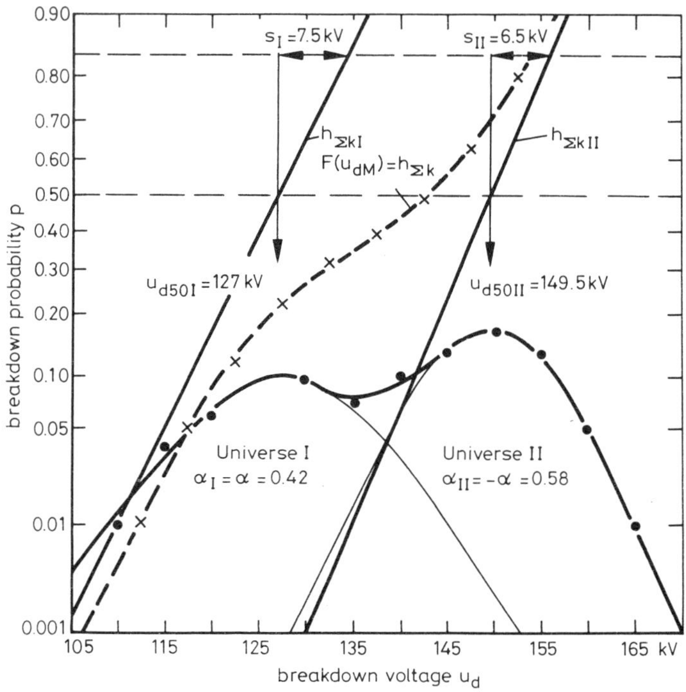

Fig. 2.26 Analysis of mixed distribution $F(u_{dM})$ from Fig. 2.25

Because of the mathematical difficulties associated with mixed distributions, they should only be used when physically essential, i.e. when the physical model of the random process being investigated, as described in Example 2.3.3.1, produces a mixed distribution. One should on no account interpret an empirically-found relationship from the outset as a mixed distribution without such a model. No doubt many relationships in nature and technology intrinsically follow mixed distributions, but in the end, in the areas that matter, it is usually individual influences that dominate, so that mixed components can be ignored. For example, when applying the enlargement law (see Section 6) to the mixed distribution evaluated in Example 2.3.3.2, it would be sufficient just to consider the normal distribution associated with universe *I*.

2.4 Basics of correlation and regression

2.4.1 Concepts and principles

If several random features (e.g. partial-discharge intensity at a certain test-voltage and level of breakdown voltage) are measured simultaneously on test pieces, we are interested in whether these features are linked to

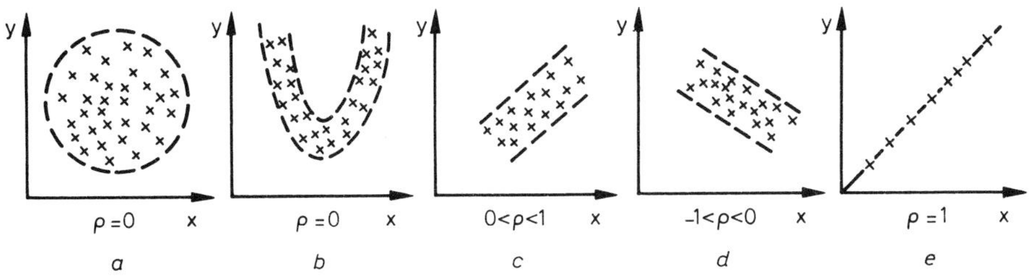

Fig. 2.27 Location of realisations of features X and Y (diagrammatic)

a Features uncorrelated
b Features linearly uncorrelated
c Positive correlation
d Negative correlation
e Complete correlation (functional relationship)

each other, how strong the link is and how it can be mathematically formulated. Correlation and regression provide a solution to these problems.

Correlation analysis first examines the question of whether there are any linear relationships between the random features X and Y being investigated, and how strong they are, by reference to a correlation coefficient ρ. It is assumed that the variates X and Y are normally distributed. The correlation coefficient can assume values $|\rho| \leqq 1$. X and Y are uncorrelated for $\rho = 0$ (Fig. 2.27*a*, *b*), i.e. there is no linear relationship between the two. The nearer $|\rho|$ is to unity, the more strongly correlated the features will be. When $\rho > 0$, X and Y increase or decrease together: this is 'positive correlation' (Fig. 2.27*c*). When $\rho < 0$, large values of Y are linked to small values of X: there is negative correlation (Fig. 2.27*d*). $|\rho| = 1$, represents complete correlation, a perfect functional dependence (Fig. 2.27*e*). The determination of estimated values r for the correlation coefficient ρ from samples with n pairs of values $(x_i\,;\,y_i)$ is dealt with in Section 2.4.2.

Regression analysis investigates, by reference to samples, the possibilities of a functional link, on the one hand between variates X and Y, and on the other between a variate and parameters (e.g. dependence of breakdown time on applied voltage). The two problems with a different content are treated mathematically in the same way. In many high-voltage engineering applications, one is particularly interested in dependence on one parameter, so we shall be considering this case in particular in the following. In the simplest case, one would represent the pairs of values $(x_i\,;\,y_i)$ graphically for the regression (x_i would then, for example, be the parameter, and y_i the realised variate). We shall here consider only the case where there is a linear relationship between the features X and Y, or one can be produced by simple transformation $X^* = f_1(X)$ and $Y^* = f_2(Y)$ (linear regression).

Example 2.4.1.1. In the investigation of the relationship between applied voltage u_d and breakdown time t_d of high-polymer insulating materials, an empirical linear relationship in the following form is obtained by

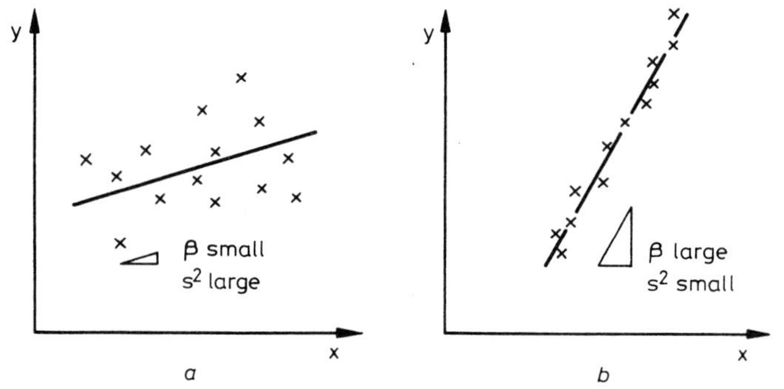

Fig. 2.28 Location of realisations of features X and Y, as a function of the strength of their relationship (diagrammatic)

a Weak relationship
b Strong relationship

β Regression coefficient
s^2 Variance

logarithmisation:

$$\log u_d = -\frac{1}{n}\log t_d + \log u_{d0}$$

(service-life law). If measured values are available, the linear regression is applied to the quantities $\log u_d$ and $\log t_d$.

In a linear regression, assuming normally distributed variates, the random features are on average linked by a linear function.[1] This regression line represents the expectation

$$\begin{aligned} EY &= \alpha + \beta EX \\ Y &= \alpha + \beta X. \end{aligned} \tag{2.111}$$

It is obtained from samples in the simplest case graphically by eye, by drawing a line of best fit through the graphically represented pairs of values $(x_i\,;\,y_i)$ (Fig. 2.28). If exact values are required, estimated values $(a;\,b)$ must be calculated for the coefficients α and β (regression coefficient) using the method of smallest error-squares (see Section 2.4.3).

If the realisations are greatly scattered about the regression line, and if the regression coefficient is consequently small, there is a weak linear relationship between features X and Y (Fig. 2.28*a*). If the scatter is slight, on the other hand, and β is large, this indicates a strong relationship (Fig. 2.28*b*).

A careful distinction needs to be drawn between a dependent and independent feature (parameter). In the estimation from y to x, the empirical

[1] Linear regression produces eminently usable results, even when the variate is only approximately normally distributed ('robust' method).

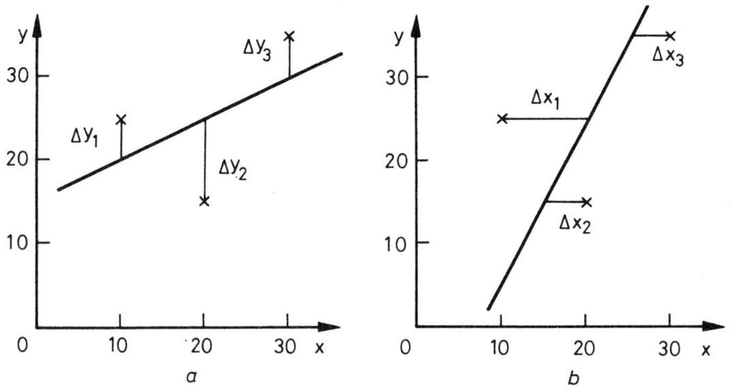

Fig. 2.29 Interchange of dependent and independent features in the regression

a Regression from y to x $(\Sigma(\Delta y_i)^2 = \min)$
b Regression from x to y $(\Sigma(\Delta x_i)^2 = \min)$

regression line

$$y = a_{yx} + b_{yx}x \tag{2.112}$$

represents an optimum fit of the empirically determined y-values (Fig. 2.29*a*). The sum of the vertical deviations is minimal, whereas in the reverse case

$$x = a_{xy} + b_{xy}y \tag{2.113}$$

the sum of the horizontal deviations is minimal (Fig. 2.29*b*). Which regression is to be executed will depend on the problem. As the correlation increases, the differences between the regression lines described by eqns. 2.112 and 2.113 disappear (Fig. 2.30). For correlation coefficients $|r| = 1$, the regression

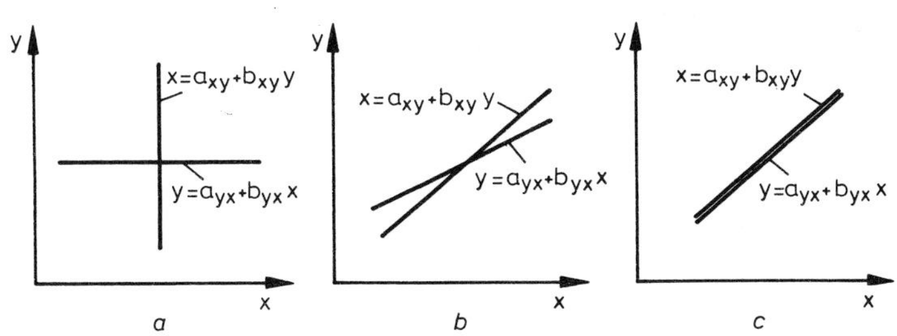

Fig. 2.30 Relationship between regression lines x and y and the correlation coefficient r (diagrammatic)

a $r = 0$
b $0 < r < 1$
c $r = 1$

lines coincide, and for the regression coefficients

$$b_{xy} = \frac{1}{b_{yx}}, \tag{2.114}$$

while generally ($|r| \leqq 1$)

$$r = \sqrt{b_{yx} b_{xy}}. \tag{2.115}$$

In the following we shall show how correlation and regression coefficients should be calculated in the simple linear case. Specialist statistical literature provides the engineer with comprehensive material for many situations well beyond this level.

2.4.2 Estimation of correlation coefficient

Take a sample consisting of n pairs of values $(x_i\,;\,y_i)$. We shall characterise them with the arithmetic means $\bar{x}$ and $\bar{y}$ (eqn. 2.41) and the mean square deviations s_x and s_y (eqn. 2.48). In addition, the empirical covariance

$$s_{xy} = \frac{1}{n-1} \sum_{i=1}^{n} (x_i - \bar{x})(y_i - \bar{y}) \tag{2.116}$$

is introduced, linking the two features and representing a mean deviation product.

From these quantities is derived the empirical correlation coefficient

$$r = \frac{s_{xy}}{s_x s_y}, \tag{2.117a}$$

which can also be advantageously calculated using

$$r = \frac{\sum_{i=1}^{n} x_i y_i - n\bar{x}\bar{y}}{\sqrt{\left(\sum_{i=1}^{n} x_i^2 - n\bar{x}^2\right)\left(\sum_{i=1}^{n} y_i^2 - n\bar{y}^2\right)}}. \tag{2.117b}$$

According to Sachs (see Bibliography), a two-way confidence region can be read off from Fig. 2.31. If the confidence region does not include the value $\rho = 0$, one can conclude that $\rho \neq 0$, i.e. a correlation exists.

Example 2.4.2.1 By determining the partial-discharge intensity (pulse charge q_i) at $u_p = 50$ kV and the breakdown voltage u_d on $n = 10$ voltage transformers, establish whether the destructive test can be replaced by non-destructive partial-discharge measurement.

Graphical representation of the measured values (Fig. 2.32) does not provide conclusive evidence. To reach a decision, the correlation coefficient and its confidence region are therefore calculated, assuming a normal distribution.

One finds

$$\bar{q}_i = 34.5\ \text{pC}; \qquad s_q = 26.0\ \text{pC}$$

$$\bar{u}_d = 98.6\ \text{kV}; \qquad s_u = 6.1\ \text{kV}$$

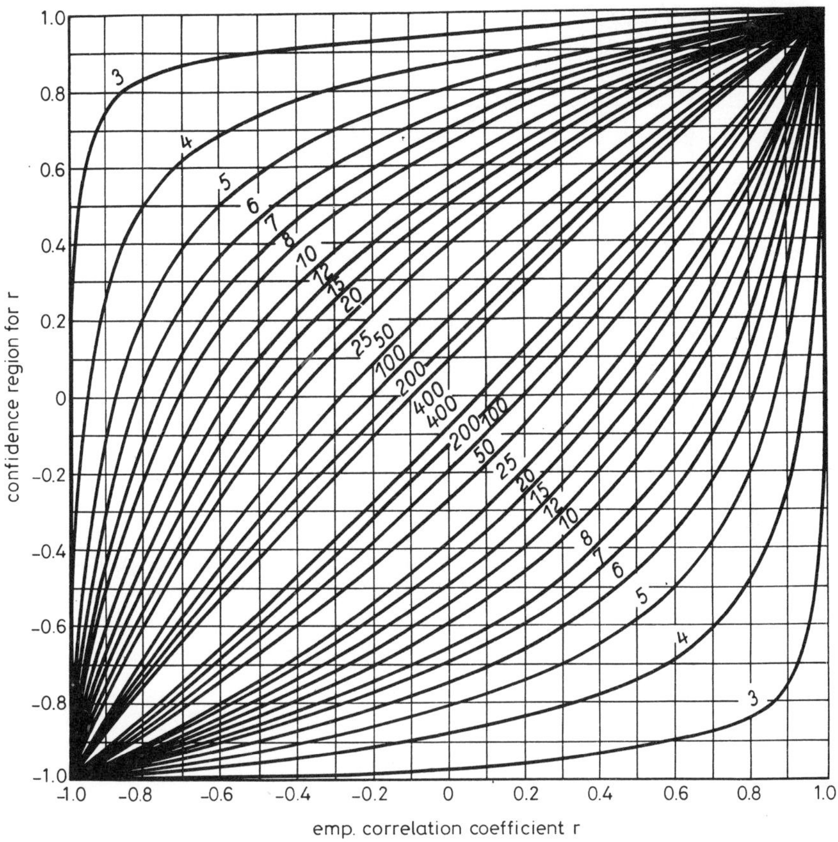

Fig. 2.31 Two-sided confidence regions for correlation coefficient with a confidence coefficient $\varepsilon = 0.95$, as a function of sample size n [2.67]

and for the covariance $s_{qu} = 85.4$ kV·pC. The correlation coefficient is consequently derived as

$$r = \frac{85.4}{26.0 \cdot 6.1} = 0.54.$$

Fig. 2.31 gives a confidence region [$g_u = -0.12$; $g_0 = +0.84$], which covers the value $\rho = 0$. It cannot therefore be regarded as certain that a correlation exists between the partial-discharge intensity and the breakdown voltage. To obtain more conclusive evidence, the sample size n needs to be increased.

2.4.3 Estimation of regression lines

Let us take a sample of n pairs of values $(x_i\,;\,y_i)$, as characterised in Section 2.4.2. In a regression from y to x (eqn. 2.112), the regression coefficient is

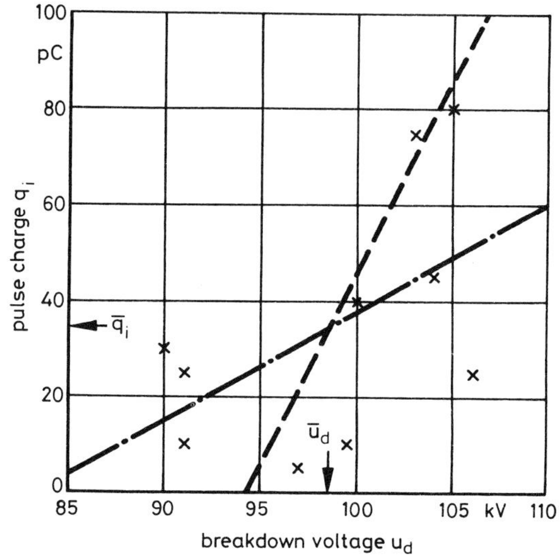

Fig. 2.32 Graphical representation of features from Example 2.4.2.1 and possible regression lines

–·–·– Regression from q_i to u_d
- - - - Regression from u_d to q_i

calculated as

$$b_{yx} = \frac{s_{xy}}{s_x^2} = r\frac{s_y}{s_x}, \tag{2.118a}$$

$$b_{yx} = \frac{\sum_{i=1}^{n} (x_i - \bar{x})(y_i - \bar{y})}{\sum_{i=1}^{n} (x_i - \bar{x})^2}. \tag{2.118b}$$

The following accordingly applies to regression from x to y (eqn. 2.113):

$$b_{xy} = \frac{s_{xy}}{s_y^2} = r\frac{s_x}{s_y}, \tag{2.119a}$$

$$b_{xy} = \frac{\sum_{i=1}^{n} (x_i - \bar{x})(y_i - \bar{y})}{\sum_{i=1}^{n} (y_i - \bar{y})^2}. \tag{2.119b}$$

The location coefficient a is obtained for the two cases as

$$a_{yx} = \bar{y} - b_{yx}\bar{x} \tag{2.120}$$

and

$$a_{xy} = \bar{x} - b_{xy}\bar{y}. \tag{2.121}$$

Example 2.4.3.1 For the data in Example 2.4.2.1, the two regression lines are to be calculated and plotted in Fig. 2.32. We obtain:

Regression from q_i to u_d:

$$b_{qu} = 2.29\frac{\text{pC}}{\text{kV}}; \qquad a_{qu} = -191.3\ \text{pC}$$

$$q_i/\text{pC} = -191.3 + 2.29 u_d/\text{kV}$$

Regression from u_d to q_i:

$$b_{uq} = 0.216\frac{\text{kV}}{\text{pC}}; \qquad a = 94.2\ \text{kV}$$

$$u_d/\text{kV} = 94.2 + 0.126 q_i/\text{pC}.$$

The two regression lines do not coincide. The regression confirms the result of Example 2.4.2.1.

The realisations are of course scattered about the regression line (Fig. 2.32). The residual variance is calculated as a measure of this dispersion:

$$s_{Ryx}^2 = \frac{1}{n-2}\sum_{i=1}^{n}(y_i - a_{yx} - b_{yx}x_i)^2 \tag{2.122}$$

and

$$s_{Rxy}^2 = \frac{1}{n-2}\sum_{i=1}^{n}(x_i - a_{xy} - b_{xy}y_i)^2. \tag{2.123}$$

Variances can also be indicated for the regression coefficient and the location coefficient. We only intend to give them here for the case of regression from y to x, which in fact occurs in all cases when regression is executed for a given parameter.

The variance of the location coefficient a is

$$s_{ayx}^2 = s_{Ryx}^2\left(\frac{1}{n} + \frac{\bar{x}^2}{(n-1)s_x^2}\right) \tag{2.124}$$

and that of regression coefficient b:

$$s_{byx}^2 = \frac{s_{Ryx}^2}{(n-1)s_x^2}. \tag{2.125}$$

It is possible, using these two variances, to calculate two-sided confidence regions for the location coefficient a

$$\left.\begin{matrix} a_o \\ a_u \end{matrix}\right\} = a_{yx} \pm t_{n-2;(1+\varepsilon)/2}\, s_{ayx} \tag{2.126}$$

and for the regression coefficient b

$$\left.\begin{matrix} b_o \\ b_u \end{matrix}\right\} = b_{yx} \pm t_{n-2;(1+\varepsilon)/2}\, s_{byx} \tag{2.127}$$

$t_{m;q}$ is the quantile of the F distribution with $m = n - 2$ degrees of freedom and order $q = (1 + \varepsilon)/2$. In addition, in [2.67, 2.2], descriptions can be found of the calculation of confidence regions for the entire regression line. Comparisons between two regression lines can be made using statistical tests.

As has already been mentioned, a linear regression can also be used when the linearity between the features investigated is only produced by a transformation. As well as being used in service-life investigations on high polymers (Example 2.4.1.1), it is also important in determining performance functions of insulation in constant-voltage tests (see Section 3.2.4). If a particular theoretical distribution function is assumed for the performance function (e.g. a normal distribution $\Phi(x)$), then linearisation will be achieved by applying the appropriate inverse function $[\Phi^{-1}(h_i) = \psi(h_i)]$.

Example 2.4.3.2 In a constant-voltage test (see Section 3.1.1), a test piece is stressed $n = 10$ times at each of $m = 5$ different voltages $u_{di}(i = 1, 2, \ldots, 5)$. The relative breakdown frequencies h_i are used as estimates for the breakdown probability. The relationship between the applied voltage and breakdown probability (i.e. the performance function) can be assumed to be a normal distribution. The empirical performance function is to be determined by linear regression. In the present case, the voltage u_{di} is a prescribed parameter; the breakdown probability estimated by h_i is a self-adjusting variate assumed as being normally distributed. A regression from h_i to u_{di} therefore needs to be performed. The problem (Fig. 2.33*a*), which is not

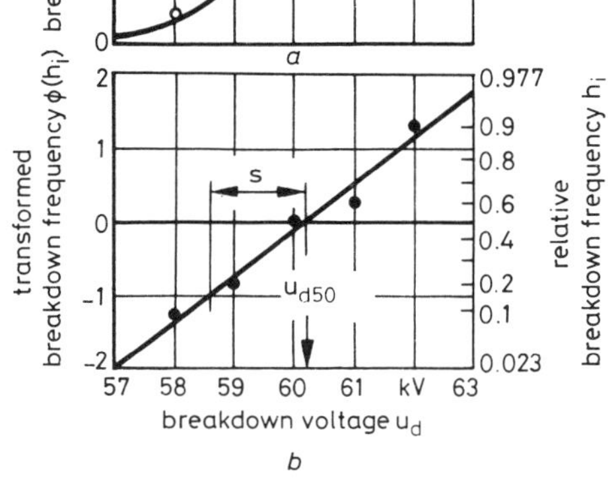

Fig. 2.33 Linear regression, Example 2.4.3.2: determination of a normally distributed performance function

a Non-linear problem
b Linearised problem (probability grid)

intrinsically nonlinear, is linearised by considering the transformed variate $\psi(h_i)$ (Fig. 2.33*b*) instead of variate h_i, where $\psi(h_i)$ is the inverse function of the normal distribution (see Müller *et al.*, Table 3*d*). Note that the same transformation is used as the basis for all normal-distribution probability paper (see Section 2.5.11). In this form, the quantities required for correlation and linear regression can best be calculated in tabular form (Table 2.24).

The results in Table 2.24 give:

- the empirical correlation coefficient (eqn. 2.117)

$$r = \frac{s_{xy}}{s_x s_y} = 0.985$$

- of which the confidence region (in Fig. 2.31 it can only be read off approximately): (≈0.70; 1)
- the regression coefficient for the regression from $y(\psi(h_i))$ to $x(u_{di})$ required here (eqn. 2.118):

$$b_{yx} = \frac{s_{xy}}{s_x^2} = 0.621 \text{ kV}^{-1}$$

- the location coefficient (eqn. 2.120)

$$a_{yx} = \bar{y} - b_{yx}\bar{x} = -37.38.$$

The regression line

$$y = -37.38 + 0.621\ x/\text{kV}$$

$$\psi = -37.38 + 0.621\ u_d/\text{kV}$$

is thus established, and the residual variance s_{Ryx}^2 can be calculated (Table 2.24). This consequently gives:

- the variance of the location coefficient (eqn. 2.124):

$$s_{ayx}^2 = 14.26; \qquad s_{ayx} = 3.78$$

- and the variance of the regression coefficient (eqn. 2.125):

$$s_{byx}^2 = 0.0040 \text{ kV}^{-2}; \qquad s_{byx} = 0.063 \text{ kV}^{-1}.$$

With the quantile $t_{3;0.975} = 3.18$ (Müller *et al.* Table 5), confidence regions with a confidence coefficient $\varepsilon = 0.95$ can now be indicated:

- for the location coefficient a_{yx} (eqn. 2.126): [−49.39; −25.37]
- and for the regression coefficient b_{yx} (eqn. 2.127): [0.421; 0.821]

It is particularly interesting that the parameters of the normal distribution can easily be calculated from the coefficients of the regression lines. Because $x_{50} = x(\psi = 0)$,

$$x_{50} = u_{d50} = -\frac{a_{yx}}{b_{yx}}, \tag{2.128}$$

Table 2.24 Linear regression. Example 2.4.3.2: determination of a normally distributed performance function

i	x_i Voltage u_{di}	Breakdowns k with	Relative breakdown frequency	y_i transformed frequency $\psi_i(h_i)$	Operands				
					$(x_i-\bar{x})$	$(y_i-\bar{y})$	$(y_i-\bar{y})(x_i-\bar{x})$	$\tilde{y}_i=-37.4+0.62x_i$	$(y_i-\tilde{y}_i)^2$
	kV	$n=10$	$h_i=k/n$		$(u_{di}-\bar{u}_d)$	$(\psi_i-\psi)$	$(u_{di}-\bar{u}_d)(\psi_i-\psi)$	$\tilde{\psi}_i=-37.4+0.62u_{di}$	$(\psi_i-\tilde{\psi}_i)^2$
1	58	1	0.1	−1.28	−2	−1.16	+2.32	−1.36	0.0064
2	59	2	0.2	−0.84	−1	−0.72	+0.72	−0.74	0.0100
3	60	5	0.5	0	0	+0.12	0	−0.12	0.0144
4	61	6	0.6	0.25	1	+0.37	+0.37	+0.50	0.0625
$m=5$	62	9	0.9	1.28	2	+1.40	+2.80	+1.12	0.0256
Equations	(2.41) (2.48)			(2.41) (2.48)	(2.116)			(2.122)	
Results	$\bar{x}=\bar{u}_d=60$ kV $s_x=s_u=1.58$ kV $s_x^2=s_u^2=2.50$ kV^2			$\bar{y}=\bar{\psi}=-0.12$ $s_y=s_\psi=0.997$ $s_y^2=s_\psi^2=0.994$			Covariance $s_{xy}=S_{u\psi}=1.553$ kV	Residual dispersion: $s_{Ryx}=0.199$ $s_{Ryx}^2=0.040$	

and because $x_{84} = x(\psi = 1)$, it follows that

$$x_{84} = u_{d84} = \frac{1 - a_{yx}}{b_{yx}},$$

which gives the empirical standard deviation directly

$$s = x_{84} - x_{50} = \frac{1}{b_{yx}}. \tag{2.129}$$

In the present example, we obtain for the parameters of the normal distribution being sought: $u_{d50} = 60.2$ kV and $s = 1.6$ kV.

The regression line drawn in Fig. 2.33*b* and the parameters estimated from it are the most reliable evaluation permitted by the existing $m = 5$ measurements. Despite the very small size of the sample, the correlation is clear and very pronounced. The confidence regions for the coefficients of the regression lines are naturally quite large.

2.5 Selected test-methods

A statistical test is a method used to test whether an assumption — or an assertion — concerning a (completely or partly) unknown probability distribution is compatible with a concrete sample that has been taken, or whether the assumed distribution differs significantly (i.e. with a predetermined statistical certainty) from the existing true distribution. The assumption made is the statistical hypothesis, whose mathematical formulation is expressed by a so-called null hypothesis H_0. (An alternative hypothesis H_1 expresses a situation differing from the null hypothesis.)

If, for example, the statistical hypothesis consists of the fact that the mean value μ of a distribution is equal to an assumed (or asserted) value μ_0, then the null hypothesis H_0 reads $\mu = \mu_0$. (And if the alternative hypothesis is that the mean value μ of this distribution is greater than μ_0, then H_1 would by $\mu > \mu_0$.) The procedure for a statistical test without explicit formulation of an alternative hypothesis (termed a significance test) can be split up into the following four characteristic steps:

(i) Setting up a hypothesis H_0
In the following, we shall be dealing with hypotheses, for example, concerning the type of distribution function under consideration (Section 2.5.1), the affiliation of a sample to a population (Section 2.5.2), as well as the randomness (especially the independence) of sample selection (Section 2.5.3).

(ii) Construction of a test quantity T
To perform the test, a suitable 'test quantity' T processing the sample is formed. This is a (random) sample function whose distribution is known, assuming the validity of the null hypothesis H_0. A realisation t of this quantity is accordingly obtained for a concrete sample.

(iii) Establishing a critical region K (or critical value k_α)
The 'critical region' is a part of the range of test quantity T such that the probability of the samples obtained of test quantity t falling in that critical range when H_0 is valid is not greater than α. α is termed the significance level, which must be chosen according to preset rules. (Usually, $\alpha \leqq 0.10$; common values are, for example, $\alpha = 0.05$ and $\alpha = 0.01$.) Depending on the significance level selected, using the critical values k_α, the limits of the critical region will be obtained, for example, as $K = (k_{\alpha 1}, \infty)$ or $K = (-\infty, k_{\alpha 2})$ for a one-sided test (Fig. 2.34*a*) and $K = [(-\infty, k_{\alpha u}); (k_{\alpha s}, \infty)]$, $k_{\alpha u} < k_{\alpha 0}$ for a two-sided test (Fig. 2.34*b*). In the latter case it is usually expedient to choose two component regions of equal area.

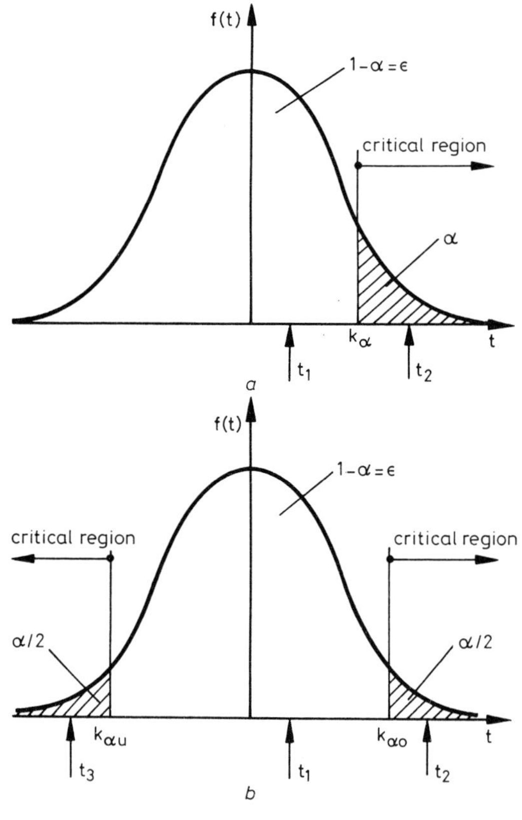

Fig. 2.34 Illustration of (a) one-sided and (b) two-sided tests

Test quantity t_1	hypothesis not rejected
Test quantity t_2; t_3	hypothesis rejected
k_α ; $k_{\alpha o}$; $k_{\alpha u}$	critical values
α	significance level

(iv) Decision rule (comparison)
If, for a concrete sample, the value t of the test quantity falls in the critical region, then H_0 should be rejected; otherwise, there is no objection to H_0 on the grounds of this test (Fig. 2.34). When the test is repeated, an erroneous decision can only occur in $(100\times\alpha)\%$ of cases by hypothesis H_0 being rejected even though true ('error of the first kind'). On this basis, the number α is also referred to as the 'error probability'.

When establishing the critical region of a test, the criterion is also important for minimising the so-called 'error of the second kind', i.e. non-rejection of a false null hypothesis. The tests are often formulated so that H_0 is rejected when $|t|>|k_\alpha|$.

Tests can be made easier for the user by using nomograms or graphical methods. A few tests relevant to high-voltage engineering are described below.

2.5.1 Distribution tests

In distribution tests, the deviations between an empirical distribution function (see Section 2.2.2) and a theoretical distribution function are investigated to determine whether they are of a random nature or significant. With regard to engineering use, it has proved advantageous to combine the test with a graphical representation of the empirical distribution function in the probability grid of the theoretical distribution function. Evaluation of the graphical representation to a certain extent provides a graphical test; we shall deal with this first, because it provides a useful visualisation of the principles.

2.5.1.1. Graphical methods (probability grids) A probability grid can be constructed for each distribution function $F(x)$ by transforming the initially linearly divided ordinate $y=p(0\leqq p\leqq 1)$ using the inverse of the distribution function being transformed $F^{-1}(y)=z$. The function values of the y scale (probability p) are then allocated to the z scale thus linearly divided.

Example 2.5.1.1.1 The normal-distribution probability grid is obtained by transformation using the inverse function of the standardised normal distribution $\psi(y)=\Phi^{-1}(y)$. Probabilities are then again allotted to the linearly divided scale $z=\Phi^{-1}(y)$.

$$z=0; \qquad y=\Phi(0)=0.500$$

$$z=1; \qquad y=\Phi(1)=0.841$$

$$z=-1; \qquad y=\Phi(-1)=-0.159$$

and so on (see Table 2.6*b*). Fig. 2.35 illustrates these relationships by reference to a normal distribution.

Co-ordinate systems can be constructed similarly for other probability grids. The ordinates of such grids are given in Fig. 2.36. The special distribution

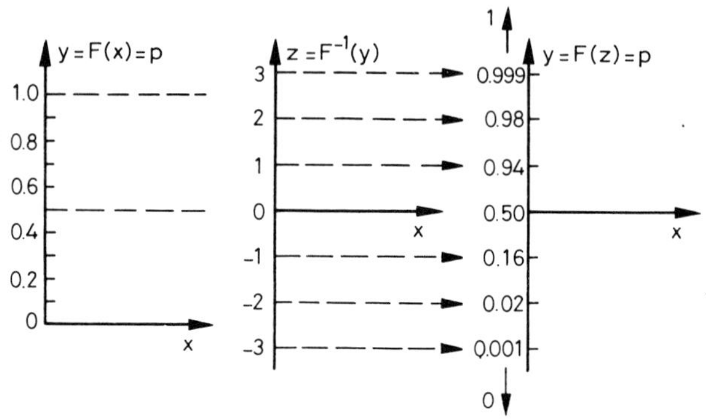

Fig. 2.35 Construction of probability paper
Example applies to normal distribution, i.e. $F(x) = \Phi(x)$ and $F^{-1}(y) = \psi(y)$

functions should have the same (equal) 50% quantile x_{50} and the same standard deviation (abscissa $x - a$). The ordinate already constructed in Fig. 2.35 is obtained for the normal distribution (c). In the case of the Weibull distributions, (d, e and f), different ordinates are obtained for the linear abscissa x as a function of the Weibull exponent δ, whereas an ordinate with arbitrary parameters exists for the double-exponential distribution (g). It will also be observed that, as δ increases, the ordinates of the Weibull distributions converge with those of the double exponential distribution. The ordinate of the double exponential distribution can be used for all

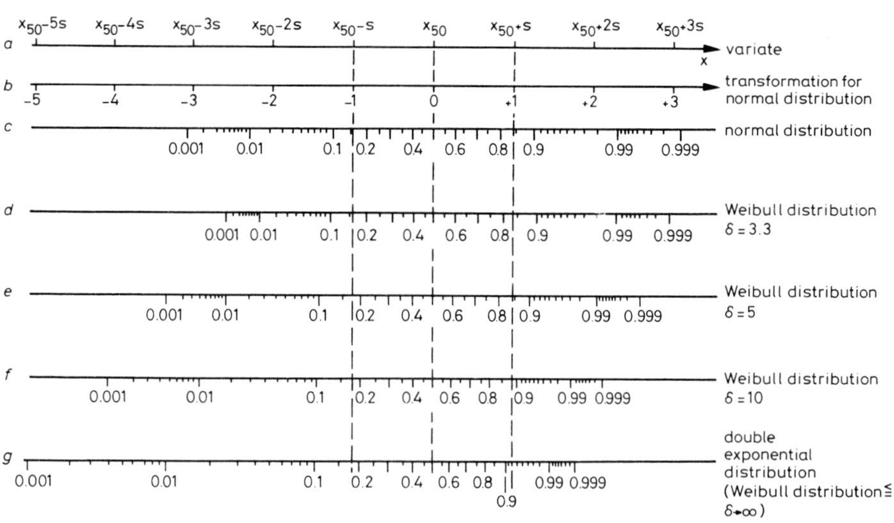

Fig. 2.36 Comparison of ordinates of various probability papers for same 50% quantiles and same standard deviation [2.53]

Table 2.25 Construction principles of probability papers

Graduation of abscissa	Graduation of ordinate	
	Inverse function of normal distribution	Inverse function of double exponential distribution
Linear	Probability grid of normal distribution	Probability grid of double exponential distribution
Logarithmic	Probability grid of lognormal distribution	Probability grid of Weibull distribution

Weibull distributions if the variate is plotted logarithmically, rather than linearly.

The ordinate of the probability grid for the normal distribution can also be applied to the log-normal distribution, provided that in this case the abscissa is also logarithmically graduated.

Table 2.25 summarises the construction principles of probability papers in common use in high-voltage engineering. There are various probability papers on the market, but it ought to be pointed out that logarithmic abscissae are better if they are constructed according to the particular requirements of the samples being evaluated.

To perform the graphical test, the empirical distribution function (preferably in the form of a cumulative frequency polygon) is reproduced on the probability grid of the theoretical distribution function being tested. A line of best fit is then drawn through the realisations, either by eye or with the aid of point estimates for parameters or linear regression (see Sections 2.3.2; 2.4.3). The deviations are either evaluated by eye (i.e. a quantitively imprecise criterion is used) or compared with tolerance limits (see Section 2.3.2.2). The evaluation is also quantitative if confidence regions can be given for the cumulative frequencies. The hypothesis for the type of distribution cannot then be rejected if the line of best fit passes through all the confidence regions (Fig. 2.37).

If one is not clear about the type of distribution function, it is necessary to reproduce it on various probability grids.

Example 2.5.1.1.2 For the sample in Example 2.2.2.1 (primary distribution table, Table 2.2), we have to test which theoretical distribution function best describes the empirical relationship. To do this, the cumulative-frequency polygon is plotted on the probability grids for normal distribution (Fig. 2.38*a*), log-normal distribution (Fig. 2.38*b*), double-exponential distribution (Fig. 2.38*c*) and two-parameter Weibull distribution (Fig. 2.38*d*), and is compared with a visually judged line of best fit. It will be observed that the best approximation is achieved with a double-exponential distribution. Here the maximum deviation between polygon and line of best fit is $d_{max} = 0.08$; with a Weibull distribution $d_{max} = 0.09$, normal distribution $d_{max} = 0.14$ and

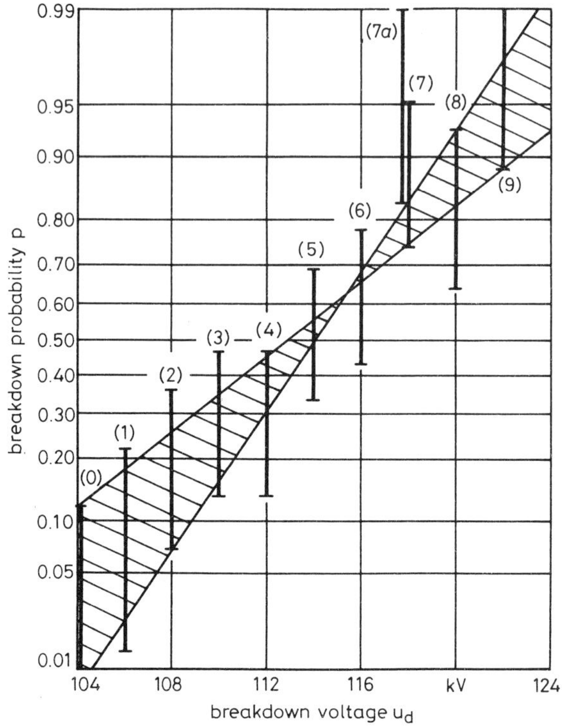

Fig. 2.37 Graphical test for normal distribution using confidence regions

Hatched area: range of possible lines of best fit
Confidence region 7: test not rejected
Confidence region 7a (instead of 7): test rejected

log-normal distribution $d_{max} = 0.17$. The double exponential distribution also provides the best approximation in the range of 'small' quantiles, which are particularly important in high-voltage engineering with respect of withstand voltage (see Section 3.4.2) and the 'enlargement law' (see Section 6). The great advantage of the graphical method lies in the evaluation, at least qualitative, of the entire fitting process.

Fig. 2.38 Graphical test of primary distribution table in Example 2.2.2.1 (Table 2.2) using probability grids

a for normal distribution ($u_{d50} = 201$ kV; $s = 10$ kV)
b for log-normal distribution ($x_{50} = 2.302$; $s = 0.020$)
c for double exponential distribution ($u_{d63} = 206$ kV; $\gamma = 9$ kV)
d for Weibull distribution ($u_{d0} = 0$; $u_{d63} = 206$ kV; $\delta = 22.5$)

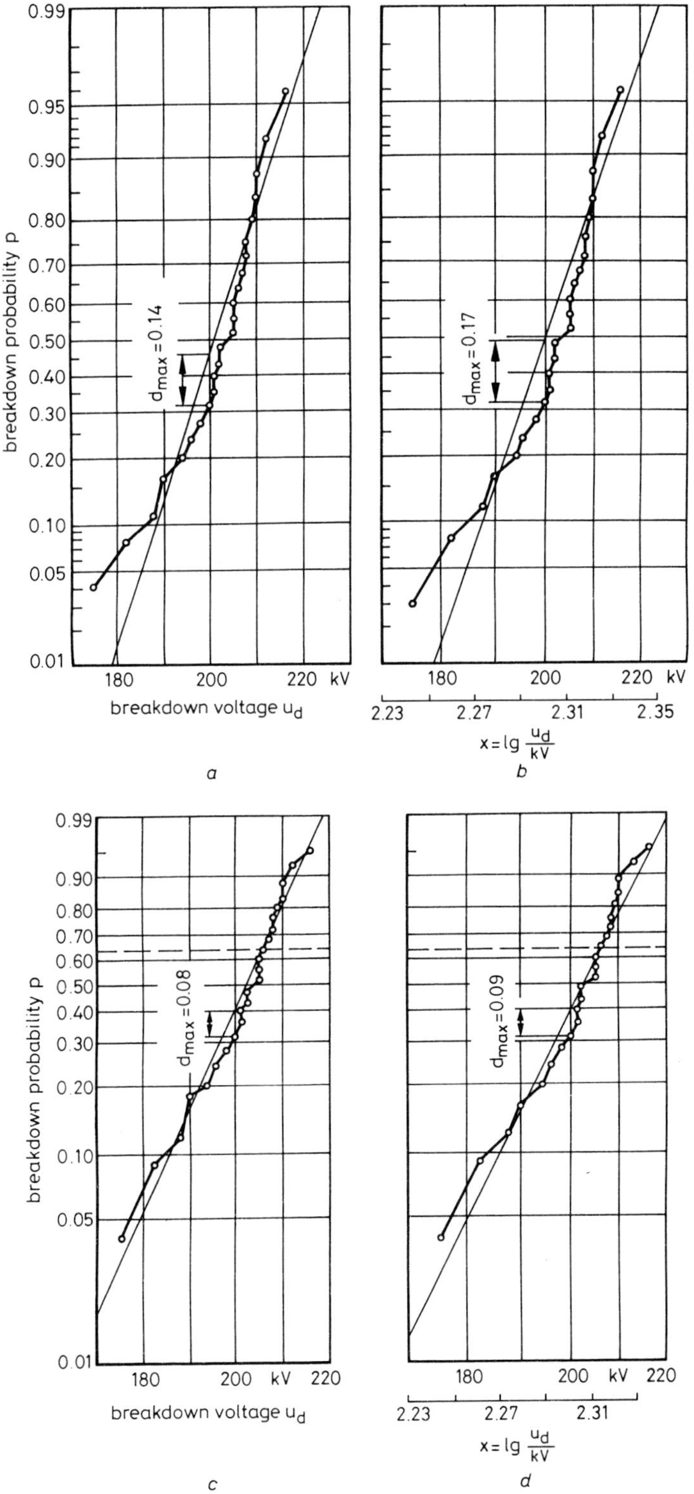

0.99
0.95
0.90
0.80
0.70
0.60
0.50
0.40
0.30
0.20
0.10
0.05
0.01
breakdown probability p
$d_{max} = 0.14$
180
200
220
kV
breakdown voltage u_d
a
$d_{max} = 0.17$
180
200
220
kV
2.23
2.27
2.31
2.35
$x = \lg \frac{u_d}{kV}$
b
0.99
0.90
0.80
0.70
0.60
0.50
0.40
0.30
0.20
0.10
0.05
0.01
breakdown probability p
$d_{max} = 0.08$
180
200
kV
220
breakdown voltage u_d
c
$d_{max} = 0.09$
180
200
kV
220
2.23
2.27
2.31
$x = \lg \frac{u_d}{kV}$
d

With small sample sizes ($5 \leqq n \leqq 20$), it is possible to dispense with the probability grid for the test for normal distribution, at the same time cutting down on calculation, using Martin's 'Minilot' template.[1] The template is used to plot the realisations according to the ordinate of a probability grid, the relative cumulative frequencies being determined with the template.

All probability grids additionally permit parameters to be estimated from the graphical representation by the quantile method. The requisite quantiles, which can then be read off directly from the empirical distribution function, are mentioned in connection with the theoretical distribution functions (see Sections 2.3.2.2 to 2.3.2.5).

2.5.1.2. Mathematical methods Mathematical tests of the approximation of an empirical distribution function by a theoretical one involve going through the steps explained above (hypothesis; test quantity; critical value; comparison). In the following we shall be considering the Kolmogorov test, which assesses the maximum deviation between the empirical and theoretical distribution. Pearson's more costly χ^2 test, which takes account of all deviations when determining the test quantity, will just be mentioned here in passing. It is exhaustively described in the literature.

Hypothesis for Kolmogorov test: The continuous but otherwise unknown (empirical) distribution function $F_n(x)$ produced by a sample of size n is in agreement with the postulated theoretical distribution function $F_0(x)$.

Test quantity: The test quantity is the maximum deviation between the empirical and theoretical value

$$d_{max} = \max. |F_n(x) - F_0(x)|, \; -\infty < x < +\infty, \qquad (2.130)$$

already illustrated in Fig. 2.38.

The test quantity can be determined, as in Fig. 2.38, either graphically (preferably linked to a graphical test) or by finding the maximum difference in tabular form according to point estimates for the parameters of $F_0(x)$ (by supplementing the primary distribution table with all the differences of the empirical and theoretical cumulative frequencies at the locations of the realisations).

Critical value: The critical value $k_{n,\alpha}$ for the two-sided test is available in tabular form (Table 19 of Müller *et al.*, of which an extract is reproduced in Table 2.26).

Comparison: The hypothesis is rejected if $d_{max} > k_{n,\alpha}$.

Problems: The test quantity $k_{n,\alpha}$ depends heavily on the size of the sample, so that the test is rather imprecise with a small sample. On the other hand, particularly the graphical representation of the test quantity (Fig. 2.39 'Kolmogorov limits') shows that an evaluation of the type of distribution in the range of 'small' and 'large' quantiles can only be expected with an extremely large sample, which cannot generally be achieved in high-voltage

[1] Commercially available together with instructions for use.

Table 2.26 Critical values for Kolgomorov test [2.11]

Size of sample n n	$k_{n,\alpha}$ for significance level $\alpha=0.10$	$\alpha=0.05$	$\alpha=0.02$
8	0.410	0.454	0.507
12	0.338	0.375	0.420
16	0.295	0.327	0.366
20	0.265	0.294	0.329
24	0.242	0.269	0.301
30	0.218	0.242	0.270
40	0.189	0.2101	0.235
50	0.170	0.188	0.211
75	0.139	0.154	0.173
100	0.121	0.134	0.150
$\to\infty$ (asymptotic expression)	$\frac{1.224}{\sqrt{n}}$	$\frac{1.358}{\sqrt{n}}$	$\frac{1.517}{\sqrt{n}}$

engineering. It is therefore advisable to combine the test with a graphical representation (Fig. 2.38) in order at least to obtain an indication of the trend in this range. When comparing different fits, it is also possible to consider the sum of the deviations between empirical and theoretical values (or the sum of the squares) in the range $p \leqq p_0$ (e.g. $p_0=0.1$).

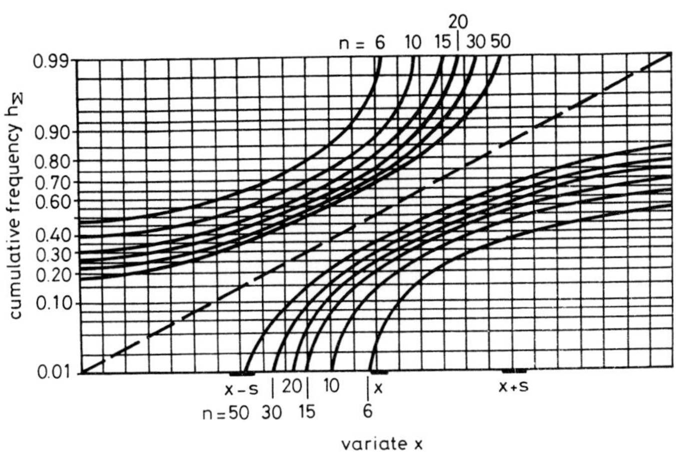

Fig. 2.39 Limits for Kolgomorov test [2.18]

Significance level $\alpha=0.10$

Table 2.27 Mathematical tests for comparing two independent samples or one sample with a known population with continuous variates

Problem	Test	Application conditions	References	Examples of high-voltage engineering applications
Is a sample with an empirical variance s^2 derived from a population with variance σ_0^2?	χ^2 scatter test	Population normally-distributed; σ_0^2 known	[2.2, S. 139] [2.11, S. 120]	Comparison of a test piece from a new batch with empirical values
Are two samples with empirical variances s_x^2, s_y^2 derived from populations with the same variance?	F test	a) Population normally-distributed; arbitrary sample sizes n_x and n_y	Section 2.5.2.1. [2.2, S. 141] [2.11, S. 137] [2.67, S. 261]	Evaluation of check measurements Clarification of test-piece dispersions
	t scatter test	b) As (a), but $n_x = n_y$	[2.67, S. 261]	
	Pillai test	c) Population roughly normally-distributed; arbitrary small sample sizes: $n_x \leqq 10$; $n_y \leqq 10$	[2.67, S. 276]	
Is a sample with an arithmetic mean $\bar{x}$ derived from a population with a mean value μ_0?	Mean-value test	a) Population normally-distributed; both parameters μ_0 and s_0^2 known	[2.2, S. 124] [2.11, S. 99] [2.67, S. 257]	Comparison of a test piece from a new batch with empirical values
	t test	b) As (a), but σ_0^2 unknown	[2.2, S. 131] [2.11, S. 130]	

Are two samples with arithmetic means $\bar{x}$ and $\bar{y}$ derived from populations with the same mean value?	Double t test	a) Population normally-distributed; dispersions unknown and in agreement (i.e. F test not rejected; if double t test not rejected, common population)	Section 2.5.3.2. [2.2, S. 136] [2.11, S. 131] [2.67, S. 266]	Evaluation of check measurements Clarification of test-piece dispersions in rising-voltage tests (same measurements on two test pieces from different batches)
	Welch test Weir test	b) As (a), but dispersions unknown and different	[2.67, S. 270] [2.67, S. 273]	
	Lord test	c) Population roughly normally-distributed; small samples: $n_x = n_y \leqq 20$	[2.67, S. 277]	
Are two samples derived from the same population regardless of their distribution function?	Mosteller test	a) Distribution-free; small sample sizes, rough approximation	[2.67, S. 284]	
	Tukey quick test	b) As (a), but the two samples of roughly the same size	[2.56, S. 289]	
	Kolgomorov-Smirnov test	c) Distribution-free for average to large sample-sizes (very precise test)	[2.67, S. 291]	
	X test (van der Waerden)	d) As (c) (also a very precise test)	[2.11, S. 224]	
	U test (Wilcoxon)	e) As (c) (less ponderous, but also less precise than (c) and (d))	Section 2.5.2.3. [2.11, S. 231] [2.67, S. 293]	

Example 2.5.1.2.1 The deviations between empirical and theoretical distribution functions in Example 2.5.1.1.2 (Fig. 2.38) are to be evaluated with the Kolmogorov test. With a significance level $\alpha = 0.10$, a critical value $k_{24;0.10} = 0.242$ is obtained for a sample of $n = 24$ from Table 2.26. All the deviations are smaller than this value, so that the hypothesis of agreement cannot be rejected. The test is therefore not precise enough to answer the question of the best approximation. The sample size n would have to be increased in the experiment to make it more precise. Had the same polygon been obtained for $n = 100$ realisations, with $k_{100;0.10} = 0.121$, the hypothesis for normal distribution ($d_{max} = 0.14$) and log-normal distribution ($d_{max} = 0.17$) would have been clearly rejected. The two extreme-value distributions consequently describe the relationship better ($d_{max} = 0.08$ and 0.09).

2.5.2 Comparison of samples in respect of common population

Comparisons between samples are often needed, to reach decisions concerning the combination of individual samples to form a large total sample, the influence of certain test parameters, or the occurrence of dependences in the course of the test (see Section 2.5.3). Such comparisons can be made, in the case of a known distribution function, by reference to their parameters (parameter tests: a hypothesis about certain parameters is tested), or in the case of an unknown distribution function, the test quantities are formed from the characteristics of the total sample[1] (distribution-free or non-parametric tests, e.g. rank tests, such as the U test).

There are a large number of tests available for handling the problem as a whole. Since we can only discuss a few selected tests here, the user is referred in the first instance to Tables 2.27 to 2.31, which contain a 'systematisation' of the problems, hints concerning possible tests and bibliographical references, as well as sample applications in high-voltage engineering:

- Table 2.27: comparison of two samples (parametric and non-parametric tests);
- Table 2.28: comparison of more than two samples;
- Table 2.29: comparison of connected samples. Such samples exist, for example, when identical measurements are made on the same test pieces before and after a particular stress (e.g. hot running, continuous-voltage running), so that a pair of realisations has been determined for each test piece. The test then tests the effect of the treatment.
- Table 2.30: comparison of relative frequencies and parameters of discrete-variate distribution functions.
- Table 2.31: correlation and regression tests.

The parameter tests have been specially developed for a normal distribution. However, when it comes to the assessment of average trends, these tests can also be applied to other distributions because of the central limiting-value set, since there are only very small deviations between different types of distribution in the range ($x_{50} \pm s$) (Fig. 2.36). Presented below are two

[1] The total sample is produced by combining all the samples being compared.

Table 2.28 Mathematical tests for comparing $k > 2$ independent samples with continuous variates (for $k = 2$, see Table 2.27)

Problem	Test	Application conditions	References	Examples in high-voltage engineering
Are the mean values of k samples $\bar{x}_1, \ldots, \bar{x}_k$ derived from populations with a common mean value μ_0?	Variance analysis (F test)	a) Populations normally-distributed; dispersions unknown but equal	[2.11, S. 138] [2.67, S. 485]	Comparison between new test-values and long-duration empirical values
	Scheffé test	b) Prerequisites as in (a); application after rejection of hypothesis in the variance analysis in order to test whether the k mean values are significantly different	[2.11, S. 138] [2.67, S. 492]	Of particular relevance to routine tests for their statistical analysis, especially for partial-discharge measurements
	Duncan test	c) Prerequisites and application as in (b), but two mean values randomly selected from k values are tested for equality after the rejected variance analysis	[2.11, S. 160]	Combining component samples to form very large samples
Are the variances of k samples $s_1^2, \ldots, s_1^2$ derived from populations with the common variance	Cochran test	a) Prerequisites as for the variance analysis, but the k samples need to be of the same size	[2.11, S. 158] [2.67, S. 480]	
	Hartley test	b) Prerequisites as in (a), but easier to execute and less precise	[2.67, S. 480]	
	Bartlett test	c) Prerequisites as in (a), but the samples can be of different sizes	[2.67, S. 483]	
Are k samples derived from the same population, regardless of their distribution function?	H test (Kruskal/Wallis)	a) Distribution-free (analogous to the U test for two samples)	[2.67, S. 302]	Combining component samples

Table 2.29 Mathematical tests for comparing connected samples

Problem	Test	Application conditions	References	Examples of applications in high-voltage engineering
Are the mean values of two connected samples derived from populations with the same mean value?	*t* test for pair differences	a) Populations normally-distributed; dispersions unknown but same	[2.2, S. 134] [2.67, S. 309]	Two measurements are made of the partial-discharge intensity on a group of test pieces both before and after long-term stress: Does the long-term stress affect the partial-discharge intensity?
	Wilcoxon test for pair differences	b) Distribution-free, nevertheless almost as precise as *t* test. It is not exactly the mean value but an average trend that is compared.	[2.67, S. 313]	
	Maximum test	c) As (b), but far less precise	[2.67, S. 314]	
	Sign test (Dixon/Mood)	d) As (b); the test can also be applied to unpaired independent samples by random pairing	[2.11, S. 220] [2.67, S. 315]	
Are the dispersions of two connected dispersions derived from populations with the same dispersion?	Variability comparison	Populations normally-distributed	[2.67, S. 311]	

Table 2.30 Mathematical tests for comparing relative frequencies and empirical parameters of discrete-variate distribution functions

Problem	Test	Application conditions	References	Examples of applications in high-voltage engineering
Is sample with a relative frequency $h = k/n$ derived from an alternative population with a probability p_0?	Testing a relative frequency	Test of empirical parameters p of a binomial distribution	[2.2, S. 154] [2.11, S. 100]	Does a check measurement of the breakdown probability agree with the empirical value?
Are two samples with relative frequencies $h_1 = k_1/n_1$ and $h_2 = k_2/n_2$ derived from an alternative population with probability p_0?	Comparison of two relative frequencies	a) Comparison of empirical parameters p of two binomial distributions	See Section 2.5.2.4. [2.2, S. 156] [2.67, S. 333]	Agreement between breakdown probabilities of two test pieces at same voltage
	Comparison of two absolute frequencies	b) Quick test for rough considerations	[2.67, S. 340]	
	Four-field chi-squared test	c) Comparison of empirical parameters p of two binomial distributions with four-field table	[2.2, S. 181] [2.11, S. 123] [2.67, S. 341]	
	Woolf's G test	d) Similar to four-field chi-squared test, but with computational advantages and better theoretical basis	[2.67, S. 346]	
Are two samples with estimated parameters derived from two Poisson distributions of the same population?	Comparison of Poisson parameters	Comparison of empirical parameters of two Poisson distributions	[2.67, S. 187]	Comparison at very low breakdown probabilities

Table 2.31 Mathematical tests for correlation and regression

Problem	Test	Application conditions	References	Examples of applications in high-voltage engineering
Does an empirical correlation coefficient r correspond to the theoretical value ρ	Test with Fisher's transformation	Normally distributed populations	[2.2, S. 158] [2.11, S. 174]	Agreement between an empirical value and a new finding
Do two empirical correlation coefficients coincide?	Test with Fisher's transformation	Normally distributed populations	[2.2, S. 161] [2.11, S. 175]	Clarification of relationship between two measured quantities
Is an empirical correlation coefficient other than zero? Are the two features investigated therefore dependent on each other or independent of each other?	t independence test	a) Normally distributed populations Important: If r only differs from zero randomly, the features being investigated can thus be regarded as independent of each other	[2.11, S. 132]	Clarification of relationship between partial-discharge intensity and amplitude of breakdown voltage
	Test with random maximum values	b) As (a)	[2.11, S. 170]	

	χ^2 independence test	c) Non-dependence is tested not with correlation coefficients but with four-field tables	[2.11, S. 123]	
Does an empirical regression coefficient b_{yx} correspond to the theoretical value b_0?	t regression-coefficient test	Dependent feature y is normally-distributed	[2.2, S. 167] [2.11, S. 133] [2.67, S. 426]	Agreement between a measured service-life characteristic and previous empirical values (a_0; b_0)
Does an empirical location coefficient a_{yx} correspond to the theoretical value a_0?	t location-coefficient test	Dependent feature y is normally-distributed	[2.2, S. 167] [2.67, S. 427]	
Do two regression lines only differ from each other by chance?	Sub-problems: a) Testing residual dispersions with F test b) Comparison of regression coefficients c) Comparison of location coefficients	Dependent features y_1 and y_2 are normally-distributed	[2.2, S. 168] [2.67, S. 428]	Agreement between two measured service-life characteristics or two performance functions

parameter tests and a non-parameteric test for comparing two samples from continuous populations, together with a test of two probabilities.

2.5.2.1. *F* test (comparison of two empirical variances)
Hypothesis: The variances σ_x^2 and σ_y^2 of two normally distributed populations represented by two samples of sizes n_x and n_y, and the empirical variances s_x^2 and s_y^2, are identical: $\sigma_x^2 = \sigma_y^2$.

Test quantity:

$$t = \frac{s_x^2}{s_y^2}, \tag{2.131}$$

where the notation x and y of the samples must be chosen so that $s_x^2 \geqq s_y^2$.

Critical value: The critical value $F_{m_1;m_2;q}$ is obtained from the F distribution (Table 2.21) as a quantile with the degrees of freedom $m_1 = n_x - 1$; $m_2 = n_y - 1$ and order $q = 1 - \alpha/2$ (two-sided test at significance level α).

Comparison: The hypothesis is rejected if $t > F_{m_1;m_2;q}$.

Example 2.5.2.1.1 Two breakdown-voltage samples are available of sizes $n_1 = 17$ and $n_2 = 21$ and with empirical variances $s_1^2 = 63.68\ \text{kV}^2$ and $s_2^2 = 32.60\ \text{kV}^2$. Because $s_1^2 > s_2^2$ we make $s_1^2 = s_x^2$, $s_2^2 = s_y^2$, $n_1 = n_x$ and $n_2 = n_y$. Using eqn. 2.131, the test quantity is obtained as $t = 1.95$. If the hypothesis is tested at significance level $\alpha = 0.05$, with $m_1 = 16$, $m_2 = 20$ and $q = 0.975$ from Table 2.21, the critical value is obtained as $F_{16;20;0.975} = 2.55$. Because $1.9 < 2.55$, the hypothesis that the two samples are derived from populations with the same variance cannot be rejected.

2.5.2.2. Double *t* test (comparison of two mean values)
Hypothesis: The mean values μ_x and μ_y of two normally distributed populations with the same variance (the test thus assumes non-rejection by the F test), which are represented by two samples of sizes n_x and n_y, mean values $\bar{x}$ and $\bar{y}$, as well as the empirical variances s_x^2 and s_y^2, are identical: $\mu_x = \mu_y$.

Test quantity:

$$t = \frac{(\bar{x} - \bar{y})\sqrt{\dfrac{n_x n_y}{n_x + n_y}}}{\sqrt{\dfrac{(n_x - 1)s_x^2 + (n_y - 1)s_y^2}{n_x + n_y - 2}}}. \tag{2.132}$$

Critical value: The critical value $t_{m;q}$ is obtained from the t distribution (Table 2.22) as a quantile with a degree of freedom $m = n_x + n_y - 2$ and order $q = 1 - (\alpha/2)$ (two-sided test at significance level α).

Comparison: The hypothesis is rejected if $|t| > t_{m;q}$.

Example 2.5.2.2.1 Two samples of breakdown voltage measurements are available, with characteristic sizes $n_x = 18$; $\bar{x} = 53.2\ \text{kV}$; $s_x^2 = 22.5\ \text{kV}^2$ and

$n_y = 24$; $\bar{y} = 48.4$ kV; $s_y^2 = 16.4$ kV^2. The test quantity is obtained from eqn. 2.132 as $t = 3.53$, while the critical value is derived from Table 2.22, with $m = 40$, $\alpha = 0.05$ and $q = 0.975$ as $t_{40;0.975} = 2.02$. Because $3.53 > 2.02$, the hypothesis is rejected. The samples are therefore not derived from a common population.

There is also a graphical method for the double t test.

2.5.2.3. *U* test (distribution-free comparison of two samples)
Hypothesis: Two populations with distribution functions $F(x)$ and $F(y)$, which are represented by two samples of sizes n_x and n_y and with realisations $x_1 \ldots x_i \ldots x_{nx}$ and $y_1 \ldots y_j \ldots y_{ny}$, are identical: $F(x) = F(y)$.

Test quantity: The $(n_x + n_y)$ sample values are jointly arranged in order of magnitude and are numbered from 1 to $(n_x + n_y)$. The numbers are referred to as 'ranks' $r(x_i)$ and $r(y_j)$, wherein the same rank (mean value of the ranks occurring) is assigned to identical realisations. The sums of the ranks are then formed for each sample.

$$R_x = \sum_{i=1}^{n_x} r(x_i) \quad \text{for } R_y = \sum_{j=1}^{n_y} r(y_j), \tag{2.133}$$

from which the test quantity

$$u = \min(u_x, u_y) \tag{2.134}$$

is obtained with

$$u_x = R_x - \frac{n_x(n_x+1)}{2}$$

and

$$u_y = R_y - \frac{n_y(n_y+1)}{2}.$$

Critical value: The critical value $U_{n_x;n_y;\alpha}$ is available in tabular form for the one-sided and two-sided test (Müller *et al.*, Table 18, of which an extract is reproduced in Table 2.32).

Comparison: The hypothesis is rejected if $u < U_{n_x;n_y;\alpha}$.

Example 2.5.2.3.1 Two samples of breakdown voltage measurements are available, for which the ranks have been determined in Table 2.33. The ranks produce $u_x = 92$ and $u_y = 40$; the test quantity is thus $u = 40$. If the hypothesis of equality of the two samples is tested bilaterally with the significance level $\alpha = 0.05$, the critical value can be obtained from Table 2.32 by interpolation, or directly from Müller *et al.* $U_{12;11;0.05} = 33$. Because $40 > 33$, the hypothesis cannot be rejected. The two samples can be assumed as stemming from a common population.

Table 2.32 Critical value $U_{n_x;n_y:\alpha}$ for U test [2.11]; two-sided test for $\alpha = 0.05$

n_y	n_x								
	8	10	12	15	20	25	30	35	40
8	13	17	22	29	41	53	65	77	89
10	17	23	29	39	55	71	87	103	119
12	22	29	37	49	69	89	109	129	149
15	29	39	49	64	90	117	143	169	196
20	41	55	69	90	127	163	200	237	274
25	53	71	89	117	163	211	258	306	354
30	65	87	109	143	200	258	317	375	434
35	77	103	129	169	237	306	375	445	515
40	89	119	149	196	274	354	434	515	596

Table 2.33 Determination of ranks and test quantity for U test (Example 2.5.2.3.1)

Sample 1 ($n_x = 12$)		Sample 2 ($n_y = 11$)	
List of individual test values x Breakdown voltage u_d/kV	Ranks	List of individual test values y Breakdown voltage u_d/kV	Ranks
221	15	198	6
226	19	231	21
228	20	225	18
190	4	207	12
203	10	189	3
208	13	221	15
200	7	221	15
204	11	201	8
232	22	186	2
202	9	195	5
222	17	184	1
235	23		
Sum of ranks $R_x = 170$		Sum of ranks $R_y = 106$	
$u_x = 170 - \frac{12(12+1)}{2} = 92$		$u_y = 106 - \frac{11(11+1)}{2} = 40$	
Test quantity (92; 40) = 40			

2.5.2.4. Comparison of two probabilities

Hypothesis: The probabilities p_1 and p_2 from two alternative populations represented by the two relative frequencies $h_1 = k_1/n_1$ and $h_2 = k_2/n_2$ are identical: $p_1 = p_2$.

Test quantity:

$$z = \frac{h_1 - h_2}{\sqrt{\dfrac{h_1 n_1 + h_2 n_2}{n_1 + n_2}\left(\dfrac{n_1(1-h_1) + n_2(1-h_2)}{n_1 n_2}\right)}}. \tag{2.135}$$

Critical value: The critical value λ_q is derived from the standardised normal distribution (Table 2.6) as a quantile of order $q = 1 - (\alpha/2)$ (two-sided test at significance level α).

Comparison: The hypothesis is rejected if $|z| > \lambda_q$.

Example 2.5.2.4.1 The relative frequencies $h_1 = 12/25 = 0.480$ and $h_2 = 11/30 = 0.367$ have been determined from two samples. To test the null hypothesis, the test quantity is obtained as $z = 0.846$ using eqn. 2.135, while the critical value with $\alpha = 0.05$ is derived from Table 2.6 as $\lambda_{0.975} = 1.960$. Because $0.846 < 1.960$, the hypothesis cannot be rejected. The fact that the two samples come from the same population can be taken as the starting point.

2.5.3 Tests of independence of realisations

The statistical evaluation of samples is only permissible if the realisations represent a random selection from the population, in the sense that they are independent of each other (see Section 2.1.3). Before any statistical evaluation, this should be verified using tests. It is advisable to employ simple graphical tests while the experiments are being conducted, and the complete list of individual test values can later be tested mathematically (see Sections 3.2.2 and 3.3.2). We shall now explain a few important independence tests (Table 2.34), both for realisations of complementary events (breakdown probabilities) and for samples of continuous variates.

2.5.3.1. Complementary events

During experiments, the following graphical control methods have proved successful in relation to independence.

The realisations of the binomially distributed variate (in the case of high-voltage engineering, usually breakdown or non-breakdown) are represented in the order in which they occur and are divided into equally powerful subgroups for which the relative frequencies are determined (Table 2.35).

If the relative frequencies of the subgroups fluctuate in a random manner about the mean value of the total sample, independence exists (Table 2.35*a*). If, on the other hand, they exhibit a systematically falling (Table 2.35*b*), rising or even periodically fluctuating tendency, then trends, i.e. dependencies, exist. The graphical check can be supplemented qualitatively by con-

Table 2.34 Mathematical tests for independence and maverick problems

Problem	Test	Application conditions	References	Examples of applications in high-voltage engineering
Should the sequence of complementary events be regarded as random?	Iteration test Stevens/Wallis/Fisz	Realisations are evaluated in respect of their chronological order (individual test values)	Section 2.5.3.1. [2.5] [2.67, S. 369]	Evaluation of test series for constant-voltage tests
Should the sequence of realisations of a continuous variate be regarded as random?	Phase test Wallis and Moore	a) Consecutive realisations are evaluated according to their difference	[2.67, S. 371]	Evaluation of test series for rising-voltage tests
	Trend test Cox and Stuart	b) Comparison of realisations at beginning and end of list of individual test values	[2.67, S. 372]	
	Neumann's trend test	c) Consecutive realisations are evaluated according to the square of their differences	[2.67, S. 367]	
	Modified iteration test	d) By comparing them with the arithmetic mean, the order of continuous realisations is re-arranged and evaluated with the iteration test	Section 2.5.3.2. [2.67, S. 369]	
Are presumably invalidated values (mavericks, especially maximum or minimum values) nevertheless derived from the same population as the other sample?	Maverick test	a) Normally distributed population. If the value is identified as a maverick, it should be removed from the sample: sample 'purification'	Section 2.5.3.2. [2.11, S. 194] [2.67, S. 279]	'Purification' of test series in rising-voltage tests of measuring and pen-recorder errors
	Thompson rule	b) As (a)	[2.11, S. 198]	
	Ranking maverick tests	c) Aa (a)	[2.11, S. 199]	

Table 2.35 Graphical check of independence for an alternative population (complementary events)

a) Sphere/plane arrangement in atmospheric air																				h_{20}	h_{100}
—	×	×	×	—	—	—	×	-	×	×	×	×	×	×	×	—	×	—	×	0.65	
×	—	×	—	—	×	×	×	×	×	—	×	×	—	×	×	×	×	×	×	0.75	
—	×	×	×	×	—	×	×	×	×	×	×	—	×	—	×	×	×	—	—	0.70	0.68
×	×	×	—	×	×	—	×	—	×	—	—	—	×	—	×	×	×	—	×	0.60	
×	×	—	×	×	×	×	—	—	×	—	×	—	×	×	×	×	×	×	—	0.70	
b) As (a), but sealed air in a vessel																				h_{20}	h_{100}
×	×	×	×	×	×	×	×	×	×	×	×	×	×	×	×	×	×	—	×	0.95	
×	—	×	—	×	×	—	×	×	—	—	×	×	×	—	×	—	—	×	—	0.55	
—	—	—	×	×	×	—	×	—	—	—	×	—	×	×	—	×	—	×	×	0.50	0.48
—	×	×	—	—	—	—	—	—	×	—	—	—	×	—	×	—	—	×	×	0.35	
—	—	—	—	—	—	×	—	—	—	—	—	—	—	—	—	—	—	—	—	0.05	

× Breakdown; — No breakdown.

ducting a mathematical comparison of the probabilities of the subgroups, among themselves and with the sample as a whole (see Section 2.5.2.4). To obtain useful information from the test, the size chosen for the subgroups n_T should not be too small. The value recommended is $n_T \geqq 20$; on no account should $n_T < 10$. If one wishes to limit the amount of calculation, it is usually sufficient to compare the first and last or particularly greatly differing subgroups.

Example 2.5.3.1.1 The sample represented in Table 2.35*a* is identified as being independent on the basis of the group values h_{ij} fluctuating about the mean value of the total sample ($h_{100} = 0.68$). Table 2.35*b* immediately suggests a falling tendency (dependence). Comparison of the probabilities (see Section 2.5.2.4) of the first ($h_{i1} = 0.95$) and last ($h_{i5} = 0.05$) subgroups produces the test quantity $z = 5.692$, which is far greater than the critical value $\lambda_{0.975} = 1.960$ at significance level $\alpha = 0.05$, so the hypothesis of equality of the relative frequencies of the two subgroups is clearly rejected.

The graphical check, with subsequent comparison, supplies information about the trend of the sample, but not about whether the two complementary events alternate in a random manner in the realisations.

Each group of similar events (i.e. each group of breakdowns or non-breakdowns) is called an iteration. If a number of iterations occur, then certain events will gather together in 'clusters', which is as strong a sign of dependence as frequent alternation. The iteration test can be used to test whether the number of iterations is sufficient to classify the sample as 'random' (i.e. the realisations as being independent of each other).

Hypothesis: The realisations are independent of each other.

Test quantity:

$$z^* = \frac{r - \dfrac{2(n-k)k}{n}}{2\dfrac{(n-k)k}{n\sqrt{n}}} \tag{2.136}$$

where n is the sample size, k the number of events being considered and r the number of iterations (groups of equal events).

Critical value: The critical value λ_q is derived from the standard normal distribution (Table 2.6) as a quantile of order $q = 1-(\alpha/2)$ (two-sided test at significance level α).

Comparison: The hypothesis is rejected if $|z^*| > \lambda_q$.

Example 2.5.3.1.2 The iteration test is applied to the sample in Table 2.35*a*. With a sample size $n = 100$, number of breakdowns $k = 68$ and number of iterations $r = 47$, the test quantity is obtained as $z^* = 0.800$. At significance level $\alpha = 0.05$, the critical value is $\lambda_{0.975} = 1.960$. Because $0.800 < 1.960$, the hypothesis cannot be rejected. The sample can be regarded as random.

With their information concerning trends and random sequence, the graphical check and the iteration test supplement each other. They ought both to be used together (see Section 3.2.2). Further methods are given in Table 2.34 or are described by Paderta (see Bibliography), with reference to high-voltage-engineering examples, using a derived test (F test of variances, χ^2 test, comparison of confidence regions).

2.5.3.2. Continuous variates Here again a graphical method is advisable in the first instance: the realisations are represented in a diagram in the order of their occurrence, as a function of the number of the individual test (Fig. 2.40). Again a visual assessment can be made during the experiments: if the realisations fluctuate in a random manner about a mean value, then there is no objection to the assumption of independence (Fig. 2.40, test series at SF_6 pressures $p_{20} = 0.15$, 0.25 and 0.40 MPa). If, on the other hand, there is a falling (Fig. 2.40, $p_{20} = 0.1$ MPa), rising or periodically fluctuating tendency, a dependence must be assumed.

The graphical check can in its turn be quantified using comparisons (F test, t test, U test — see Section 2.5.2) of subgroups of the total sample. In particular, the first and last subgroups should be tested for a common

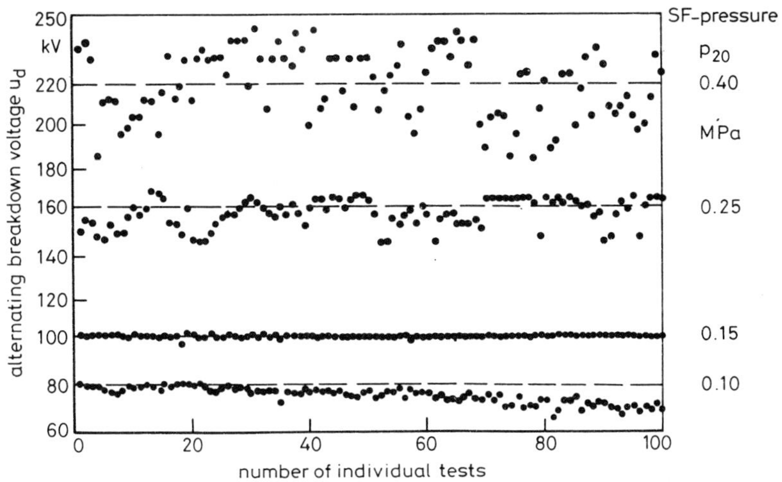

Fig. 2.40 Graphical check of independence in the case of a continuous population

Coaxial cylinders in SF_6; test series for $p_{20} = 0.15$, 0.25, 0.40 MPa independent; test series for $p_{20} = 0.10$ MPa dependent

population to assess the question of independence. There should be at least ten realisations in each subgroup.

It is expedient to reduce the test of independence of the continuous realisations to an iteration test. To do so, the realisations are compared with the arithmetic mean: if they are larger, a plus sign is inserted; if they are smaller, a minus sign. This sequence of complementary events is tested in the iteration test (see Section 2.5.3.1). The quantity k is formed from the plus sign; should the mean value happen to coincide with a realisation, a plus sign is inserted for this.

Example 2.5.3.2.1 An independence test is to be carried out for the sample of size $n = 14$ represented in Table 2.36. To do so, the realisations are compared with the mean value $\bar{u}_d = 104.8$ kV, and the iteration test is applied to the sign: with $k = 7$ plus signs and $r = 8$ iterations, the test quantity is obtained from eqn. 2.136 as $z^* = 0.535$. With a significance level $\alpha = 0.05$, a critical value $\lambda_{0.975} = 1.960$ is obtained. Because $0.535 < 1.960$, the hypothesis is not rejected. There is consequently no objection to the sequence of realisations being regarded as independent.

A common problem is that of a value in the sample appearing to be invalidated in some way (e.g. by measuring instrument errors or logging errors). This presumably invalidated and thereby dependent value would not then be representative of the population and would have to be removed from the sample as a so-called maverick or outlier (cleaning-up sample).

A maverick test can be undertaken to test whether a maverick is present (Table 2.34).

Table 2.36 Application of iteration test to a sequence of continuous realisations (Example 2.5.3.2.1)

No.	Breakdown voltage u_{di}/kV	Sign of $u_{di}-\bar{u}_d$ ($\bar{u}_d = 104.8$ kV)	Number of plus signs k	Iterations r	Test quantity
1	103	−		1	
2	107	+	1	2	
3	101	−		3	eqn. 2.136
4	109	+	2	4	
5	104	−			$z^* = 0.535$
6	102	−		5	
7	106	+	3		
8	105	+	4		
9	108	+	5	6	
10	102	−			
11	100	−			
12	99	−		7	
13	110	+	6		
14	111	+	7	8	

Table 2.37 Critical value $W_{n;\alpha}$ for maverick test [2.11]

Size of sample n	$W_{n;\alpha}$ for significance level $\alpha = 0.02$	$\alpha = 0.05$	$\alpha = 0.10$
10	2.540	2.414	2.294
15	2.800	2.638	2.494
20	2.959	2.779	2.623
30	3.156	2.958	2.792
40	3.281	3.075	2.904
50	3.370	3.160	2.987
$\rightarrow \infty$	$W_{n,\alpha} \approx u\sqrt{\frac{2(n-1)}{2n-5+u^2+(3+u^2+2u^4)\frac{1}{6(2n-5)}}}$		
(asymptotic relationship)	$u = \psi\left(1-\frac{0.02}{2n}\right)$	$u = \psi\left(1-\frac{0.05}{2n}\right)$	$u = \psi\left(1-\frac{0.10}{2n}\right)$

$\psi(q) = \Phi^{-1}(q)$ is the inverse function of the normal distribution (see Table 2.6 and especially [2.11])

Hypothesis: The value x_E belongs to a normally distributed population represented by a sample with an arithmetic mean $\bar{x}$ and variance s^2; x_E is not a maverick.

Test quantity:

$$t^* = \frac{x_E - \bar{x}}{s}. \tag{2.137}$$

Critical value: The critical value $W_{n;\alpha}$ is available in tabular form (Müller *et al.*, Table 14*c*); an extract has been reproduced in Table 2.37 for the two-sided test at significance level α).

Comparison: The hypothesis is rejected, i.e. X_E is a maverick, if $|t^*| > W_{n;\alpha}$.

Example 2.5.3.2.2 A series of tests has produced the individual values reproduced in Table 2.36. The sample has a mean value $\bar{u}_d = 104.8$ kV and a mean square deviation $s = 3.83$ kV. The size of the sample is to be increased to $n = 15$ by additional measurements. In these the value $u_{dE} = 89$ kV is determined. There is a suspicion that this value is a maverick. The total sample is extended by u_{dE}, and the parameters $\bar{u}_d = 103.7$ kV and $s = 5.50$ kV are obtained. Using eqn. 2.137, the test quantity for the maverick test is obtained as $t^* = -2.681$. The critical value $W_{15;0.05} = 2.638$ is obtained from Table 2.37 at significance level $\alpha = 0.05$ (by linear interpolation). Because $|-2.681| > 2.638$, the hypothesis is rejected: u_{dE} should be regarded as a maverick and must not be included in the sample.

Chapter 3

Planning, execution and evaluation of measurements

The mathematical/statistical fundamentals gathered together in Chapter 2 constitute the tools for planning, performing and evaluating high-voltage engineering measurements. It is already apparent from the examples in Chapter 2 how the relationships and methods explained can be used, but there they were arranged according to mathematical criteria. In this Chapter, we shall consider their application to engineering and present their relationships in a technically logical sequence.

3.1 Selection of measuring methods, experimental parameters and test equipment

3.1.1 Stressing insulation systems

The experimentally determined distribution function of a variate describing the insulating capacity (breakdown voltage, maximum breakdown field strength, electric strength, etc.) should provide information directly about the insulation used under practical conditions. The stresses implemented when determining the distribution function must therefore be equivalent to stresses occurring in actual service. The insulation systems, for example, in an electrical-engineering installation or in standard tests (see Chapter 4), are stressed with voltages whose amplitude and time characteristics are clearly determined — even if not always in a known manner. It is therefore not the stress but the behaviour of the insulation that is of a random nature: does breakdown (or partial discharge) occur? Repeated stressing with identical voltages produces the problem dealt with for the binomial distribution, of how often an event (breakdown) occurs within a series of n tests (see Section 2.3.1.3). In the result of such a test of size n, one obtains an estimate of the breakdown probability at the given voltage. The same procedure will generally then be repeated at other voltages, e.g. at Δu higher on each occasion (Fig. 3.1), thereby producing a correlation of relative breakdown frequencies as estimates for breakdown probabilities at the given voltage values. It has become customary to refer to the overall procedure illustrated by Fig. 3.1 for several voltages (each constant) as a 'constant-voltage test'. Following the example of Hochrainer, the relationship between stress and breakdown probability (or inception probability of partial discharges) obtained in the constant-voltage test is called the 'performance function'. For a particular voltage, the value of the performance function is $V(u) = p$.

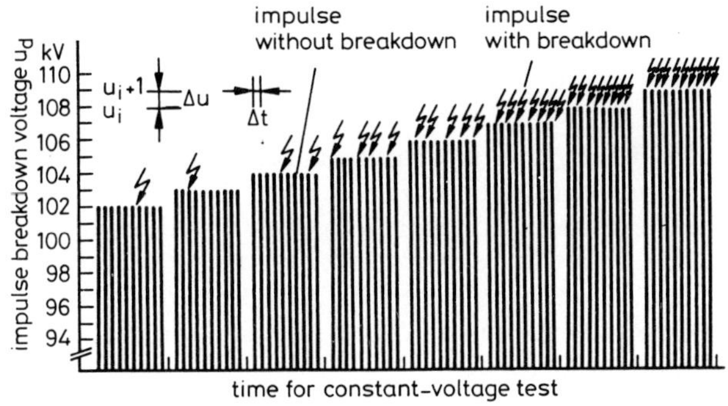

voltage step	m = 1	2	3	4	5	6	7	8
voltage applications per step	n = 10	10	10	10	10	10	10	10
breakdowns per step	k = 1	1	2	4	5	7	8	9
breakdown frequency	h = 0.1	0.1	0.2	0.4	0.5	0.7	0.8	0.9
breakdowns, total	g = 1	2	4	8	13	20	28	37

Fig. 3.1 Procedure for a constant-voltage test (diagrammatic)

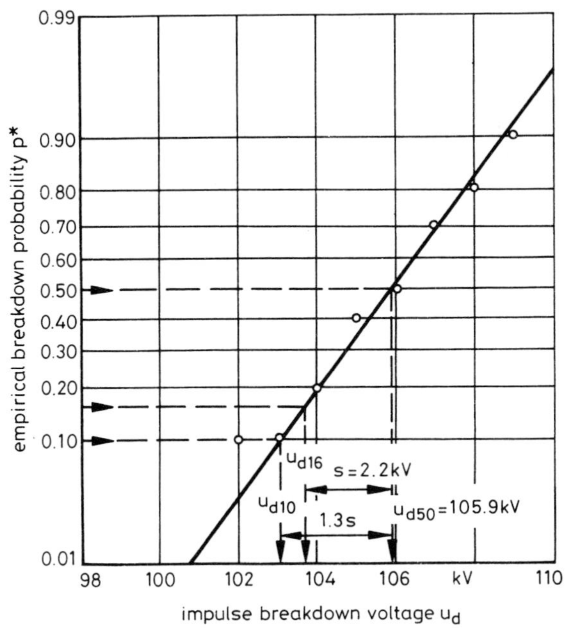

Fig. 3.2 Representation of a performance function in a normal-distribution probability grid

(diagrammatic: as Fig. 3.1, but read 1.3s instead of 1.3σ)

Like distribution functions, performance functions are reproduced on probability grids (see Section 2.5.1) (Fig. 3.2), even though their properties, described below (Section 3.1.2), can differ considerably from them.

Example 3.1.1.1 The graphical representation of the relative breakdown frequencies obtained in Fig. 3.1 for the $m=8$ voltage steps demonstrates as point estimates for the breakdown probabilities in the probability grid (Fig. 3.2), that the performance function can, like an empirical distribution function, be approximated by a normal distribution. The parameters can be determined from quantiles (see Section 2.3.2.2) or by linear regression (see Example 2.4.3.2).

With a view to the practical applicability of test results, the aim is always to perform constant-voltage tests. However, in tests where the voltage is not applied as an impulse, but has to be raised slowly (direct and alternating voltage), as well as in the case of expensive test-pieces, constant-voltage tests are difficult to achieve, either for technical or for economic reasons. One would often like to obtain more precise information from a test piece than a mere yes-no decision. In such cases the voltage is raised in a well defined manner during a test until breakdown occurs. The voltage involved may be a steadily raised direct voltage, an alternating voltage being steadily increased in respect of peak values, or an impulse voltage that is raised through Δu after each voltage application (Fig. 3.3).

Breakdown is then certain as an event: what is random is the amplitude of the breakdown voltage (or of a quantity derived from it). Such a 'rising-voltage test', consisting of repeated voltage rises, produces a sample of n realisations of the selected continuous variate. The sample is evaluated in

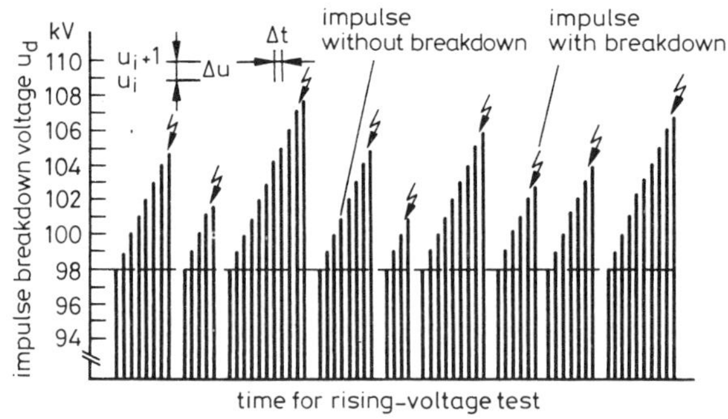

individual test	n = 1	2	3	4	5	6	7	8	9
breakdowns per individual test	k = 1	1	1	1	1	1	1	1	1
cumulative frequency	h_Σ= 0.6	0.2	0.9	0.5	0.1	0.7	0.3	0.4	0.8
breakdowns, total	g = 1	2	3	4	5	6	7	8	9

Fig. 3.3 Procedure for a rising-voltage test, using impulse voltages with a fixed initial voltage u_0

(diagrammatic)

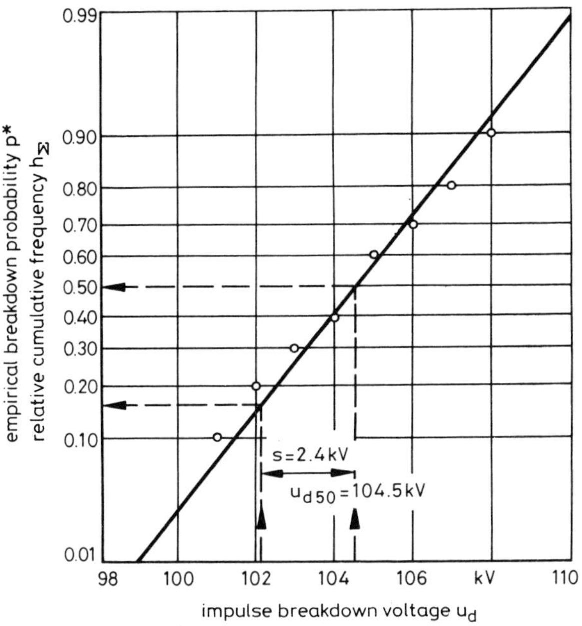

Fig. 3.4 Representation of a cumulative frequency function on a normal-distribution probability grid

(diagrammatic; according to Fig. 3.3)

precisely the manner described for continuous variates (see Sections 2.2.2 and 2.5.1) and produces an empirical distribution function which is termed a 'cumulative frequency function', as distinct from a performance function (Fig. 3.4). It should, of course, be noted that the parameters of the rising-voltage test (rate of rise of voltage v_u or step height Δu; initial voltage u_0) affect the characteristic of the cumulative frequency function. Compared with the constant-voltage test, the rising-voltage test provides more information for a smaller number of voltage applications resulting in breakdown (cf. Figs. 3.1 and 3.3, with a total of 37 and 9 breakdowns, respectively), but the results are dependent on the test parameters and cannot be directly applied to practical application of the insulation. We intend, therefore, first of all to take a closer look at the properties of performance functions and cumulative frequency functions.

3.1.2 Performance function and cumulative frequency function

The monotonic rise of a performance function from zero to unity, as illustrated in Fig. 3.2, is not conclusive. It is possible, for example, with an electrode arrangement in a gas (in the region of the so-called anomaly areas — for point/plane arrangements or coaxial cylinders in air, in the sparking-distance range of a few centimetres), for electrical breakdown to occur at lower voltages from a streamer discharge, and at higher voltages from glow

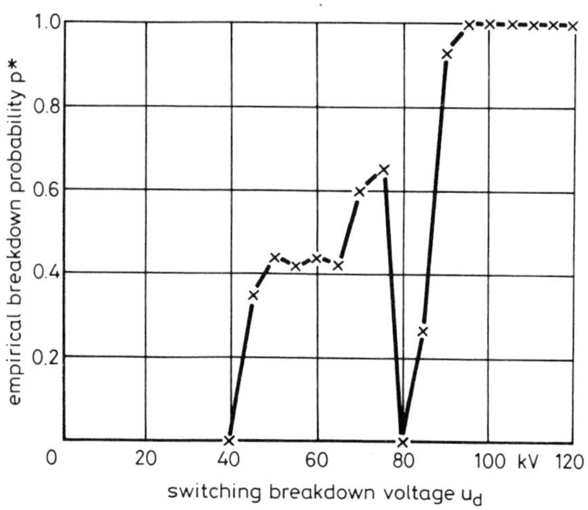

Fig. 3.5 Empirical performance function, which can be approximated by a theoretical distribution function [3.5]

Coaxial cylinders in air; $d_i = 7.5$ mm; $d_a = 90$ mm; switching voltage + 190/18600

discharge. The result of this voltage-dependent change in the discharge mechanism is that, as the voltage rises, the breakdown probability is first reduced and then increases again. However, this deviation from a monotonic rise should not be regarded as being typical of performance functions. In the overwhelming majority of cases, in the relatively narrow range of the spread of the realisations, no voltage-dependent change in the discharge mechanism will occur. Most performance functions consequently rise monotonically, like distribution functions (Fig. 3.2). There is consequently no objection to approximating performance functions by theoretical distribution functions of continuous variates (see Section 2.3.2) and to working with them as with distribution functions (see Section 3.2). If there is a suspicion in particular cases that the performance function does not rise monotonically, this should be checked by confidence estimates for the breakdown probability (Fig. 3.6). If the suspicion is confirmed, approximation with a theoretical distribution function must naturally not be used. In such rare cases one should continue to work with the empirically determined performance function.

If the performance function $V(u_d)$ of an insulation arrangement is known for a particular impulse voltage (or for a direct or alternating voltage of a given amplitude and duration), the cumulative frequency function can be calculated for any step-height Δu. The probability of breakdown occurring at the initial voltage u_0 is derived directly from the performance function as $p_0 = V(u_0)$.[1] After the first step $(u_0 + \Delta u)$, the breakdown probability at this voltage is naturally $V(u_0 + \Delta u)$, but the first breakdown at this step presupposes that no breakdown has taken place at the previous step.

[1] u_0 is naturally chosen so that breakdown is extremely improbable, i.e. $V(u_0) \approx 0$.

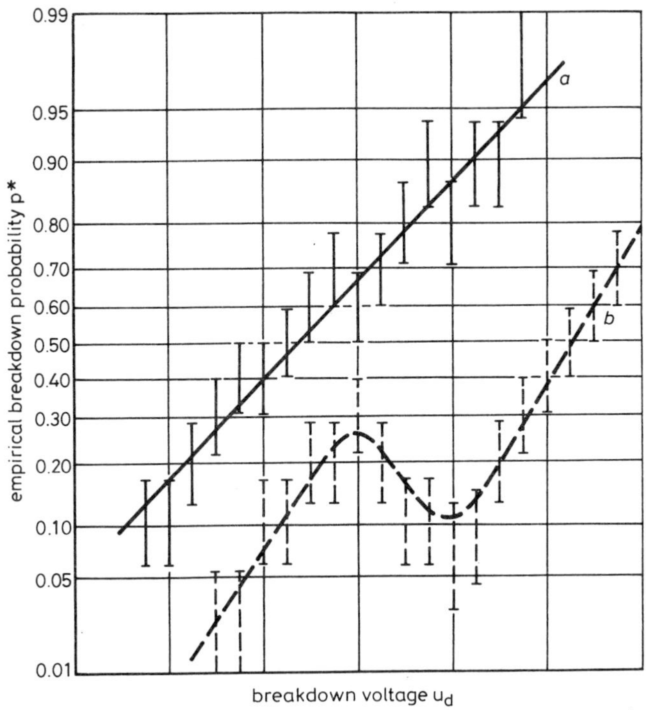

Fig. 3.6 Checking a performance function (diagrammatic)

a Approximation by distribution function possible;
b Approximation by distribution function impermissible

Consequently, for breakdown at the first step:

$$p_1 = V(u_0 + \Delta u)[1 - V(u_0)].$$

Similarly, for breakdown at the second step $(u_0 + 2\Delta u)$, it is assumed that no breakdown has occurred at the two previous steps, and for breakdown at the second step:

$$p_2 = V(u_0 + 2\Delta u)[1 - V(u_0)][1 - V(u_0 + \Delta u)].$$

For breakdown at the kth step, we similarly obtain:

$$p_k = V(u_0 + k\Delta u) \prod_{j=0}^{k-1} [1 - V(u_0 + j\Delta u)].$$

The value of the cumulative frequency function at $(u_0 + k\Delta u)$ is the sum of the probabilities of breakdowns at the individual steps,[1] where the number of steps and the absolute amplitude of the voltage at the steps is determined

[1] Probabilities p_0 to p_k supply the empirical density function.

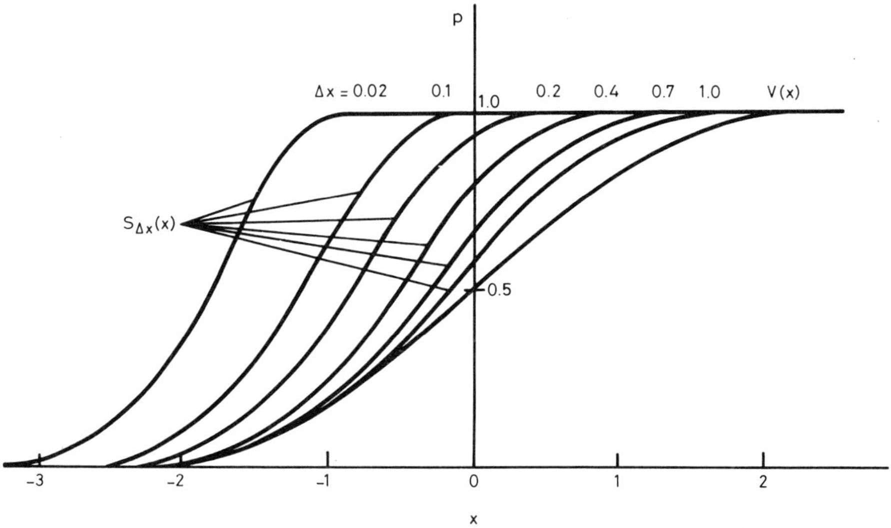

Fig. 3.7 Calculation of cumulative frequency functions $S_{\Delta x}(x)$ for various step-heights Δ_x from a performance function $V(x)$ assumed as being $N(0;\,1)$-distributed.

$$\Delta x = \frac{\Delta x}{\sigma}; \qquad x = \frac{u-\mu}{\sigma}$$

by the step height Δu and it is assumed that $V(u_0 - \Delta u) = 0$:

$$S_{\Delta u}(u_0 + k\Delta u) = \sum_{i=0}^{k} \left\{ V(u_0 + i\Delta u) \prod_{j=0}^{i-1} [1 - V(u_0 + j\Delta u)] \right\}. \qquad (3.1)$$

With a given performance function $V(u_d)$, $S_{\Delta u}(u_d)$ will differ from this more with a smaller chosen step height Δu, i.e. more steps to reach a particular voltage value. In this connection, Fig. 3.7 shows the result of a calculation on a computer. A standardised normal distribution $V(x) = N(0;\,1)$ is adopted as the performance function, and the cumulative frequency functions $S_{\Delta x}(x)$ are calculated for the standardised step-heights $\Delta x = \Delta u/\sigma$, using eqn. 3.1. It is observed that the cumulative frequency functions always lie on the 'sure' side of the performance function, i.e. indicating excessively high breakdown probabilities. For $\Delta x \geqq 1$ $(\Delta u \geqq \sigma)$, the difference between $V(x)$ and $S_{\Delta x}(x)$ is still small.

The procedure for determining the cumulative frequency function can also be modified so that several voltage applications can take place at each voltage step. The relationship between performance function and the cumulative frequency functions that are then possible is derived in a similar way to eqn. 3.1. It is apparent that such methods are better suited for determining small quantities of the breakdown voltage, which we shall consider in greater detail in Section 3.4.2.

3.1.3 Questions of statistical test planning

High-voltage engineering large-number measurements must be planned with great awareness of the attendant responsibility, since the cost of test pieces, the expense of test-plant use, and the time cost of test personnel are generally considerable. When determining the test parameters, particular attention must therefore be paid to ensuring

- comparability of the results with other laboratory tests and with practical stresses
- the general applicability of the results, by appropriate variation of the test parameters and by sufficiently large sample sizes.

To achieve this, one should be clear from the outset as to the statistical aim of the test (e.g. estimation of the average trend of a dependence or of the complete distribution function). Known models of the relationships being investigated (e.g. streamer theory of gas breakdown, damage accumulation in solid insulation) should be used when determining the test parameters.

Considerable expenditure on tests must always be expected when there are numerous test parameters and when the influencing tendency of the parameters is unknown, at least quantitatively. It is here that the methods of statistical test planning can be advantageously employed. Whereas in conventional test planning, provision is made for changing one variable in stages while the others are kept constant, in statistical test planning all the variables are often changed simultaneously within a test plan. By evaluating a relatively small number of tests, it is then possible to separate essential variables (main influences) from non-essential, and also to assess the interaction between different variables (parameters). The method of statistical test planning is attributable to R.A. Fisher. There is now a very extensive and, in mathematical terms, in most cases extremely exacting literature relating to such problems. Scheffler has produced a useful introduction for the engineer, which dispenses with mathematical details and is based on the work of Box, but it relates mainly to metallurgy and chemistry. The linear method described by Scheffler is easy to understand: we shall therefore explain it in more detail as an example. These explanations are only intended to point the reader in the direction of further specialist literature (see Bibliography) and of more complicated applied examples from high-voltage engineering, and to make that path easier.

A random process is being investigated as a function of several influencing factors (variables, parameters) x_1 to x_k, where the trend and strength of the influences are unknown. Only two settings of each influencing factor are investigated: the lowest value (initial state, characterised by '−') and the highest value (final state, characterised by '+') of these variables. Test planning is based on all possible parameter-combinations. With k influencing factors, $N = 2^k$ combinations are obtained and, by the same token, N individual tests. In Table 3.1, these parameter combinations are tabulated in a so-called plan matrix for $k = 4$. From the table, the plan matrices can be obtained for $k = 2$ ($N = 4$ tests), $k = 3$ ($N = 8$ tests) and $k = 4$ ($N = 16$ tests). In the answer matrix, the results (answers) y_{nj} of the N tests are represented.

Table 3.1 First-order test plan with $k = 4$ parameters and $2^k = 16$ test settings (diagrammatic)

	Plan matrix (influencing factors) x_l				Matrix of independent variables (sign z_{ij})																Answer matrix y_{nj}			
					0 $i =$	1	2	12	3	13	23	123	4	14	24	124	34	134	234	1234				
					Main influencing factors	■	■		■				■											
					2-factor interaction			■		■	■			■	■		■							
					3-factor interaction							■				■		■	■					
j	x_1	x_2	x_3	x_4	4-factor interaction															■	y_{1j}	y_{2j}	y_{3j}	. . .
1	−	−	−	−	+	−	−	+	−	+	+	−	−	+	+	−	+	−	−	+				
2	+	−	−	−	+	+	−	−	−	−	+	+	−	−	+	+	+	+	−	−				
3	−	+	−	−	+	−	+	−	−	+	−	+	−	+	−	+	+	−	+	−				
4	+	+	−	−	+	+	+	+	−	−	−	−	−	−	−	−	+	+	+	+				
5	−	−	+	−	+	−	−	+	+	−	−	+	−	+	+	−	−	+	+	−				
6	+	−	+	−	+	+	−	−	+	+	−	−	−	−	+	+	−	−	+	+				
7	−	+	+	−	+	−	+	−	+	−	+	−	−	+	−	+	−	+	−	+				
8	+	+	+	−	+	+	+	+	+	+	+	+	−	−	−	−	−	−	−	−				
9	−	−	−	+	+	−	−	+	−	+	+	−	+	−	−	+	−	+	+	−				
10	+	−	−	+	+	+	−	−	−	−	+	+	+	+	−	−	−	−	+	+				
11	−	+	−	+	+	−	+	−	−	+	−	+	+	−	+	−	−	+	−	+				
12	+	+	−	+	+	+	+	+	−	−	−	−	+	+	+	+	−	−	−	−				
13	−	−	+	+	+	−	−	+	+	−	−	+	+	−	−	+	+	−	−	+		(Test results	—answers—)	
14	+	−	+	+	+	+	−	−	+	+	−	−	+	+	−	−	+	+	−	−				
15	−	+	+	+	+	−	+	−	+	−	+	−	+	−	+	−	+	−	+	−				
$N = 16$	+	+	+	+	+	+	+	+	+	+	+	+	+	+	+	+	+	+	+	+				
	Effect matrix →			y^*_{1i}	$\bar{y}_1$																			
				y^*_{2i}	$\bar{y}_2$	(Mean value formation of answers taking sign into account)																		
				y^*_{3i}	$\bar{y}_3$																			
				⋮	⋮																			

Only one quantity (e.g. $y_{1j} = u_d$ breakdown voltage) can be evaluated, but several quantities ($y_{2j} = u_i$ partial-discharge inception voltage; $y_{3j} = u_e$ partial-discharge cessation voltage, etc.) can also be recorded in the answer matrix and be evaluated later in the effect matrix. To be in a position to perform this assessment, the matrix of independent variables is formed.

The influences on the result (effects) caused by the main influencing factors x_1 (in Table 3.1, $i = 1$; 2; 3; 4), by the two-factor interactions (12; 13; 23; 14; 24; 34), by three-factor interactions (123; 124; 134; 234), by four-factor interactions (1234), and so on (up to k factor interactions), are separated from each other by performing suitable sign-bearing averagings. The matrix of the independent variables now contains the signs with which the results y_{nj} are to be combined in the averaging. As a reference quantity, the arithmetic mean (column 0: only positive sign) is also determined for all the columns as a 0-factor interaction.

The effect matrix contains numerical values which are calculated from the responses y_{nj} using

$$y^*_{ni} = \frac{2}{N} \sum_{j=1}^{N} (z_{ij} y_{nj}) \qquad (3.2)$$

where N is the number of tests and z_{ij} is the sign at point ij of the matrix of independent variables (i is the value of the main influencing factors or interactions; see Table 3.1). In addition, the effect matrix also contains the arithmetical mean values of the answers $\bar{y}_{nj}$. The numerical values of these mean values are the largest values in the lines of the effect matrix, with which the numerical values (effects) determined using eqn. 3.2 can be compared. The value of the effects calculated for the individual influencing factors or interactions is a measure of the influence on the process being investigated.

It is generally sufficient to consider main effects (Table 3.1: $i = 1$; 2; 3; 4) and two-factor interactive effects (Table 3.1: $i = 12$; 13; 23; 14; 24; 34). Higher-order interactive effects can be neglected. The significance of the influencing factors concerned, including their interactions, can be assessed on the basis of the effects, thereby providing indicators for the optimum selection of influencing factors. Let us further illustrate the relationships with an example.

Example 3.1.3.1 A coaxial cylinder arrangement in SF_6 is to be experimentally optimised. (In practice, of course, theoretical optimisation is also possible using the model representation of SF_6 breakdown.) Preliminary consideration has revealed that the insulating gas pressure can be varied in the range 0.25–0.40 MPa, the radius of the inner electrode in the range 3–6 cm, and the maximum roughness-depth of the inner electrode in the range 4–10 μm. If we start with the smallest and largest values of the influencing factors, which are characterised by the signs '−' and '+' in the plan matrix (Table 3.2), then $N = 2^3 = 8$ possible combinations are obtained, which for the sake of clarity are marked with the letters $j = a$ to h. The matrix of independent variables, which contains the signs for the averagings, can be derived directly from Table 3.1 for $k = 3$ or $N = 8$. The aim of

Table 3.2 First-order test plan for coaxial cylinders in SF_6. Outer radius $r_a = 15$ cm fixed; inner radius r_i, insulating gas pressure p_{20} and roughness of inner electrode ρ variable

	Influencing factors			Matrix of independent variables									Result
	① Pressure p_{20}	② Radius r_i	③ Roughness ϱ	0	$i =$	1	2	12	3	13	23	123	Mean value of alternating breakdown voltage u_d
	— (0.25 MPa)	— (3 cm)	— (4 μm)		Main influencing factors	●	●		●				
	+	+	+		2-factor interactions			●		●	●		
	(0.40 MPa)	(6 cm)	(10 μm)		3-factor interactions							●	kV
a	−	−	−	+		−	−	+	−	+	+	−	815
b	+	−	−	+		+	−	−	−	−	+	+	1230
c	−	+	−	+		−	+	−	−	+	−	+	735
d	+	+	−	+		+	+	+	−	−	−	−	1400
e	−	−	+	+		−	−	+	+	−	−	+	780
f	+	−	+	+		+	−	−	+	+	−	−	1150
g	−	+	+	+		−	+	−	+	−	+	−	885
h	+	+	+	+		+	+	+	+	+	+	+	1310
	Plan matrix			1038		469	89	76	−14	71	44	49	Answer matrix
			Effect matrix										Answer

optimisation is to give the arrangement as high a breakdown voltage as possible. The answer matrix therefore contains a single quantity in the form of the breakdown voltages measured at the eight parameter-settings (answer matrix with one column, Table 3.2). For evaluation purposes, the effect matrix is formed (eqn. 3.2), which contains the arithmetical mean value $\bar{u}_d = 1038$ kV (eqn. 2.41) as a reference quantity. Among the main influences, that for $i = 1$ with $u^*_{d1} = 469$ kV, caused by the pressure, is by far the strongest, while the radius ($i = 2$) produces a clearly lesser effect ($u^*_{d2} = 89$ kV). Finally, increasing the roughness ($i = 3$) causes a light negative effect ($u^*_{d3} = -14$ kV), i.e. the breakdown voltage is reduced. Among the interactions, those connected with the insulating gas pressure ($i = 12$ or 13; conditionally also $i = 123$) are relatively strong. It can thus be concluded that the optimum range lies at high pressure, large internal radius and low electrode roughness (case $j = d$). Let us illustrate and critically check this conclusion.

For $k = 3$, the test series can be represented by a cube whose corners constitute the test settings $j = a$ to h (Fig. 3.8). For the pressure, the lower face of the cube, for example, represents the initial state (−); the upper, the final state (+). The left and right side faces represent the radius, the front and rear the roughness. The main effect is now the difference between the mean value of the results in the final state (i.e. for the pressure, from *b, d, f, h*) and the mean value in the initial state (for the pressure, from *a, c, e, g*). A two-factor interactive effect is, for example, for $i = 13$, half the difference between effect 1 on the final state from influencing factor 3 (roughness), i.e. the difference from *f, h* and *e, g*, and effect 1 on the initial state from influencing factor 1, i.e. the difference from *b, d* and *a, c*. The relationships can also be represented in conventional diagram form (Fig. 3.9). The main effect caused by the pressure is derived from the assumed linear change in the resultant breakdown-voltage magnitude at the transition

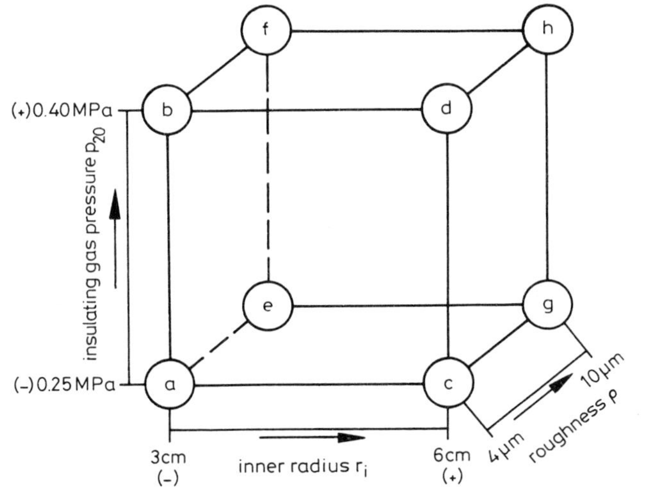

Fig. 3.8 In connection with Example 3.1.3.1: illustration of statistical test-planning

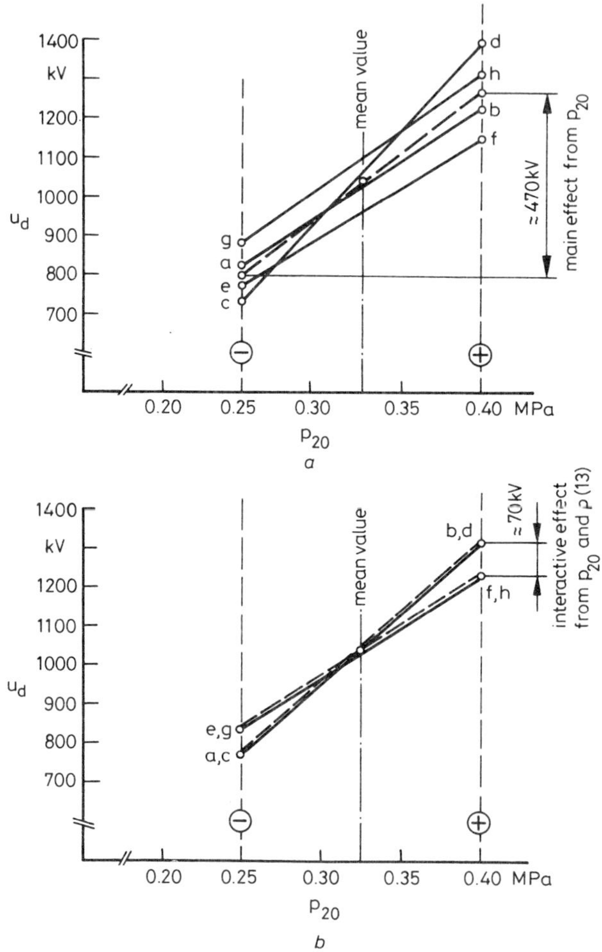

Fig. 3.9 In connection with Example 3.1.3.1: graphical representation of
a main effect 1 (from p_{20});
b interactive effect 13 (from p_{20} and ρ).

from the initial state (*a*, *c*, *e*, *g* in Fig. 3.9*a*) to the final state (*b*, *d*, *f*, *h*), as an average self-adjusting change. Interaction 13 due to pressure and roughness is derived (Fig. 3.9*b*) from the effect of the roughness on the final state (line *e*, *g*–*f*, *h*) and of the roughness on the initial state (line *a*, *c*–*b*, *d*). If the two lines are moved in parallel, so that they intersect at the mean value $\bar{u}_d = 1038$ kV, the interactive effect 13 can be read off at the final state of the pressure (+) (Fig. 3.9*b*).

With its linear assumptions, the method presented in the example is essentially limited to qualitative statements, which can of course be very important for subsequent test design. The relationships represented in Fig.

3.9 in fact permit a mathematical formulation, but their information value is problematical, especially when the assumption of linear relationships is questionable. In the example chosen, this is the case, for example, in the dependence of the breakdown voltage on the inner radius r_i, which for coaxial systems actually passes through a maximum ($r_a/r_i = e = 2.72, \ldots$). Since for r_i the initial and final states (3 cm, 6 cm) are too far apart, linear statistical test planning claims an optimum at the largest possible inner radius.

One solution here is to approximate the relationship between influencing factors and target quantity (breakdown voltage) using quadratic terms. If the statistical test-planning is conducted with such quadratic terms (in addition to the linear and interactive terms), they not only permit qualitative evaluation of the individual influencing factors undertaken in the example, but also quantitative optimisation of the system. Statistical test planning becomes an extremely useful tool for the engineer whenever systems with a large number of parameters have to be optimised, and when experimental or computer-related expenditure has to be minimised.

Statistical test planning has, rather like reliability theory, developed into an independent branch of mathematical statistics. To go any further here would be to exceed the scope of the present book. Reference should again be made to the Bibliography.

3.1.4 High-voltage test equipment for large-number measurements

When planning large-number measurements, great care must be devoted to the selection of the high-voltage test equipment. Naturally one has to take the equipment present in the particular test plant as the initial basis, but the planning or reconstruction of a test plant ought to take account of future measurements under statistical criteria.

To reduce subjective influences — reduction of operator involvement and rational utilisation of test plants — a largely automated test procedure is recommended. In many high-voltage laboratories, efforts are being made to automate test procedures, and test equipment manufacturers also offer complete measuring and control systems. In such a system (Fig. 3.10) for investigations using alternating voltage (and in a similar form for direct voltage), the program is fed into the control equipment (e.g. preselection of rate of rise of voltage and the number of individual tests), and the random values of the breakdown voltage are measured digitally. These data are either printed out or fed for evaluation into a computer. The same system can also be used for partial-discharge measurements (Fig. 3.10), where the voltage is now specified, and the investigated characteristic partial-discharge quantities (e.g. apparent charge, impulse frequency, etc.) occur as random variables, which are displayed as analogue signals or digitally.

It should be stressed here that the digital output of test data is extremely important in the context of statistical evaluation, since the data can then be directly processed numerically. Analogue signals are frequently misread, and the use of calibration curves and subsequent rounding-off (e.g. to

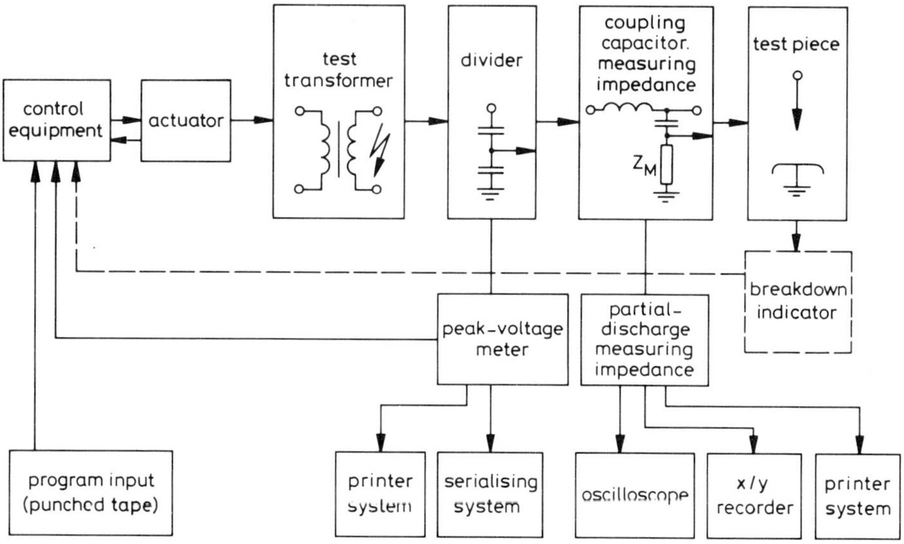

Fig. 3.10 Measuring and control system for alternating voltage

Basic circuit diagram system used at Hermann Matern VEB Transformer and X-ray Works, Dresden (TuR)

whole-number kilovolt values) causes further errors, which may result in certain values accumulating in lists of original test values. It is consequently essential to think in terms of digital-display measuring instruments.

It is more or less impossible to perform statistical large-number measurements with impulse voltages using 'classical' manually operated test equipment: the use of modern control equipment or, better still, automation of the entire test procedure, is a prerequisite for statistical evaluation. The measuring and control system for impulse voltages (Figs. 3.11 and 3.12) can be constructed similarly to that for alternating voltage, but, besides the peak value of the impulse voltage, information on the breakdown/non-breakdown event, charging voltage and, when required, auxiliary quantities such as weather data or times of day, can also be recorded.

Digitalisation of the analogue data supplied by the voltage divider (Fig. 3.13) enables all the relevant variate realisations describing the breakdown process (e.g. peak value of breakdown voltage, instantaneous value of breakdown voltage, breakdown time) to be determined and stored on a microcomputer. Before the data are displayed, they are at least prepared on a microcomputer. Often the micro itself will be capable of performing the statistical evaluation.

It is readily apparent that the need for statistical large-number measurements in high-voltage engineering has to a large extent determined the development of high-voltage test equipment and its automation over recent years. Anyone interested in large-number investigations should therefore

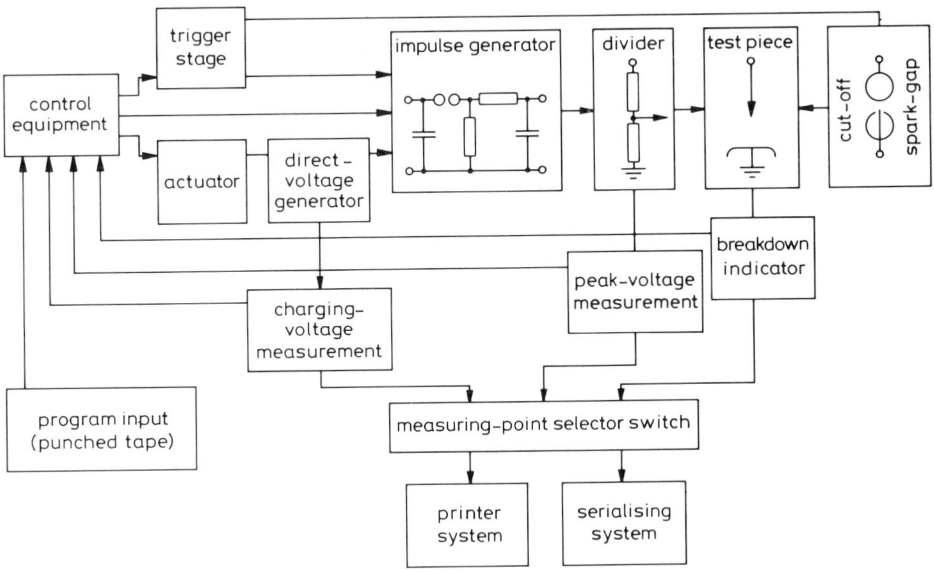

Fig. 3.11 Measuring and control system for impulse voltage

Manufacturer: Hermann Matern VEB Transformer and X-ray Works, Dresden (TuR)

Fig. 3.12 Measuring and control system for impulse voltage

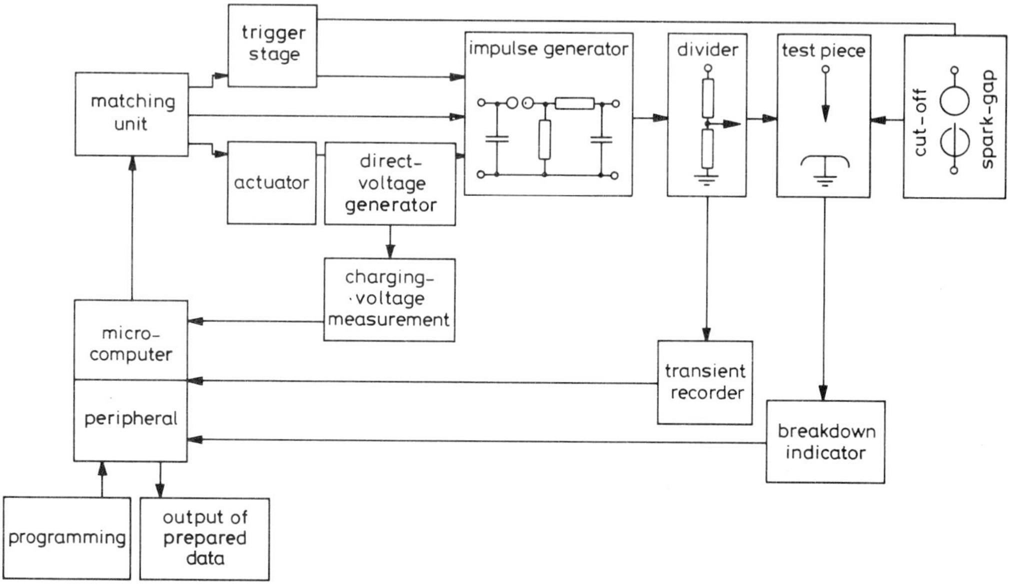

Fig. 3.13 Measuring and control system for impulse voltage, using microcomputer

pay constant attention to the latest developments in test equipment. Engineers' own developments have proved extremely satisfactory in laboratories.

3.2 Constant-voltage tests to determine the performance function

Constant-voltage tests can be performed using all kinds of voltage. For impulse voltages, they represent the 'classical' method of determining breakdown probability: the individual stress generally consists of one voltage impulse, but several impulses can theoretically also be evaluated as an individual stress. (If breakdown occurs in one of them, then 'breakdown' is evaluated as the result for the individual stress.) In the case of continuous voltages (alternating and direct voltage), the individual stress consists of a stress of a given amplitude and duration. For example, with regard to the standard test methods, a stress duration of 1 minute is employed. In respect of the long-term insulating capacity, an individual stress can also be of a duration of ten, a thousand or a few thousand hours. In the following, the relationships are explained by reference to examples of impulse voltages (one impulse per individual stress). They can, however, be accordingly applied directly to all constant-voltage tests using the most varied kinds of voltage.

3.2.1 Test parameters and size of sample

Each constant-voltage test must be so planned and executed as to ensure the independence of the individual tests in relation to each other, the reproducibility and general applicability of the test results. To this end, all the influences on the test procedure must be carefully weighted and recorded as test parameters. Test parameters are, of course, the characteristic quantities of the test voltage used, the ambient conditions (correction to a reference atmosphere has to be undertaken for air insulation), and the parameters of the measuring instruments used. The test parameters of the constant-voltage test in a narrower sense are obtained from the test procedure (Fig. 3.1):

- number of voltage steps (m)
- number of voltage applications per step (n)
- voltage difference between two steps (Δu)
- interval between two voltage applications (Δt).

The total number of individual tests is

$$z = mn \tag{3.3}$$

which is affected, on the one hand, by the dispersion of the random process being investigated, and on the other by accuracy requirements and by economic considerations (test-piece costs, test duration, personnel expenses). An engineering decision is consequently called for that takes into account the risk of faults in the insulation concerned in subsequent laboratory tests, high-voltage tests and especially in service. It is generally important that a new test piece should be used for each individual test, each produced under the same conditions. The interval Δt will then be derived solely from the time taken to disconnect and reconnect the test pieces and to make the voltage available. If, of course, it is manifest — as in the case of insulation arrangements in atmospheric air — that the insulating capacity has been fully restored between two stresses, then all the individual tests can be performed on the same test piece. The interval Δt, and also the total number of voltage applications z, must then be determined in the interest of the mutual independence of the results (see Section 3.2.2).

The number of voltage steps m and the voltage difference Δu are connected. One must first ascertain, from empirical values or preliminary tests, in which voltage range Δu_R the breakdown probability is greater than zero and less than unity ($0 < p(u_d) < 1$). Bearing in mind the voltage graduation Δu_p that can be achieved with the test equipment, the possible number of voltage steps is obtained as

$$m_T = \frac{\Delta u_R}{\Delta u_P}. \tag{3.4}$$

The number of voltage steps used for the test, $m \leqq m_T$, should where possible be greater than $m = 10$; in any case, $m \geqq 5$ must be ensured to determine a complete performance function. In special cases, when only the 50% breakdown voltage has to be estimated, it will be possible to be content with $m \geqq 2$.

The voltage difference is then obtained as

$$\Delta y \approx \frac{\Delta u_R}{m}, \tag{3.5}$$

where it should again be noted that, at all m steps, the breakdown probability must be greater than zero and less than unity, so that $1 \leqq k \leqq (n-1)$ applies to the number of breakdowns k. It is important, however, to choose steps in which $p(u_d)$ is close to zero or unity.

Determining the number of voltage applications per step n has a most emphatic effect on the accuracy of the result. If, for example, the breakdown probability is in practice not estimated accurately for two steps by an extremely large number of voltage applications, then a normal distribution describing the performance function, or a different two-parameter distribution, has already been determined. n can be established on the basis of the confidence limits for unknown probabilities (see Section 2.3.1.3, Fig. 2.11). The width of the confidence region is reduced as the sample size is increased. Fig. 3.14 illustrates this for $p=0.50$. $\Delta p = p_0 - 0.5 = 0.5 - p_u$ can then be specified as a permissible error (e.g. $\Delta p = 0.1$), and the number of voltage applications n (size of sample) be determined from there (e.g. $n=90$).

The method described is suitable for a certain probability, i.e. a relevant quantile of the breakdown voltage. For probabilities $p \neq 0.5$, the requisite nomograms (as in Fig. 3.14) can easily be derived from Fig. 2.11 or be calculated using eqns. 2.58–2.60. One can also, however, start with the complete performance function and treat it as a distribution function. In Section 3.3.1, sample sizes for rising-voltage tests are derived for mathematically exact distribution functions. Hoppadietz has modified this method based on tolerance limits for constant-voltage tests. Reference should be made to his original publication (see Bibliography) and to Section 3.3.1. It

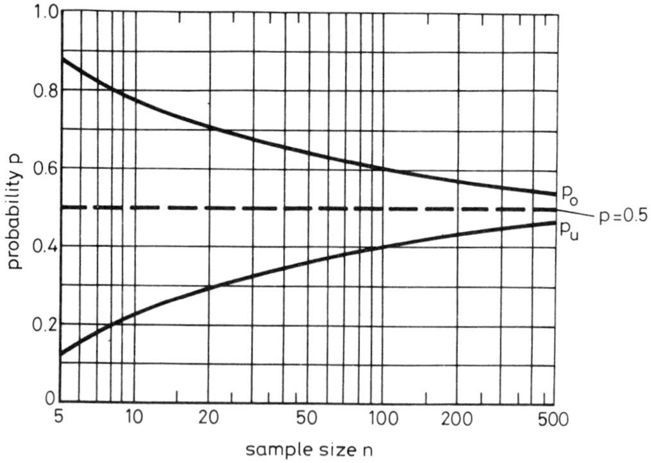

Fig. 3.14 Confidence interval with limits p_0 and p_u for a measured relative frequency $h_n = 0.5$, as a function of sample size n (for confidence coefficient $\beta = 0.95$)

should be borne in mind, however, that the sample size then determined should not be used for the total number z of individual tests, but rather for the number of voltage applications n per step.

Example 3.2.1.1 The performance function of an air insulating arrangement is to be determined in an automated switching-voltage test. The statistical parameters of the constant-voltage test are to be so determined that the error for the breakdown probability $\Delta p < 0.1$. Fig. 3.14, with $p_0 = \Delta p + 0.5 = 0.6$, produces a voltage-application number $n^* = 90$; $n = 100$ voltage applications per step has been chosen. In a preliminary test, it was established that above 1140 kV breakdowns always occur, and below 1065 kV no breakdowns occur. We are therefore interested in a voltage range with a width $\Delta u_R = 75$ kV. The voltage graduation capable of being set on the impulse generator is $\Delta u_p = 5$ kV, so that the possible step number $m_T = 15$ is obtained from eqn. 3.4. Because of the relatively large number of voltage applications per step, the step number $m = 12$ is considered sufficient, so that, using eqn. 3.5, the voltage difference can be established as $\Delta u \approx 6$ kV. According to eqn. 3.3, the total number of individual tests is $z = 1200$. The first step should be at 1065 kV (if, in $n = 100$ voltage applications, no breakdown occurs, this step is not counted). Since the insulating capacity is relatively quickly restored in outdoor air insulation, the gap between two pulses required by the generator, $\Delta t = 30$ s, is sufficiently long as an interval.

This example concerns measurements with a relatively high accuracy. In rough estimates one can manage with considerably fewer individual tests. For accurate measurements, on the other hand, considerably more time and effort has to be devoted. If the performance functions are to be determined down to very small quantiles, then extremely large sample voltage applications per step are required (e.g. for ranges where $p(u_d) < 0.03$ or $p(u_d) > 0.97$, $n \geqq 400$ should be provided). Such large-number tests take a long time, so great attention has to be paid to the constancy of all test parameters and to preventing conditioning effects (especially due to electrode erosion or electrode coatings).

3.2.2 Ensuring independence

If a new test piece is not used for each individual test, the mutual independence of the realisations has to be checked. Dependences of the results of an individual test on the preceding tests are caused by physical changes to the test piece, which in their turn are associated with the test piece itself (insulating material, electrodes), the type of voltage, the energy conversion in preceding partial discharges or breakdowns, as well as with the intervals chosen.

In the case of insulation using atmospheric air, the practically unlimited gas volume ensures a continuous replacement, so that, with intervals in the second range, effects of previous arcs can no longer be detected. Dependences are usually due to changes in the electrodes, produced by partial discharges or breakdowns. Electrodes with small radius of curvature (point electrodes) are particularly prone to erosion; those having small surface areas are prone to roughening or to the formation of non-metallic coatings.

If any such vulnerable electrode arrangement is to be investigated, steps must first be taken in the tests to counter the changes (e.g. limitation of the energy converted in a breakdown, by using series resistors, parallel spark gaps, etc.), and secondly, the independence of the realisations should be checked during the tests (see Section 3.3.2).

With compressed-gas insulation, in addition to the electrode problems for air insulation, there are changes which occur in the enclosed volume of gas. First, it should be borne in mind that discharges produce chemical reactions and thereby changes in the composition of the gas. Secondly, breakdowns in particular leave behind space charges, which recombine comparatively slowly. If the next voltage application occurs after a breakdown before the space charge has recombined, there will be a different field-strength distribution, which will either accelerate or inhibit the development of discharge compared with the case free from space charge. If there are solid dielectric boundary layers in the insulation, due, for example, to supporting elements, firmly adhering surface charges can be produced there — this also happens in air — which will certainly produce dependences. The intervals need to be chosen with great care here; favourable effects are also achieved by using 'interval stresses'. In the interval between two (usually impulse-shaped) stresses, an alternating voltage of limited amplitude (roughly 10–20% of the impulse voltage) is applied; this speeds up recombination of the space charge or surface charge. It is still essential to check independence during the test.

The conditions with liquid insulation are similar to those with compressed-gas insulation, but chemical changes due to discharges (sooting up, gas formation) are more intensive and the space charges are less mobile. Constant-voltage tests with acceptable intervals can only be performed in special cases: one should use rising-voltage tests, in which independence can be better ensured (Section 3.3.2).

In the case of solid insulation there is practically no prospect of a test piece being used for several individual tests. Discharges cause irreversible destruction of the structure of the solid material, and certain structural changes must even be expected before discharges are initiated. A separate test piece has to be used for each individual test. In this instance, dependences can only be produced if the test pieces themselves exhibit dependences (e.g. due to changes in the production process). The test pieces must represent a genuine random selection and, where appropriate, be selected using random-number tables.

Possible ways of checking independence are described in Section 2.5.3.1. Since dependences can be expressed both in the trend of the sample (e.g. increase in breakdown probability in the course of voltage applications), and also in the sequence of the two complementary events (breakdown and non-breakdown), both have to be checked in tabular form during the tests (Table 2.35). In addition, the probabilities of subgroups of the sample can be tested for equality after the tests have been concluded (see Section 2.5.2.4). The test of whether a sufficient number of alternations between breakdowns and non-breakdowns are occurring is performed with an iteration test, also after conclusion of the tests.

Table 3.3 Graphical check of independence. Part test of step 1 = 7 of constant-voltage test; Example 3.2.2.1

$i-j$	Events: × Breakdown – Non-breakdown										h_{ij}	h_n
1–10	–	–	×	–	–	–	×	×	–	–	0.30	
11–20	–	×	–	–	×	–	–	–	–	–	0.20	
21–30	×	–	–	×	–	–	–	×	×	–	0.40	
31–40	–	–	×	–	–	–	–	–	–	×	0.20	
41–50	–	–	–	–	×	–	×	×	–	–	0.30	0.29
51–60	–	–	×	–	–	×	–	–	×	–	0.30	
61–70	–	–	–	×	–	–	–	×	–	–	0.20	
71–80	×	×	×	–	–	×	–	–	×	–	0.50	
81–90	–	–	×	–	–	–	–	–	–	–	0.10	
91–100	×	–	–	–	×	–	×	–	–	×	0.40	
Voltage applications $n = 100$	Iterations $r = 48$										Breakdowns $k = 29$	

Iteration test: test quantity acc. to equation (2.136): $z^* = 1.66$; critical value for $\alpha = 1.960$ } Random sample

Example 3.2.2.1 Independence has to be checked at each voltage step for the constant-voltage test planned in Example 3.2.1.1. The procedure is explained at step $1 = 7$ ($u_d = 1100$ kV) (Table 3.3). Before the test, a table is prepared for the $n = 100$ individual tests, consisting of 10 subgroups each of 10 individual tests, on which the realisations are entered (original list of individual test values). Not less than 10 individual tests should be provided per subgroup. The table is completed during the test. No clusters are apparent, and the breakdown frequencies of the subgroups h_{ij} are randomly dispersed about the value of the total sample h_n. There is thus no reason to abort the test prematurely and to repeat it with different parameters.

The list of test values as a whole demonstrates that there is no systematic trend. Neither comparison of the two extreme probabilities ($h_{71\text{-}80} = 0.50$ and $h_{81\text{-}90} = 0.10$) with the test quantity in accordance with eqn. 2.135 $z = 1.952$, nor with the critical value at significance level $\alpha = 0.05$ ($\lambda_{0.975} = 1.960$), contradicts the hypothesis that all the subgroups come from the same population: a trend is thus absent. An adequate number of alternations is confirmed by the iteration test (eqn. 2.136); the test quantity $z^* = 1.66$ is below the critical value at significance level $\alpha = 0.05$ ($\lambda_{0.975} = 1.960$). There are consequently no grounds for not regarding the sample as independent, either from the trend or the alternations.

The list of test values needed to assess independence does not, of course, have to be filled out by hand: automatic test apparatus (see Section 3.1.4) can print out equivalent lists or even conduct an evaluation of the list of test values.

3.2.3 Empirical performance function

The relative breakdown frequency has to be determined for all m voltage steps, as described in Example 3.2.2.1 for one voltage step. The table, which assigns the breakdown frequencies to the voltages, should be understood as being an empirical performance function.

Example 3.2.3.1 For the constant-voltage test planned in Example 3.12.1.1, the results are recorded in Table 3.4 in the guise of an empirical performance function. The voltage difference between the individual steps is not exactly $\Delta u = 6$ kV as planned, but fluctuates between $\Delta u = 4$ kV and $\Delta u = 9$ kV, since the impulse generator used does not allow an absolutely precise setting. The value of the empirical performance function is not affected by the not-quite-fixed voltage difference. For the relative frequencies, it is advisable to calculate confidence regions for the breakdown probabilities (see Section 2.3.1.3) or, more simply, to read them off from Fig. 2.11. At the same time, the result of the independence test described in the previous Section should also be noted in the table. (The plus sign indicates that the hypothesis of independence is not rejected.) The graphical representation of Table 3.4 is then an illustration of the empirical performance function (Fig. 3.15). The confidence regions show that the slight reduction of breakdown probability between 1083 kV and 1089 kV is random and not significant. The

Table 3.4 Result of the constant-voltage test (Example 3.2.3.1). Insulation arrangement in air, $n = 100$ shots per voltage step

Voltage step	Voltage	Number of breakdowns	Relative breakdown frequency	95% confidence limits as in Fig. 2.11		Independence test	
l	u_d kV	k_l	$h_l = k_l/n$	p_u	p_o	Trend	Change
1	1065	2	0.02	0	0.07	+	+
2	1071	3	0.03	0.01	0.08	+	+
3	1075	5	0.05	0.02	0.12	+	+
4	1083	11	0.11	0.07	0.19	+	+
5	1089	10	0.10	0.06	0.17	+	+
6	1094	21	0.21	0.14	0.29	+	+
7 (Table 3.3)	1100	29	0.29	0.21	0.39	+	+
8	1107	48	0.48	0.39	0.59	+	+
9	1111	56	0.56	0.46	0.65	+	+
10	1120	88	0.88	0.81	0.93	+	+
11	1128	98	0.98	0.93	1	+	+
$m = 12$	1135	99	0.99	0.95	1	+	+

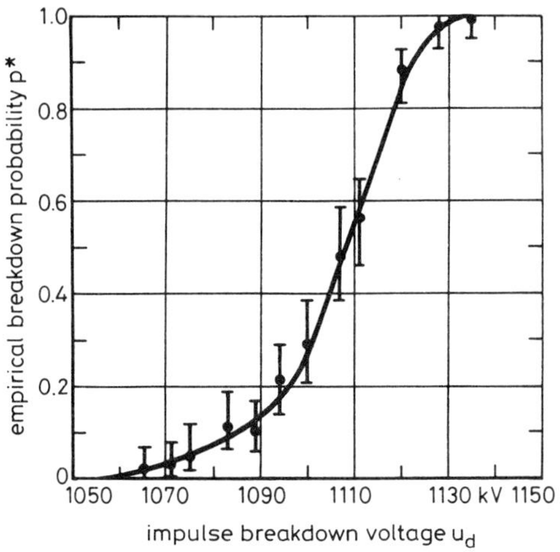

Fig. 3.15 Empirical performance function of an air spark-gap (Example 3.2.3.1).

performance function can therefore be taken as rising monotonically, and there is no objection to its being approximated by a theoretical distribution function (see Section 3.2.4).

In the constant-voltage test, the information content of an individual test (breakdown or non-breakdown) is relatively small. To increase the information content, besides the event in the test, one tries to obtain further information by recording and evaluating the instantaneous value of the breakdown voltage u_{md} or breakdown time t_d for the given peak value. A normal empirical distribution function $F(t_d)$ is obtained for the recorded breakdown times, as it also is for the instantaneous breakdown voltages $F(u_{md})$, if related not to the total number of voltage applications n, but to the number of breakdowns k. $F(t_d)$ and $F(u_{md})$ should be taken as being cumulative frequency functions (Fig. 3.16), from which interesting physical conclusions can be drawn. $F(t_d)$ should preferably be used for lightning voltage, $F(u_{md})$ for switching voltage, since in this case breakdown is generally initiated in the front region of the impulse. The peak value $\hat{u}_d$ should be used as the instantaneous value, both for breakdowns at the peak and on the tail.

According to a suggestion by Ouyang, one can relate to the total number of voltage applications n, so that one obtains a function $V^*(u_{md}\,;\,\hat{u}_d)$ ending at breakdown probability p of the constant-voltage test, this function being closely related to the performance function $V(u_d)$ (Fig. 3.16). If there is no influence of the shape of the impulse (i.e. rate of rise of voltage) on the switching breakdown voltage, the functions can be converted into each other. This is not generally the case, however. In this instance, $V^*(u_{md}\,;\,\hat{u}_d)$ can assist in creating, for the very vague conditions, a statistical tool for the dimensioning of air insulation for extra high voltages and for insulation coordination: the voltage/time characteristics of the voltages used are recorded, and instantaneous values producing the same breakdown probability

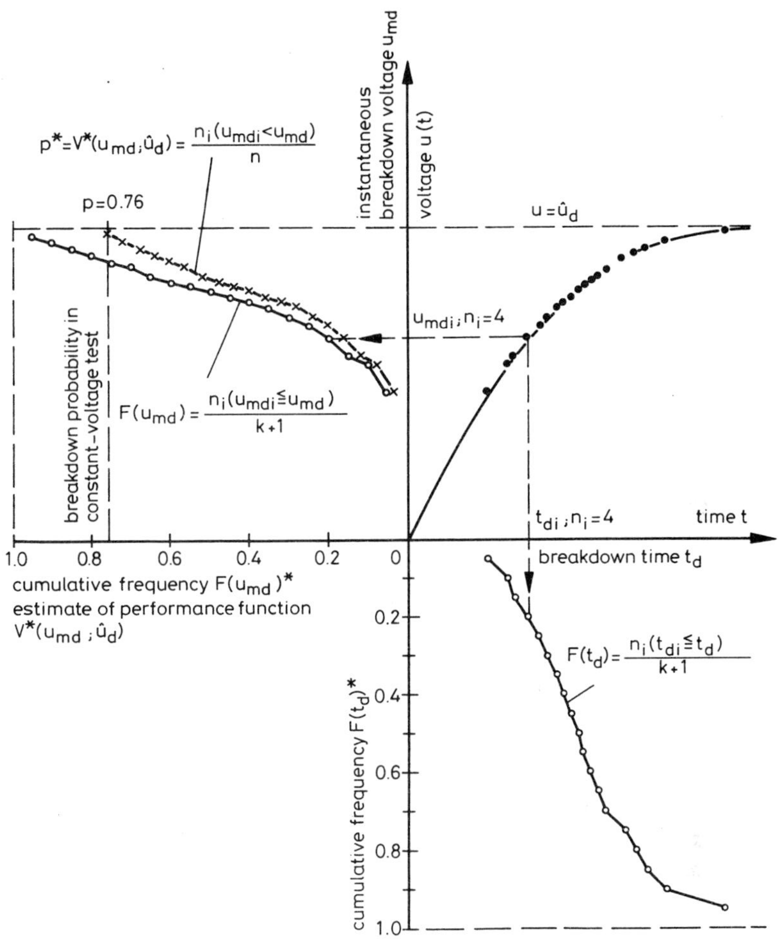

Fig. 3.16 Determination of distribution functions of the instantaneous breakdown voltage $F(u_{dm})$ and breakdown time $F(t_d)$ in a constant-voltage test using switching voltage $\hat{u}$ and $n=25$ individual tests with $k=19$ breakdowns

The points plotted on the voltage characteristic indicate the breakdown instants

are joined together by 'contour' lines (Fig. 3.17). If the process is repeated for a wide variety of impulse waveforms, families of contour lines are obtained (Fig. 3.18). The envelope for selected probabilities is referred to a 'minimum contour line' (in Fig. 3.18, broken line for $p_{min}=0.50$ and $p_{min}=0.023$). The minimum contour lines (Fig. 3.19) are then a very convenient tool for evaluating the performance of an air-insulation arrangement over the possible wide range of switching voltages. Contour lines can, of course, also be obtained by determining performance functions for very varied impulse waveforms and then evaluating them in the form of Fig.

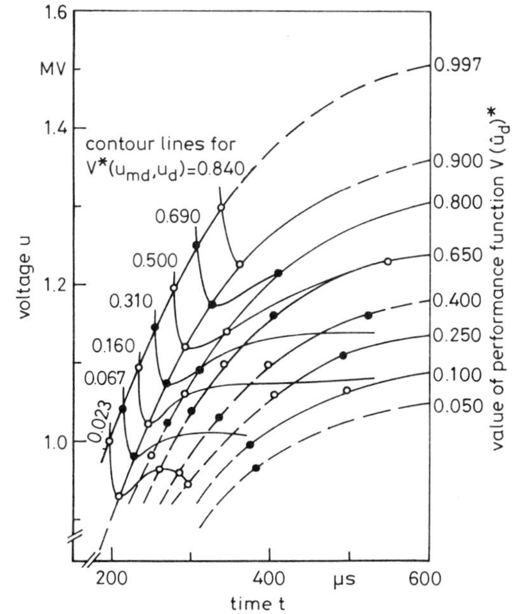

Fig. 3.17 Determination of contour lines [3.41]

Constant-voltage tests with eight different peak values of waveform 680/5200; post insulator in air

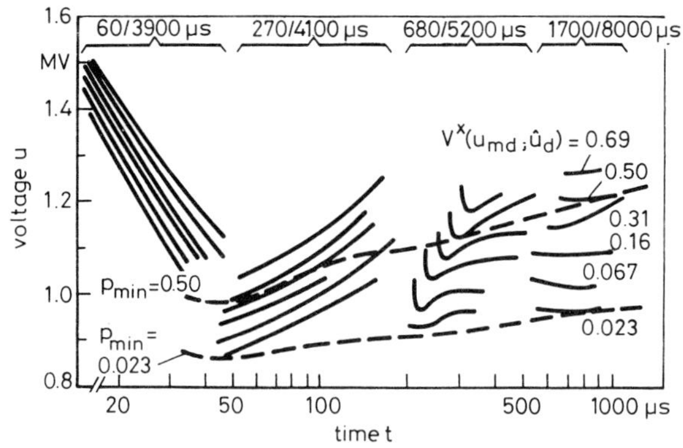

Fig. 3.18 Contour lines for four different switching voltages and determination of minimum contour line [3.41]

Post insulator in air

p_{min} probability of minimum contour line

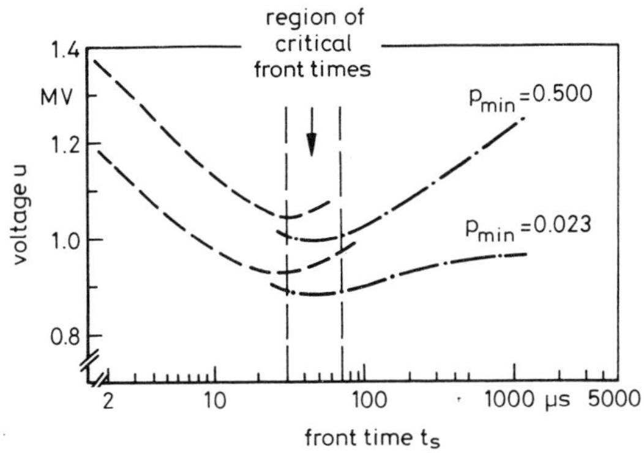

Fig. 3.19 Selected quantiles of breakdown voltage, as a function of the front time of the switching voltage, determined by the minimum-contour-lines method [3.41]

Post insulator in air

3.19. The method attributable to Ouyang and Carrara has, for more complex measurements (not just measuring the peak value, but also the instantaneous value), the advantage that it involves less time and effort. Constant-voltage tests (and not only those on air insulation) ought therefore to be more comprehensively planned, in order to permit the additional evaluation of all-embracing physical or technically interesting information.

3.2.4 Matching performance functions with theoretical distribution functions

When the performance function increases monotonically, as in Fig. 3.15, there are no objections to approximating it by a theoretical distribution function. It should be borne in mind that no probability-theory situation should then be created, but only a mathematically prepared formalism — in fact the theoretical distribution function — appears. In content, the relationship between breakdown probability and voltage (or other specifiable quantities) is still a performance function, as required for the technical evaluation of insulation systems (see Section 3.1.1).

The following fundamentals must be borne in mind when matching performance functions by theoretical distribution functions.

The distribution function must approximate to the empirical performance function with sufficient accuracy. Criteria for the quality of the approximation are obtained on the one hand from the physical or technical objective, while on the other hand statistical distribution tests can be used (see Section 2.5.1).

The theoretical distribution function (see Section 2.3) should correspond in content as far as possible to the model assigned to the physical process being investigated. Such relationships exist, for example, between the ignition processes of gas discharges and the model of an extreme-value distribution. The model of a normal distribution can be advantageously applied to air breakdown developing from partial discharges. Further advice will be found both in the treatment of theoretical distribution functions in Section 2.3.2 and also in Section 5, in the description of insulating capacity.

Finally, the theoretical distribution function used to approximate a performance function should also confer an advantage to the engineer. Complicated approximations should be rejected.

The relationship between the empirical performance function available as a table (Table 3.4) and the theoretical distribution function is usually reproduced on a probability grid (see Section 2.5.1.1), but is better represented by a linear regression (see Section 2.4.3). Matching with a normal distribution has already been dealt with in Example 2.4.3.2. Nothing can be added to that evaluation here. The extensive statistical instruments available for the normal distribution (Section 2.3.2.2) permit a rough estimation of confidence regions and tolerance limits. The number of voltage applications n per voltage step should be used as the sample size. Confidence limits can be calculated for the breakdown probabilities, and also for any quantiles of the performance function matched by a normal distribution. Evaluation of a constant-voltage test can be performed in a probability grid, dispensing with a regression calculation, but regression calculation using pocket calculators is very simple and should be given preference.[1]

Besides a normal distribution, empirical performance functions can also be approximated using other distribution functions for continuous variates. Extreme-value distributions are preferably used. The process will be explained in greater detail below for matching by a double exponential distribution.

Example 3.2.4.1 The empirical performance function given in Table 3.4 and reproduced in Fig. 3.15 is to be investigated with regard to matching by a double-exponential distribution. The point estimates and confidence estimates are first plotted on the probability grid for a double-exponential distribution (Fig. 3.20). It is readily apparent that a line can be drawn through all the confidence regions. The best line ought, however, to be determined by linear regression. To do so, the ordinate is linearised using the inverse function of the double-exponential distribution

$$y = \ln[-\ln(1-p)] \qquad (3.6)$$

and the problem, which is now linear, is dealt with in Table 3.5. With the transformed breakdown probability y and breakdown voltage $u_d = x$, the regression line

$$y = a_{yx} + b_{yx}x = -88.7 + 0.0797x,$$

is obtained, from which the parameters of the double exponential distribution

$$u^*_{d63} = \frac{-a_{yx}}{b_{yx}} \qquad (3.7)$$

[1] The confidence limits of the regression line as a whole are confidence limits for any quantiles of the performance function.

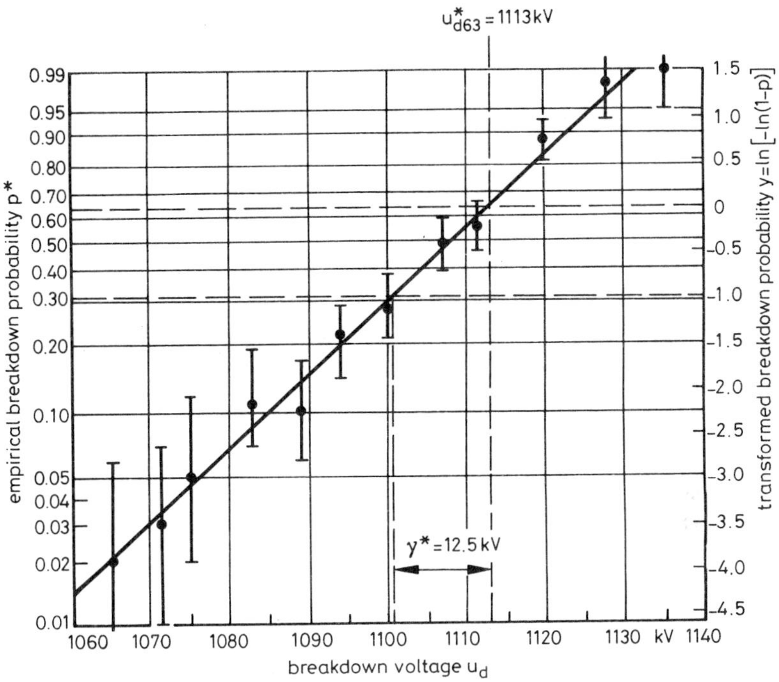

Fig. 3.20 Approximation of empirical performance function of a slightly non-uniform insulation arrangement in air by a double exponential distribution (Example 3.2.4.1)

and

$$\gamma^* = \frac{1}{b_{yx}} \tag{3.8}$$

can easily be calculated (cf. normal distribution, Example 2.4.3.2, eqns. 2.128 and 2.129). In the present case, $u^*_{d63} = 1113$ kV and $\gamma^* = 12.5$ kV. The double-exponential distribution plotted in Fig. 3.20 with these parameters passes through all the confidence regions of the breakdown probabilities and produces a good approximation to the empirical performance function. This can be confirmed, for example, by tagging-on a Kolmogorov test (see Section 2.5.1.2).

3.2.5 Evaluation chart and use of computers

The individual steps in the execution of a constant-voltage test will now be reviewed by reference to Fig. 3.21. After preparation and execution of the tests, lists of test values are available for each of the m voltage steps for test evaluation. These should be tested for independence and be further evaluated if independence is confirmed. After the breakdown frequencies associated with the voltage steps have been determined, the empirical performance

Table 3.5 Evaluation of a constant-voltage test using regression (Example 3.2.4.1)

i	x_i Voltage u_{di} kV	Relative frequency h_i	y_i Transformed frequency $y_i = \ln[-\ln(1-h_i)]$	Operands $(x_i-\bar{x})$	$(y_i-\bar{y})$	$(x_i-\bar{x})(y_i-\bar{y})$
1	1065	0.02	−3.902	−33	−2.714	89.6
2	1071	0.03	−3.491	−27	−2.303	62.2
3	1075	0.05	−2.970	−23	−1.782	41.0
4	1083	0.11	−2.150	−15	−0.962	14.4
5	1089	0.10	−2.250	−9	−1.062	9.6
6	1094	0.21	−1.445	−4	−0.257	1.0
7	1100	0.29	−1.072	2	0.116	0.2
8	1107	0.48	−0.425	9	0.763	6.9
9	1111	0.56	−0.197	13	0.991	12.9
10	1120	0.88	0.752	22	1.94	42.7
11	1128	0.98	1.364	30	2.552	76.6
$m=12$	1135	0.99	1.527	37	2.715	100.5
	$\bar{x}=1098$ kV $s_x=22.8$ kV $s_x^2=521.8$ kV^2		$\bar{y}=-1.188$ $s_y=1.839$ $s_y^2=3.381$	Covariance: $s_{xy}=41.6$ Correlation coefficient: $r=0.992$ Regression coefficient: $b_{yx}=0.0797$ Location coefficient: $a_{yx}=-88.7$	Regression line $y=-88.7+0.0797x$	

function is available and with a monotonically rising characteristic, it can be matched by a theoretical distribution function. After the match has been checked and confirmed, the parameters of the performance function and quantiles of technical interest can be indicated. The test results should be compiled in a suitable manner to enable them to be interpreted. The mathematical formulae for all the steps described are included in this book; the figures given in parentheses in Fig. 3.21 refer to the relevant Sections.

Since evaluation consists of a sequence of mathematical or logical steps, there is nothing to prevent processing on a computer, particularly since the test values from modern test equipment are supplied directly for computer input. In many high-voltage laboratories, computer programs are available for 'off line' statistical evaluation. For the individual mathematical operations, programs have been published and are available in most computer centres. Compiling programs thus presents no fundamental difficulties; it is rather a matter of taking account of the peculiarities associated with a particular computer and programming language. For this reason, we have not included programs here.

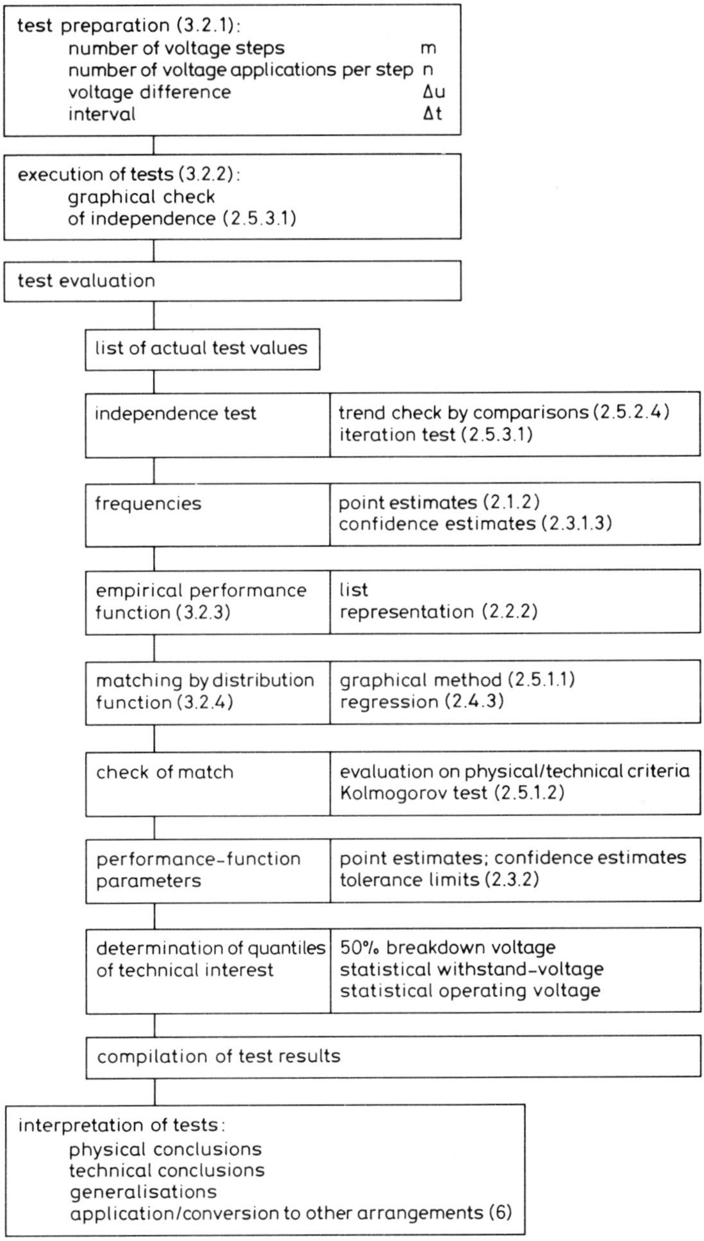

Fig. 3.21 Evaluation chart for a constant-voltage test

The advent of microcomputers in automated test plants has made it possible in future to carry out statistical evaluation at the test plant on-line. The first attempts have already been made (Fig. 3.13).

3.3 Rising-voltage tests to determine the cumulative-frequency function

Rising-voltage tests can be performed using all kinds of voltage. Direct voltages and the amplitudes of alternating voltages are often steadily increased (Fig. 3.22*a*), if one disregards the 'discretisation' dictated by the

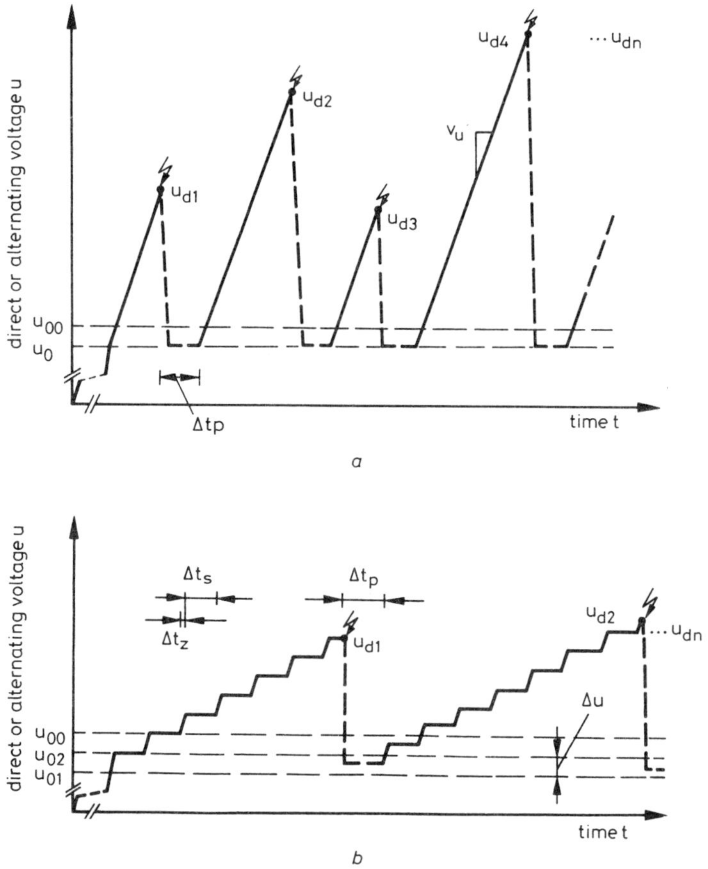

Fig. 3.22 Procedure for a rising-voltage test using continuous voltages (diagrammatic)

a continuous voltage-rise
b voltage rise in steps, with variation of initial voltage u_0

test-plant actuators. The voltage rise can also take place in steps of amplitude Δu (Fig. 3.22*b*), with the direct or alternating voltage kept constant for a given step duration Δt_s (Δt_s should be selected according to the respective requirements and may be between a few tens of seconds and — in endurance tests — several hundred hours). The time Δt_z taken for the rise to the next step should be negligible compared with the step duration Δt_s ($\Delta t_z \ll \Delta t_s$). If a breakdown occurs during the voltage rise, it ought to be established it is assigned to the step from which it was raised. Impulse voltages in rising-voltage tests are always raised in steps (Fig. 3.3), the initial voltage u_0, step amplitude Δu and interval Δt_p being the most important parameters.

In constant-voltage tests, the instantaneous values of the breakdown voltages u_{md} and the breakdown times t_d can also be evaluated as cumulative-frequency functions in the context of a rising-voltage test (Fig. 3.16). Details of this have already been discussed in Section 3.2.3.

3.3.1 Test parameters and sample size

As was the case with the constant-voltage test (see Section 3.2.1), independence and 'generality' of the test results must also be ensured in the rising-voltage test. The general test parameters (test voltage, ambient conditions and measuring equipment) must be kept constant. In a rising-voltage test with a continuously increasing voltage (Fig. 3.22*a*), test parameters in the statistical sense are:

- the rate of rise of the voltage v_u
- the number of individual tests n (identical to the number of breakdowns; realisations; size of sample)
- the interval between two individual tests Δt_p and
- the initial voltage u_0

In rising-voltage tests with the voltage increased in steps these parameters again apply (Figs. 3.3 and 3.22*b*), but with

- the step amplitude Δu

instead of the rate of rise of the voltage, and in the case of continuous voltage,

- the step duration Δt_s and
- the rise time $\Delta t_z \ll \Delta t_s$

in addition.

A fresh test piece generally has to be used for each individual test. Only when it is possible to guarantee with certainty that the insulating capacity is completely restored can all the individual tests be performed on the same test piece. As has already been explained for the constant-voltage test (Section 3.2.1), the mutual independence of the realisations must not be in question when determining the intervals Δt_p and the number of individual tests n. All the test parameters therefore need to be chosen with care.

In all rising-voltage tests, the initial voltage u_0 can be chosen greater than zero. In order not to affect the test result, however, u_0 must be lower than an upper limit u_{0_0} (to be considered in more detail later: see Section 3.3.4), which must not be exceeded by u_0 ($0 \leqq u_0 \leqq u_{0_0}$). As a rule of thumb, the lowest of all breakdown voltages $u_{d\min}$ measured should be at least 10% higher than u_0 ($u_{d\min} > 1.1u_0$). Preliminary tests should be arranged for this purpose. When the voltage is raised in stages, u_0 must be chosen at a lower level for smaller step amplitude Δu. It is expedient to vary the value u_0 between a lower value u_{01} and an upper value $u_{02} = u_{01} + \Delta u$ in steps equivalent to the lowest voltage capable of being set (or the accuracy of reading of the measuring instrument) (Fig. 3.22*b*, which is also applicable to impulse voltages: cf. Fig. 3.3 with fixed value of u_0). In this way, realisations of the variate breakdown voltage are not restricted to a small number of discrete voltages: what is produced rather is a continuum, as in the case of a steadily increasing voltage. In a rising-voltage test, the voltage is normally disconnected at breakdown, and the actuator lowered to a position corresponding to u_0. The voltage u_0 is reapplied after an interval Δt_p, which is timed from the instant of breakdown. Overvoltages are generally produced. These must, of course, neither cause breakdown nor a change in the test piece affecting the breakdown process. Should such a danger exist, u_0 should be specified so that overvoltages do not affect the tests.

At the bottom end of the scale, the step amplitude Δu is limited by the voltage graduation Δu_p which can be achieved with the test plant. At the top end, Δu should not be greater than the mean square deviation s of the breakdown voltage: one should be aiming at $\Delta u \approx 0.5s$. With direct or alternating voltage, the step duration Δt_s will depend on the physical or technical purpose of the test. In tests of short duration, $\Delta t_s = 1$ minute is recommended, since this is the standard duration for proving rated withstand voltages (see Section 4.2). The rise time Δt_z should be as short as possible and must be negligible compared with Δt_s ($\Delta t_z \ll \Delta t_s$). In any case one must ensure that the subsequent step is not overrun during Δt_z due to the high rate of rise of the voltage.

The rate of rise of the voltage v_u for a steady rise can be related to the stepwise rise by the expression

$$v_u \approx \frac{\Delta u}{\Delta t_s}. \tag{3.9}$$

Like Δu, v_u depends on the aim of the test. In many cases the actuators of high-voltage test plants work in a nonlinear manner; sliding transformers in particular have this disadvantage. One therefore needs to be absolutely aware of the actual voltage/time characteristic, as supplied by the test plant during the rising-voltage test. Major deviations from the ideal characteristic (Fig. 3.22*a*) can produce test-result dependences. On the other hand, the breakdown voltages of slightly nonuniform gas-insulation arrangements, for example, are more or less independent of the rate of rise of voltage selected. In any case, the performance of the test plant should be experimentally compared with the reaction of the test piece. Only such a comparison can show whether the rate of rise of voltage that can be achieved with the test plant meets statistical requirements. The number of individual tests, i.e. the sample size n, is determined by the test-result accuracy required (cf. Section 3.2.1). The accuracy of an estimation is of course also affected by the dispersion of the random process being investigated. n can be

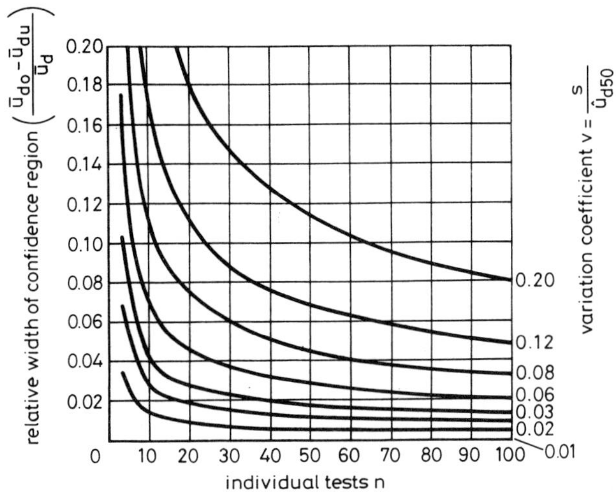

Fig. 3.23 Relative width of confidence region with $\beta = 0.95$, as a function of n and variation coefficient v

established on the basis of the relative width of the confidence region for the mean value, taking the variation coefficient into account (Fig. 3.23). Where economics allow, a width of

$$\frac{\bar{u}_{do} - \bar{u}_{du}}{\bar{u}_d} < 0.02.$$

should be aimed at, but $n \geqq 20$ should be ensured in all cases. Measurements with a smaller sample-size can provide a reference for the mean value, but not sufficient information concerning the distribution function.

It is also possible to determine sample size n using distribution-free tolerance limits. Fig. 3.24 shows the ensured order of the quantile belonging with a confidence coefficient $\beta = 0.95$ to the lowest measured value (lower tolerance limit τ_u) and highest measured value (upper tolerance limit τ_0). For example, with $n = 100$ and $\beta = 0.95$, the smallest realisation $u_{d\min}$ is not less than the 3% quantile u_{d03}. With $n = 28$, $u_{d\min}$ can be regarded as the 10% quantile u_{d10}. The sample size can be selected even more simply using Fig. 3.25, since the graphical representation of the tolerance limits on a double-logarithmic scale produces more or less straight lines. Let us illustrate the procedure with an example:

Example 3.3.1.1 The alternating breakdown voltage, with a steadily increased voltage, is to be measured on a sphere/plane arrangement in air. Preliminary tests have shown that the rate of rise of the voltage does not affect the amplitude of the breakdown voltage in the range $5\,\text{kV/s} \leqq v_u \leqq 35\,\text{kV/s}$. Adopting a value $v_u = 20\,\text{kV/s}$ provides sufficient assurance that deviations by $\pm 15\,\text{kV/s}$ caused by the test plant (i.e. by the position of the

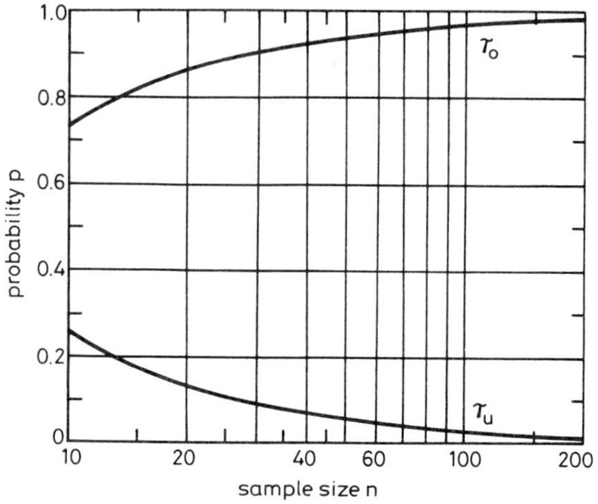

Fig. 3.24 Ensured order of the quantile belonging to the highest measured value (upper tolerance limit τ_0) and lowest measured value (lower tolerance limit τ_u) of a sample, with a confidence coefficient $\beta = 0.95$

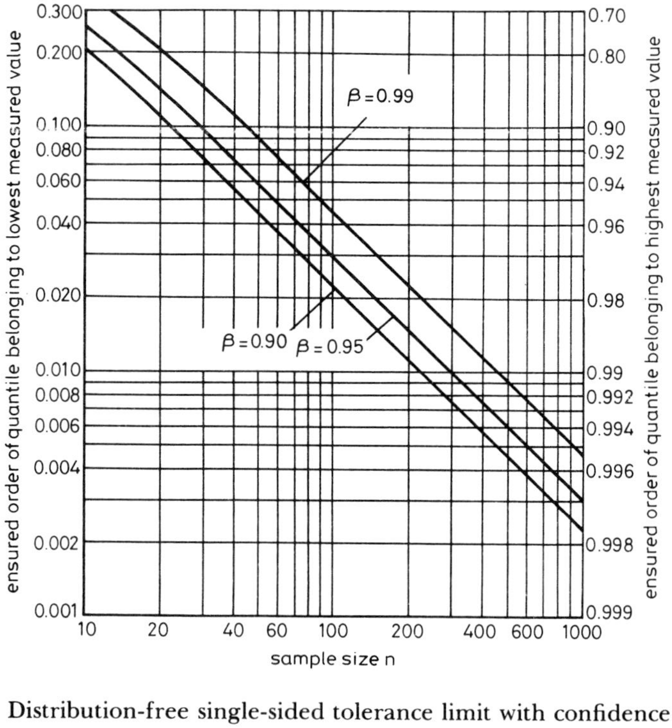

Fig. 3.25 Distribution-free single-sided tolerance limit with confidence coefficient β [2.11]

sliding transformer) do not invalidate the test result. The interval between two individual tests can be set at $\Delta t_p = 0$, since the initial voltage $u_0 = 0$ was chosen because of possible overvoltages, and the normal test procedure (voltage-free return of the actuator and voltage rise) between two breakdowns produces periods of about 1 minute, which are quite sufficient for regeneration of insulating capacity. The sample size n is determined using Fig. 3.25. With a confidence coefficient $\beta = 0.90$, it should be possible to regard the lowest measured value as 5% quantile. Fig. 3.25 then gives $n = 44$. $n = 49$ is chosen to obtain a 'smooth' reference quantity $(n+1) = 50$.

3.3.2 Ensuring independence

In rising-voltage tests too, ensuring independence depends on how the insulation recovers its insulating capacity after breakdown. How this happens for various insulation systems is covered in detail in connection with constant-voltage tests (see Section 3.2.2).

The independence of individual tests can generally be ensured better in the case of rising-voltage tests than it can for constant-voltage tests, because the voltage is always reduced after a breakdown (Figs. 3.3, 3.22). There are usually considerable intervals between two breakdowns, which both assist recovery and foster low electrical stresses in the region of the initial voltage. The latter have an action similar to the interval stresses in constant-voltage tests. It is generally true that independence can be ensured more easily for bipolar stresses (e.g. by alternating voltage) than unipolar (e.g. by direct voltage). With air insulation, compressed-gas insulation and liquid insulation, rising-voltage tests can generally be planned so that dependences are avoided.

Methods for checking independence are given in Section 2.5.3.2. As with the constant-voltage test, one must test both the trend of the sample by assessing the list of test values (graphical representation as in Fig. 2.41; comparison of first and last subgroups using suitable tests: Section 2.5.2), and also the variations of the realisations by feedback to the iteration test. The list of test values should be represented graphically straight away during the test. The other tests are not possible until the complete sample is available. Let us explain the procedure in closer detail with an example.

Example 3.3.2.1 The rising-voltage test planned in Example 3.3.1.1 is performed and the list of test values graphically represented (Fig. 3.26). The graph reveals no dependences. The realisations apparently fluctuate in a random manner about the mean value u_{d50} in a range of roughly $\pm 2\ s$ (s = standard deviation). To assess the trend, the first ten and last ten values of the sample are compared with each other. Taking the mean and the variance for the first part-sample ($\bar{u}_{d1} = 949.6$ kV; $s_1^2 = 288.3$ kV^2) and for the last part-sample ($\bar{u}_{d1} = 952.5$ kV; $s_1^2 = 203.8$ kV^2), the test quantity for the F test is obtained as $t_F = 1.414$ (see Section 2.5.2.1) and for the t test as $t_t = 0.413$ (see Section 2.5.2.2). Both quantities are smaller than the respective critical values ($F_{9;9;0.975} = 4.03$; $t_{18;0.975} = 2.101$), so that there is no objection to a common population and, by the same token, to the independence of the sample. The requisite number of variations is obtained by comparison

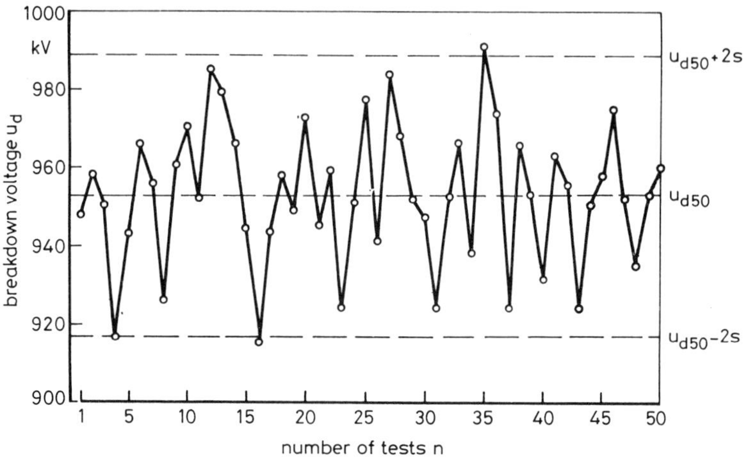

Fig. 3.26 Graphical representation of values from rising-voltage test (Example 3.3.2.1)

with the mean value (see Section 2.5.3.2), whereby the number of iterations can also be read off directly from Fig. 3.26. All realisations above u_{d50} are given a positive sign, those below a negative sign. The intersections with the lines for u_{d50} give the number of iterations as $r=30$ for $k=25$ 'positive' signs and a sample size $n=49$. The test quantity for the iteration test (eqn. 2.136) consequently has the value $z^*=1.574$ and is smaller than the critical value at significance level $\alpha=0.05$ ($\lambda_{0.975}=1.960$) derived from the normal distribution. The sample illustrated in Fig. 3.26 thus has enough fluctuations to be regarded as independent. The modified iteration test also contains information concerning the trend: if a trend occurs, comparison with the mean value at the start of the sample will cause particular signs to cluster together, with the opposite signs clustering together at the end. If this test is performed, the common-population tests can generally be omitted.

3.3.3 Empirical cumulative-frequency functions and their matching by theoretical distribution functions

The determination and representation of empirical cumulative-frequency functions is identical to the procedure explained in Section 2.2.2 and 2.5.1 for empirical distribution functions:

Example 3.3.3.1 For the rising-voltage test planned in Example 3.3.1.1, the list of test values is provided by Fig. 3.26. The primary distribution table (see Section 2.2.2) derived from these is reproduced in Table 3.6. The primary distribution table is shown to best advantage if all the realisations are graphically represented. Fig. 3.27 shows the empirical cumulative-frequency function with a linear probability scale.

Table 3.6 Primary distribution-table for rising-voltage test (Example 3.3.3.1); $n = 49$

Breakdown voltages u_{di} kV	Tally	Relative frequency h	Relative cumulative frequency h_Σ
915	I	0.02	0.02
917	I	0.02	0.04
924	IIII	0.08	0.12
926	I	0.02	0.14
931	I	0.02	0.16
935	I	0.02	0.18
938	I	0.02	0.20
941	I	0.02	0.22
943	II	0.04	0.26
944	I	0.02	0.28
945	I	0.02	0.30
947	I	0.02	0.32
948	I	0.02	0.34
949	I	0.02	0.36
950	I	0.02	0.38
951	II	0.04	0.42
952	IIII	0.08	0.50
953	I	0.02	0.52
955	I	0.02	0.54
956	I	0.02	0.56
958	III	0.06	0.62
959	I	0.02	0.64
960	I	0.02	0.66
961	I	0.02	0.68
963	I	0.02	0.70
965	I	0.02	0.72
966	III	0.06	0.78
968	I	0.02	0.80
970	I	0.02	0.82
972	I	0.02	0.84
973	I	0.02	0.86
975	I	0.02	0.88
977	I	0.02	0.90
979	I	0.02	0.92
984	I	0.02	0.94
985	I	0.02	0.96
991	I	0.02	0.98

The empirical cumulative-frequency function can now be approximated by a theoretical distribution function. The principles already dealt with for the constant-voltage test apply (see Section 3.2.4): good matching, correspondence between physical and mathematical model, and convenience of matching.

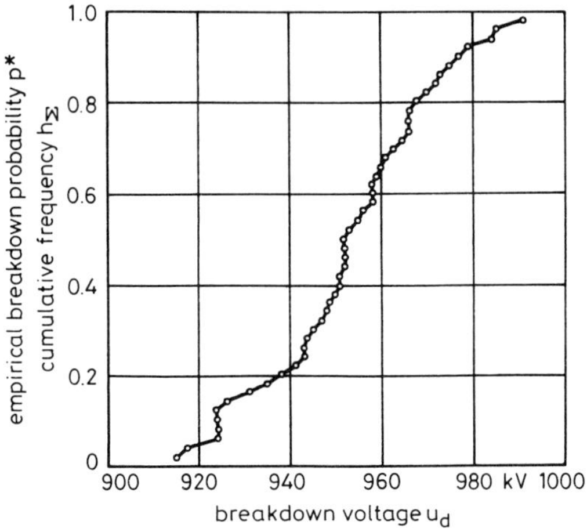

Fig. 3.27 Empirical cumulative-frequency function of a spark-gap in air (Example 3.3.3.1)

Matching of the theoretical distribution function is through parameter estimations (see Section 2.3.2), representation of the empirical distribution function in a probability grid (see Section 2.5.1.1), or by mathematical distribution tests (see Section 2.5.1.2).

Example 3.3.3.2 We know that the breakdown voltages are generally normally distributed for the rising-voltage test on a sphere/plane arrangement in air planned in Example 3.3.1.1. The empirical cumulative-frequency function (Fig. 3.27) should therefore be approximated by a normal distribution. The point estimation of the parameters (see Section 2.3.2.2) produces $m = \bar{u}_d = 953$ kV for the mean value and $s = 18.2$ kV for the standard deviation. The normal distribution with these parameters is represented as straight lines on a normal-distribution probability grid (Fig. 3.28), at the same time as the empirical cumulative-frequency function. It is observed that the normal distribution $N(953;\ 18.2^2)$ is a good approximation to the relationship obtained. This impression is confirmed by the Kolmogorov test (see Section 2.5.1.2): the test quantity $d_{max} = 0.07$ (Fig. 3.28) is clearly below the Kolmogorov limit $k_{50;0.05} = 0.188$. Calculation of the lower tolerance limit (see Section 2.3.2.2, broken line in Fig. 3.28) shows even more clearly that approximation by a normal distribution is still sufficiently certain in the technically relevant range of very small probabilities. On this basis, there are no further objections to regarding the breakdown voltage investigated as being normally distributed. It is now advisable to calculate confidence limits for the parameters of this normal distribution. The formulae in Table 2.7 are used to obtain the two-sided confidence estimates for the confidence coefficient $\varepsilon = 0.95$ for the mean value (947 kV; 959 kV) and for the standard deviation (15.0 kV; 22.8 kV).

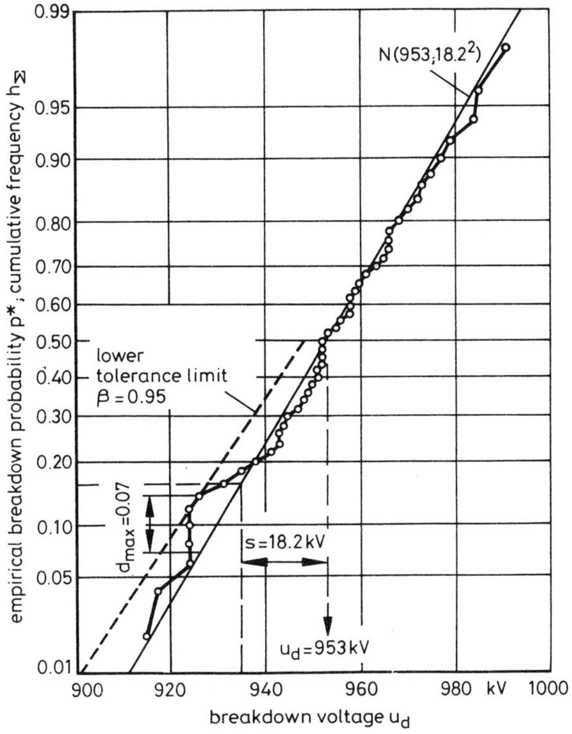

Fig. 3.28 Empirical and approximate cumulative-frequency function on a normal-distribution probability grid (Example 3.3.3.2)

The customary method of parameter estimation described above produces the most accurate estimate and, with simultaneous graphical representation, also a very vivid estimate. If parameter estimation presents difficulties, e.g. with a Weibull distribution, after representation on a probability grid for the theoretical distribution function provided, the line of best fit can also be obtained by linear regression (see Section 2.4.3 and Example 3.2.4.1). It should be emphasised that all the realisations available should be used for the regression. The parameters can then be determined from the lines of best fit as quantiles.

3.3.4 Determination of performance functions from cumulative-frequency functions

Between cumulative-frequency functions and performance functions there is a relationship determined by the test parameters selected (see Section 3.1.2). The possible cumulative-frequency functions $S\Delta_u(u)$ (Fig. 3.7) can be calculated for a known performance function $V(u)$, using eqn. 3.1.

Compared with constant-voltage tests, the rising-voltage tests producing the cumulative-frequency function offer advantages in testing terms, but practical conclusions concerning the performance of insulation systems when particular stresses occur can in practice only be derived from performance functions (see Section 3.1.1). If a cumulative-frequency function is used as a performance function, no technical risk is of course involved, since cumulative-frequency functions are always on the 'safe' side of performance functions (Fig. 3.7). It is nevertheless advisable to give closer consideration to the determination of performance functions from cumulative-frequency functions.

The relationship between the two functions is best illustrated for a voltage increasing in stages. If, in the case of impulse voltages (Fig. 3.3) or continuous voltages (Fig. 3.22*b*), the voltage rise starts from a fixed initial voltage u_0, the same voltage values will be repeated in all the individual tests. The breakdown probability can then be estimated for each voltage value, after a whole test has been completed, by relating the occurrence of the event 'breakdown' to the number of voltage applications at that value. In Fig. 3.3, for example, $k_1^* = 1$ breakdown is obtained for $u_d = 102$ kV with $m_1^* = 8$ voltage applications (second individual test), and $k_2^* = 2$ breakdowns for $u_d = 105$ kV with $m_2^* = 5$ voltage applications (first and fourth individual tests). The estimates obtained therefrom for the breakdown probability $p_1^* = 0.125$ and $p_2^* = 0.40$ belong to the performance function being sought. Modification of the evaluation plan consequently produces the estimate of the performance function directly, but such large quantiles are of course estimated very inaccurately.

The estimate can be improved by basing calculation on approximation of the cumulative-frequency function by a theoretical distribution function. The method of calculation described in the literature and described in the following only presupposes a cumulative-frequency function $S_{\Delta u}(u_d)$ which is determined using the fixed step amplitude Δu but is otherwise arbitrary.

If eqn. 3.1 is transformed and if the abbreviations

$$a_i = V(u_0 + i\Delta u); \qquad i = 0;\ 1;\ 2;\ 3\ldots \quad \text{and}$$

$$b_k = S_{\Delta u}(u_0 + k\Delta u); \qquad k = 0;\ 1;\ 2;\ 3\ldots,$$

are used, eqn. 3.1 assumes the form

$$b_k = \sum_{i=1}^{k} a_i \prod_{j=0}^{i-1} (1 - a_j).$$

The following applies, moreover:

$$\begin{aligned} b_{k+1} &= b_k + a_{k+1} \prod_{j=0}^{k} (1 - a_j). \\ &= b_k + a_{k+1}\left[1 - a_0 - \sum_{i=1}^{k} a_i \prod_{j=0}^{i-1} (1 - a_j)\right] \\ &= b_k + a_{k+1}(1 - a_0 - b_k). \end{aligned}$$

$a_0 = V(u_0)$ can be equated to zero, because the breakdown probability ought to be more or less zero at the initial voltage. It therefore follows that

$$a_{k+1} = \frac{b_{k+1} - b_k}{1 - b_k}$$

or (after the abbreviations have been replaced by the original relationships) the desired general expression for calculating performance functions from cumulative-frequency functions:

$$V(u_d) = \frac{S_{\Delta u}(u_d) - S_{\Delta u}(u_d - \Delta u)}{1 - S_{\Delta u}(u_d - \Delta u)}. \tag{3.10}$$

If a cumulative-frequency function is sufficiently well known, there will be no difficulty in calculating the performance function for the point using the values of Δu used. Using eqn. 3.10, however, the associated performance function can be calculated by substitution in that formula, even in the case of mathematically simple distribution functions with which $S_{\Delta u}$ is approximated. If a double exponential distribution with the parameters u_{d63S} and γ is assumed for $S_{\Delta u}(u_d)$, the 'closed' solution

$$V(u_d) = 1 - \exp\left\{-\exp\frac{u_d - \left\{u_{d63S} - \gamma \ln\left[1 - \exp\left(-\frac{\Delta u}{\gamma}\right)\right]\right\}}{\gamma}\right\}. \tag{3.11}$$

will be obtained, for example. The performance function is again double-exponentially distributed, and only the 63% quantile (mode) will change:

$$u_{d63V} = u_{d63S} - \gamma \ln\left[1 - \exp\left(\frac{-\Delta u}{\gamma}\right)\right]. \tag{3.12}$$

The difference between u_{d63S} and u_{d63V} is less than 2% if $\Delta u \geqq \gamma$ is chosen. If the step amplitude Δu is sufficiently large, the conversion can be dispensed with and $S_{\Delta u}(u_d)$ be interpreted straight away as performance function $V(u_d)$, since the error is on the technically safe side.

If the cumulative-frequency functions are approximated by mathematically more complicated functions, then, after substitution in eqn. 3.10, such handy solutions are not often obtained for the performance function. Complicated relationships are produced if a normal distribution is used for the cumulative-frequency function. A transformation method based on computer simulation of the relationships has proved suitable for solution of the problem.

The breakdown voltages of an insulation arrangement in a rising-voltage test and in a constant-voltage test are both approximated as being normally distributed (variates U_d and $U_{d\Delta u}$; parameters μ; σ^2 and $\mu_{\Delta u}$; $\sigma^2_{\Delta u}$). Both the variates are standardised with μ; σ^2:

$$X = \frac{U_d - \mu}{\sigma}, \tag{3.13}$$

$$X_{\Delta x} = \frac{U_{d\Delta u} - \mu}{\sigma}. \tag{3.14}$$

The performance function $V(u_d)$ is thereby transformed into a standardised normal distribution $\Phi(x)$ with the parameters $\mu=0$; $\sigma^2=1$. The cumulative-frequency function $S_{\Delta u}(u_d)$ becomes a normal distribution $F_{\Delta x}(x)$ with the parameters $\mu_{\Delta x}$ and $\sigma^2_{\Delta x}$. The step amplitude Δu is likewise standardised:

$$\Delta x = \frac{\Delta u}{\sigma}. \tag{3.15}$$

The variates X and $X_{\Delta x}$ are linked by

$$X_{\Delta x} = \sigma_{\Delta x} X + \mu_{\Delta x} \tag{3.16}$$

The relationship between $\Phi(x)$ and $F_{\Delta x}(x)$ is given by eqn. 3.1. The expressions used have been compiled in a transformation plan (Fig. 3.29). If the performance function $V(u_d)$ is to be determined from $S_{\Delta u}(u_d)$, it is also possible to make use of the fact that $\Phi(x)$ is also produced by the standardisation

$$X = \frac{U_{d\Delta u} - \mu_{\Delta u}}{\sigma_{\Delta u}} \tag{3.17}$$

It follows from eqns. 3.16 and 3.17 that

$$X_{\Delta x} = \frac{U_{d\Delta u} - \mu_{\Delta u}}{\sigma_{\Delta u}} \sigma_{\Delta x} + \mu_{\Delta x}. \tag{3.18}$$

Comparing the coefficients of eqns. 3.14 and 3.18 produces the desired

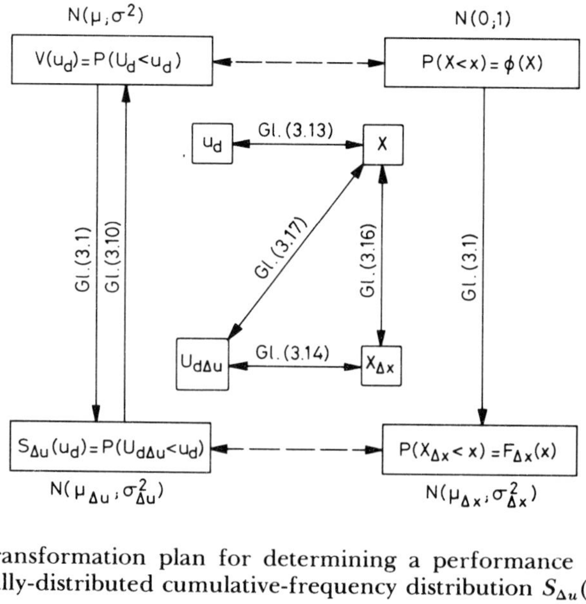

Fig. 3.29 Transformation plan for determining a performance function $V(u_d)$ from a normally-distributed cumulative-frequency distribution $S_{\Delta u}(u_d)$

parameters of the performance function:

$$\sigma = \frac{\sigma_{\Delta u}}{\sigma_{\Delta x}}, \tag{3.19}$$

$$\mu = \mu_{\Delta u} - \mu_{\Delta x}\frac{\sigma_{\Delta u}}{\sigma_{\Delta x}}. \tag{3.20}$$

The standardised step-amplitude (eqn. 3.15) is now:

$$\Delta x = \frac{\sigma_{\Delta x}}{\sigma_{\Delta u}}\,\Delta u. \tag{3.21}$$

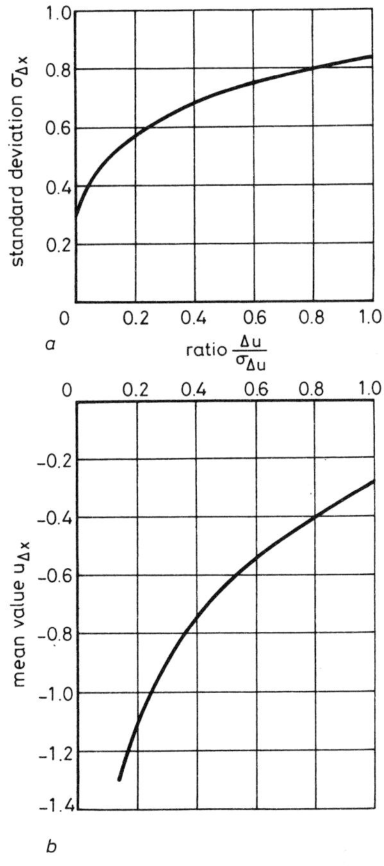

Fig. 3.30 Nomograms for determining the parameters of a transformed normally-distributed cumulative-frequency function

a Standard deviation $\sigma_{\Delta x}$
b Mean value $\mu_{\Delta x}$

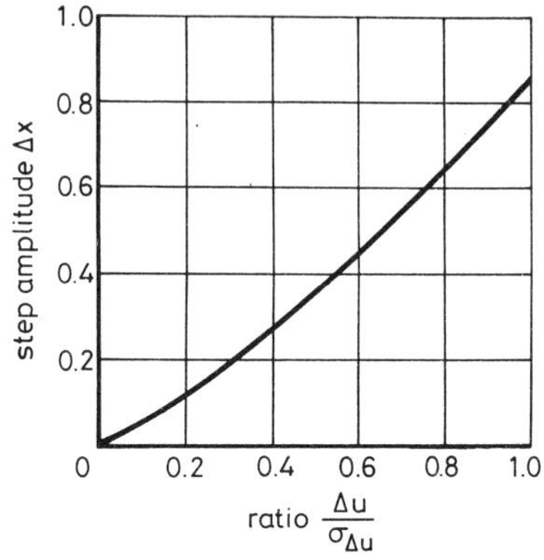

Fig. 3.31 Standardised step-amplitude Δx and invariable ratio $\Delta u/\sigma_{\Delta u} = \Delta x/\sigma_{\Delta x}$

The ratio

$$\frac{\Delta u}{\sigma_{\Delta u}} = \frac{\Delta x}{\sigma_{\Delta x}}$$

is thus invariable in respect of linear transformations. It is familiar in the result of the rising-voltage test.

Calculation of σ and μ also calls for $\sigma_{\Delta x}$ and $\mu_{\Delta x}$, which are determined by computer simulation as a function of $(\Delta x/\sigma_{\Delta x}) = (\Delta u/\sigma_{\Delta u})$. These quantities required for the transformation are available in diagram form (Fig. 30), as is the standardised step-amplitude (Fig. 3.31).

Simulation of the relationships also provides details of the upper permissible limit u_{0_0} of initial voltage u_0 (cf. Section 3.3.1). It is exceeded whenever the cumulative-frequency function is a function of this initial voltage. The requisite initial voltage is determined (Fig. 3.32) as a function of step

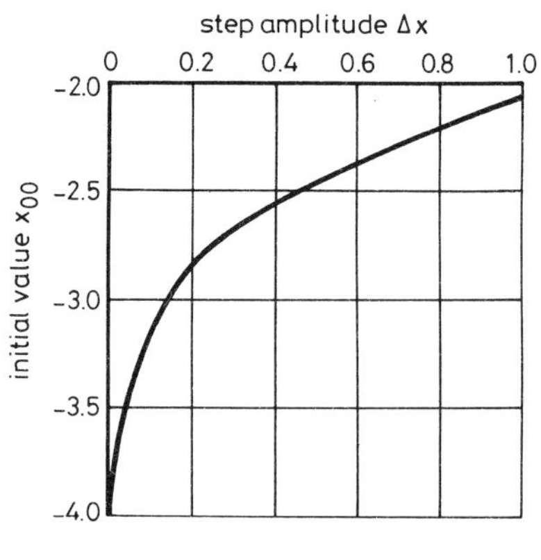

Fig. 3.32 Upper limit x_{0_0} of initial voltage u_0 for rising-voltage tests

Assumption: normally-distributed cumulative-frequency function amplitude Δx (Fig. 3.31), using the standardisation

$$x_{0_0} = \frac{u_{0_0} - \mu_{\Delta u}}{\sigma_{\Delta u}} \tag{3.22}$$

The smaller the step amplitude, the lower the initial voltage has to be.

Example 3.3.4.1 The performance function is to be determined for a point/plane arrangement in oil (electrode gap $d = 50$ mm). Experience has shown direct determination to be inappropriate, hence the cumulative-frequency function method is selected. Before the tests are started, the mean value of the cumulative-frequency function is estimated from empirical values as

$$u^*_{d50S} = \mu^*_{\Delta u} = 150 \text{ kV}$$

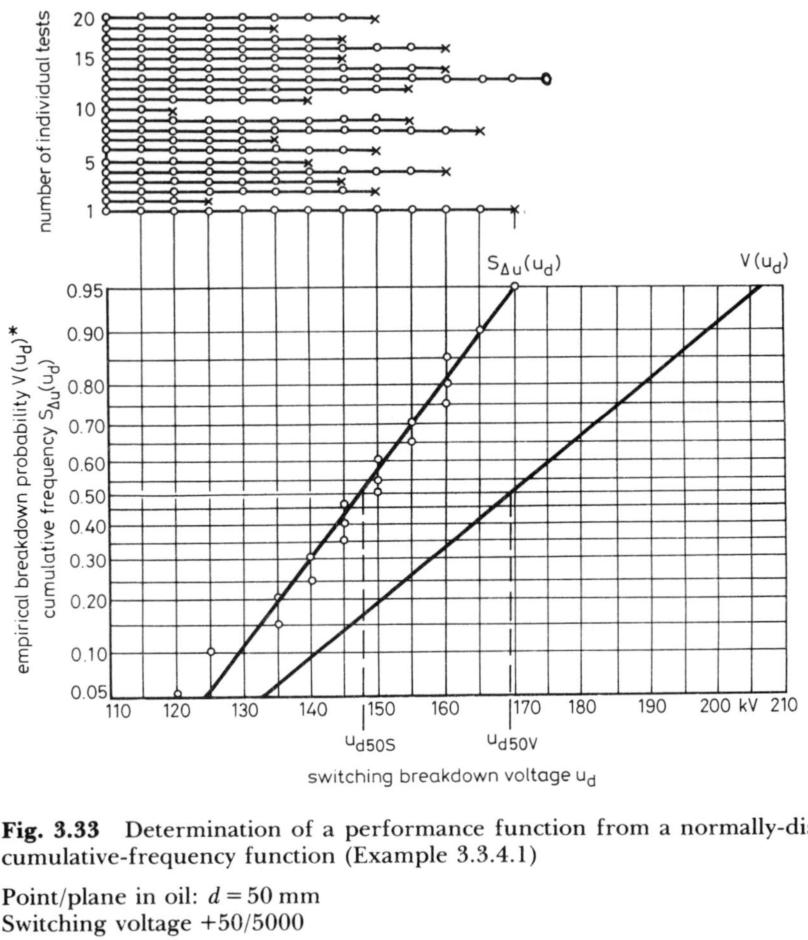

Fig. 3.33 Determination of a performance function from a normally-distributed cumulative-frequency function (Example 3.3.4.1)

Point/plane in oil: $d = 50$ mm
Switching voltage +50/5000

and the standard deviation as

$$\sigma^*_{\Delta u} = 0.08\mu^*_{\Delta u} = 12\ \text{kV}.$$

The step amplitude is selected as $\Delta u = 5$ kV. An estimated value of 0.4 is thus obtained for $\Delta u/\sigma^*_{\Delta u}$, with which, according to Fig. 3.31, a $\Delta x \approx 0.25$ is associated. The standardised upper limit of the initial voltage is then obtained as $x_{0_0} = -2.7$ (Fig. 3.32), so that

$$u_{0_0} = \mu^*_{\Delta u} + x_{0_0}\sigma^*_{\Delta u} = 150 - 2.7 \cdot 12 \approx 118\ \text{kV}$$

$S_{\Delta u}(u_d)$ is determined accurately with a selected initial voltage $u_0 = 110\ \text{kV} < u_{0_0}$. Although u_0 should be varied, it is kept constant here, for the sake of clarity. The number of individual tests is determined as $n = 20$. The test result is illustrated in Fig. 3.33.

The parameters of the cumulative-frequency function can now be estimated as $\mu^{**}_{\Delta u} = u_{d50S} = 148$ kV and $\sigma^{**}_{\Delta u} = s_{\Delta u} = 15$ KV. With the invariable ratio $\Delta u/s_{\Delta u} = 0.33$, $\sigma_{\Delta x} = 0.65$ is obtained from Fig. 3.30*a* and $\mu_{\Delta x} = -0.88$ from Fig. 3.30*b*. Eqns. 3.19 and 3.20 are used to calculate the parameters $s = 23$ kV and $u_{d50V} = 169$ kV of the performance function (normal distribution), plotted in Fig. 3.33. Were the transformation not to be performed, the far-too-low voltage $u_{d50S} = 148$ kV would be the representative mean value.

According to the results of experimental checks, the transformation method is quite powerful. The performance function $V(u_d)$ was determined for an air insulating clearance (Fig. 3.34), via the cumulative frequency

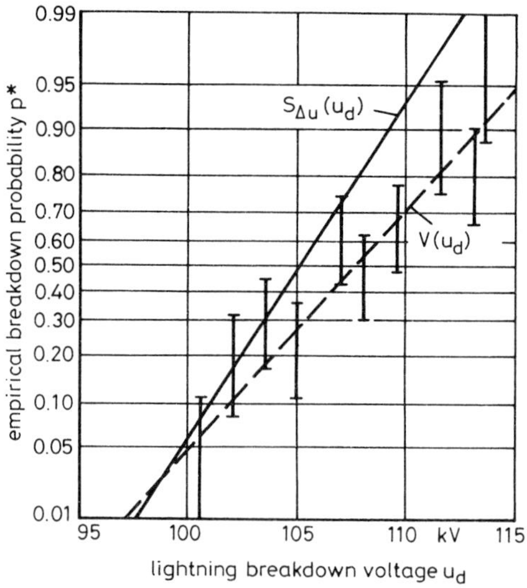

Fig. 3.34 Cumulative-frequency function and performance function of an air insulating clearance

Point/plane: $d = 150$ mm
Lightning voltage 1.2/50

function $S_{\Delta u}(u_d)$ as in the above example (measured values omitted for the sake of clarity). Direct determination of the performance function in the constant-voltage test produced confidence regions which were a good match for the performance function calculated from $S_{\Delta u}(u_d)$. The method is thus usable and, since there are also no dependences in the constant-voltage test, is checkable.

In a similar check on an oil insulating gap, similar agreement was only achieved when intervals of $\Delta t_p = 300$ s were used in the constant-voltage test (Fig. 3.35). With short intervals ($\Delta t_p = 30$ s), dependences invalidating the result occur in the constant-voltage test. These are caused by space charges produced during breakdowns which cannot recombine in the short intervals. It is advisable, in the case of liquid insulation and metal-clad gas insulation, to determine the performance function via the cumulative-frequency function.

If one wishes to determine performance functions down to very small probabilities, it is advisable to conduct rising-voltage tests with several voltage applications per stage ($m > 1$). Starting at an initial voltage u_0, the voltage is increased in stages by Δu until breakdown occurs for the first time at a stage. It is unimportant at which of the m voltage applications breakdown occurs. After breakdown, the test is interrupted — regardless of how many voltage applications were still due at that step — and the voltage again reduced to u_0 for the next test. The probability of no breakdown occurring

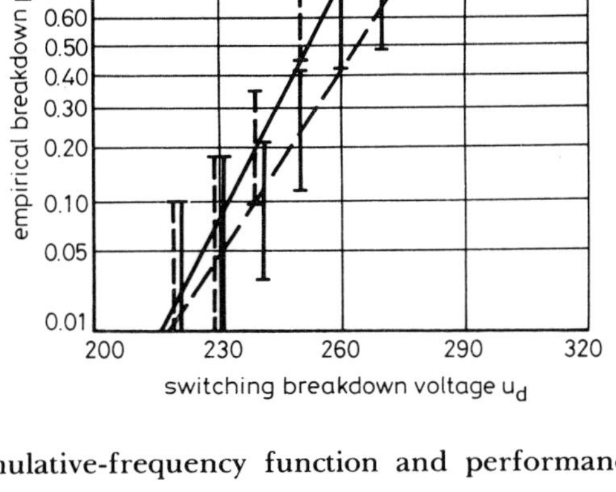

Fig. 3.35 Cumulative-frequency function and performance function of an oil insulating gap

Point/plane: $d = 70$ mm
Switching voltage 200/3000

at the m voltage applications at a step is given by

$$p_{Nk} = [1 - V(u_0 + k\Delta u)]^m. \tag{3.23}$$

At least one breakdown occurs with a probability

$$p_{Dk} = 1 - p_{Nk} = 1 - [1 - V(u_0 + k\Delta u)]^m \tag{3.24}$$

the test in fact being interrupted after the first breakdown. Using these relationships, the differentiation to calculate cumulative-frequency functions from performance functions (see Section 3.1.2) can now be repeated for m voltage applications ($m > 1$). With a known performance function $V(u_0 + k\Delta u)$, the following is obtained for the value of the cumulative-frequency function at u_0:

$$S_{\Delta u}(u_0 + k\Delta u) = \sum_{i=1}^{k} \left\{ \left(1 - [1 - V(u_0 + i\Delta u)]^m \right) \prod_{j=0}^{i-1} [1 - V(u_0 + j\Delta u)]^m \right\}. \tag{3.25}$$

For $m = 1$, eqn. 3.25 is transformed into eqn. 3.1. Conversely — by analogy to the differentiation leading to eqn. 3.10 — with a known cumulative-frequency function $S_{\Delta u}(u_d)$ (determined for $m > 1$ voltage applications per voltage step), the performance function $V(u_d)$ can be calculated:

$$V(u_d) = 1 - \sqrt[m]{1 - \frac{S(u_d) - S(u_d - \Delta u)}{1 - S(u_d - \Delta u)}}. \tag{3.26}$$

For the case of $m = 1$ already investigated, eqn. 3.26 is identical to eqn. 3.10. With a rising-voltage test procedure with m voltage applications per step (in most cases $m = 10$–100), the performance function in the range of extremely small probabilities can be determined using eqn. 3.26. The total number of voltage applications required is naturally very large (see Section 3.4.2).

In conclusion, let us follow up the exhaustive treatment of the problems with a stepwise voltage rise with a few remarks on conditions with a continuous voltage rise (Fig. 3.22*a*). The quantity 'time' has to be introduced here in addition to the voltage. With a linearly increasing voltage, voltage and time are proportional to each other, as are the realisations of breakdown voltage and breakdown time.

$$u_d = v_u t_d. \tag{3.26a}$$

We now intend to calculate the performance function for a stress duration T_0, which is short compared with the total duration of the rising-voltage test, and which is physically or technically significant. T_0 can, for example, be equal to the duration of an alternating-voltage half-cycle. According to unpublished reports by Speck, the continuous voltage rise should now be transformed into a step function with a step duration Δt_s and a reference time T_0. If the deliberations leading to eqn. 3.1 or 3.10 are applied to this model, if independence is assumed, if the time (eqn. 3.26*a*) is used instead of the voltage, and if the boundary transition $\Delta t_s \rightarrow dt_d$ is executed, then:

$$V(t_d;\, T_0) = 1 - \exp\left\{\frac{T_0\left(-\dfrac{dS(t_d)}{dt_d}\right)}{1 - S(t_d)}\right\} = 1 - \exp\left\{\frac{-s(t_d)\,T_0}{1 - S(t_d)}\right\}. \qquad (3.26b)$$

$S(t_d)$ is calculated, using $t_d = u_d/v_u$, from the cumulative-frequency function $S_v(u_d)$ determined experimentally with v_u. The quantity $dS(t_d)/dt_d$ is the rise of the cumulative-frequency function, i.e. the density function $s(t_d)$, at the point t_d under consideration.

If a two-parameter Weibull distribution (parameters u_{d63}; δ) is assumed for the cumulative-frequency function $S_v(u_d)$, then, using eqns. 3.26*a* and *b*, one obtains the performance function of the breakdown voltage

$$V(u_d,\, T_0) = 1 - \exp\left\{-\left[\frac{u_d}{u_{d63}\left(\dfrac{u_{d63}}{v_u T_0 \delta}\right)^{1/(\delta-1)}}\right]^{\delta-1}\right\}. \qquad (3.26c)$$

If $S_v(u_d)$ is double-exponentially distributed (u_{d63}, γ), the double-exponential distribution for $V(u_d, T_0)$ is

$$V(u_d,\, T_0) = 1 - \exp\left\{\frac{-v_u T_0}{\gamma} \exp\left(\frac{u_d - u_{d63}}{\gamma}\right)\right\}. \qquad (3.26d)$$

The use of other theoretical functions for $S_v(u_d)$ is, of course, possible, but this produces less convenient expressions.

Eqns. 3.26*b*, *c* and *d* theoretically permit a comparison, for example, between the alternating-breakdown-voltage cumulative-frequency function obtained for a steadily rising voltage on an electrode arrangement, and the switching-breakdown-voltage performance function of the same arrangement. Although other applications suggest themselves, e.g. service-life investigations, no examples of such applications have so far been published. It is recommended that such problems be tackled.

3.3.5 Evaluation chart and use of computers

Let us review the individual steps involved in performing and evaluating a rising-voltage test, referring to Fig. 3.36. After preparation and execution of the tests, lists of test values with n realisations are available for test evaluation. The mutual independence of the realisations has to be tested: when independence has been confirmed, evaluation can be proceeded with. The empirical cumulative-frequency function is determined from the test values. It is then possible either to approximate this function, using a theoretical distribution and converting to a performance function using this matched function, or by converting the empirical performance function point by point. Whereas the first possibility often results in a certain matching of the performance function (e.g. according to Section 3.3.4, a double-exponential distribution is obtained, and, in the case of the transformation method, a normal distribution), then — particularly in the case of the second route — the empirical performance function should be matched by a

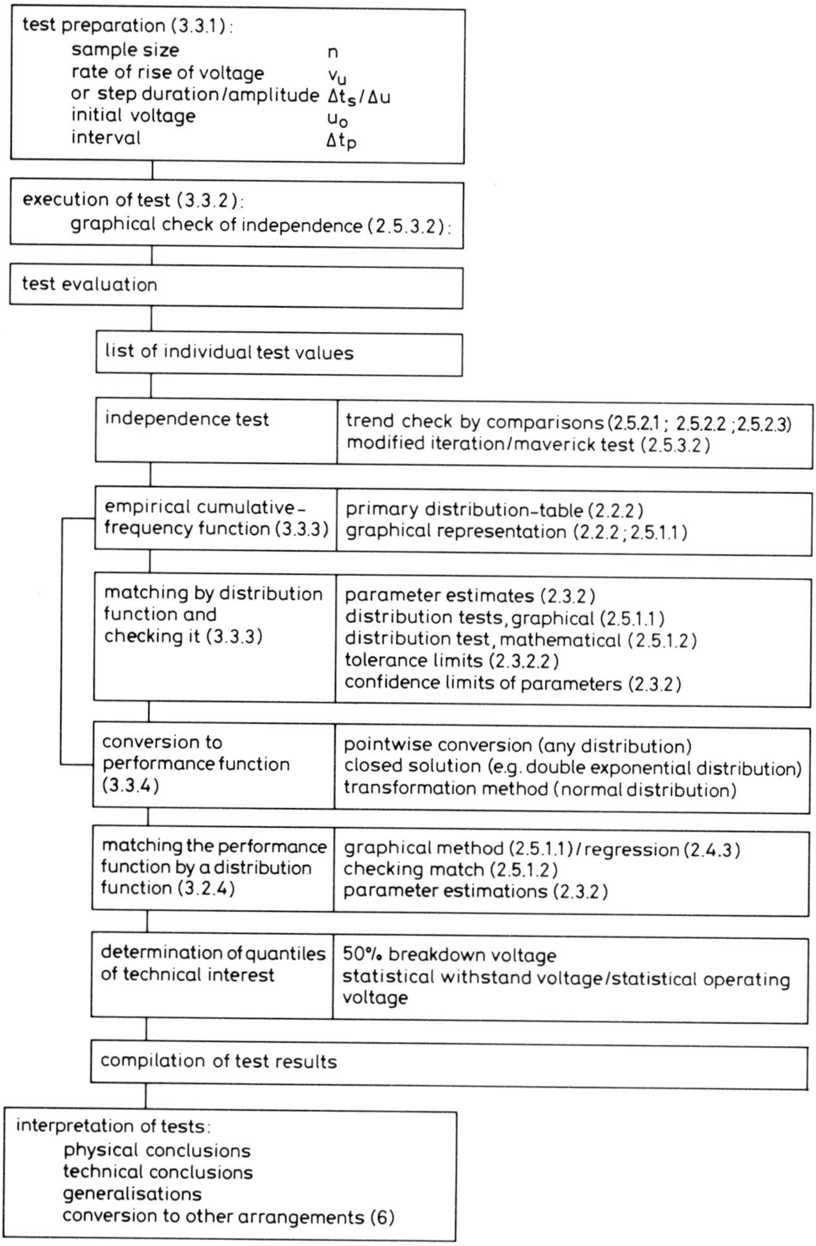

Fig. 3.36 Evaluation chart for a rising-voltage test

theoretical distribution function. The quantiles of technical interest can be derived from the performance function. The test results are finally compiled in a suitable form and interpreted. All the mathematical operations needed in the chart (Fig. 3.36) are dealt with in the sections indicated in parentheses after the key words.

As in the evaluation of constant-voltage tests (see Section 3.2.5), the procedure outlined in Fig. 3.36 can be adopted to evaluate rising-voltage tests using a digital computer. For the reasons already put forward (see Section 3.2.5), we have refrained from including computer programs and will refer the reader to the relevant literature. Through the use of microcomputers to control high-voltage test plants, there are prospects of performing at least simple statistical evaluation operations directly during the tests online.

3.4 Methods for determining selected quantiles

3.4.1 Up-and-down method

The up-and-down method of Dixon and Mood permits a quite reliable estimation of the 50% breakdown voltage when the breakdown voltage is normally distributed. In this case the method also provides an estimate for the standard deviation. If the breakdown voltage cannot be assumed to be normally distributed, indication of the dispersion should be abandoned. However, the up-and-down method then also supplies usable estimates u^{*}_{d50} for the 50% breakdown voltage, corresponding to the control limiting-value expression (see Section 2.3.2.2).

In the execution of the method, the voltage is initially raised in steps of a fixed amplitude Δu, from an initial value u_{00} at which with certainty no breakdown occurs, until the first breakdown occurs at a voltage u_{l1} (Fig. 3.37). The voltage is then reduced by Δu. If no breakdown occurs at voltage $u_{l2} = u_{l1} - \Delta u$, the test voltage should again be raised through Δu, otherwise be reduced by Δu. The process is repeated until a predetermined number n of voltage values u_{l1}, $u_{l2} \ldots u_{ln}$ are obtained. Particularly with a large sample n, the arithmetic mean of these voltages in itself provides a preliminary estimate of the 50% breakdown voltage being sought. Closer examination shows, however, that the result is affected in particular by the step amplitude Δu, as is clearly shown by investigation of the breakdown probability associated with the first breakdown voltage u_{d1} (Fig. 3.38). The smaller Δu is chosen, relative to the standard deviation of the population, the lower will be the voltages u_{l1} and the probabilities p_1 at which the first breakdown occurs. Dixon and Mood therefore determine the estimates $u^{(*)}_{d50}$ and s on the basis of the event (breakdown or non-breakdown) occurring more rarely at the n voltages. This event occurs k times in the course of the n tests, and the complementary event q times ($n = k + q$). To evaluate the test, n voltage values u_l are classified according to magnitude (Fig. 3.37*c*). The index $i = 0$ is assigned to step u_i at which the (rarer) event concerned first occurs. The subsequent higher steps are given the indices

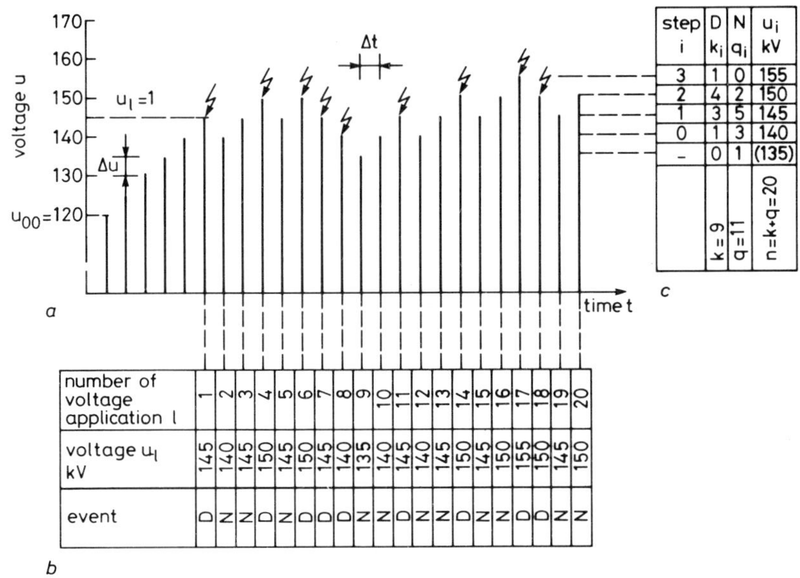

Fig. 3.37 Up-and-down method of Dixon and Mood [3.53]

a Test plan
b and ***c*** Evaluation

n Size of sample
k Number of breakdowns
D Breakdown
N Non-breakdown

$i=1, 2, \ldots, r$. The 50% breakdown voltage can now be estimated using

$$u_{d50}^{(*)}=u_0+\Delta u\left(\frac{A}{k}\pm\frac{1}{2}\right) \tag{3.27}$$

where $A=\sum_{i=1}^{r} ik_i$ (k_i is the number of events concerned at the ith step). If the events concerned are breakdowns, the sign in eqn. 3.27 is negative; in the case of non-breakdowns it is positive. The standard deviation is estimated using

$$s^{(*)}=1.62\cdot\Delta u\left(\frac{kB-A^2}{k^2}+0.029\right) \tag{3.28}$$

where $B=\sum_{i=1}^{r} i^2k_i$. According to Dixon and Mood, estimates can further be obtained for the standard deviation of the 50% breakdown voltage

$$s_m=G\frac{s^{(*)}}{\sqrt{k}} \tag{3.29}$$

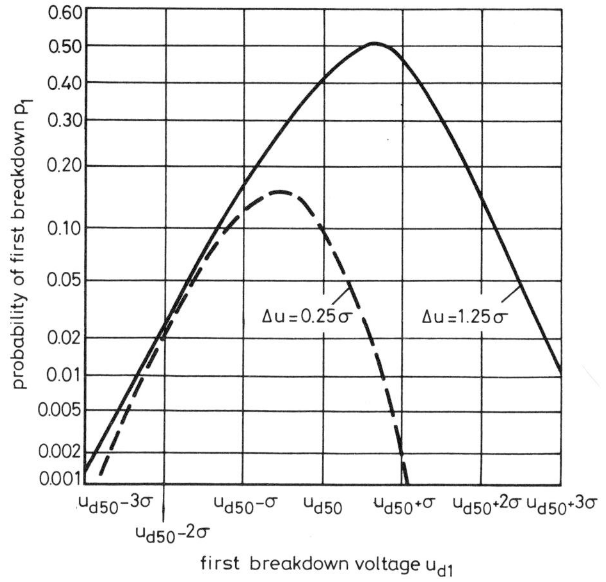

Fig. 3.38 Probability of first breakdown voltage using the up-and-down method of Kučera [3.54]

and the empirical standard deviation

$$s_s = H\frac{s^{(*)}}{\sqrt{k}}, \tag{3.30}$$

where the factors G and H can be derived from Fig. 3.39. Using s_m and s_s, Dixon and Mood give for u_{d50} confidence estimates

$$u_{d50}^{(*,0,u)} = u_{d50}^{(*)} \pm \lambda_{(1+\varepsilon)/2} s_m$$

and for other quantiles u_{dq}

$$u_{dq}^{(*,0,u)} = u_{dq}^{(*)} \pm \lambda_{(1+\varepsilon)/2}\sqrt{s_m^2 + \lambda_q^2 s_s^2}$$

($\lambda_{(1+\varepsilon)/2}$ and λ_q are quantiles of normal distribution $N(0;\ 1)$).

Simulation of the up-and-down method has shown, however, that eqns. 3.27 to 3.30 (and by the same token the confidence estimates as well) should be handled with a degree of caution. Particular attention should be paid to the influence of the number of voltage values n and the step amplitude $\Delta u(u_{l1} \ldots u_{ln})$. The deviations from the true value will be greater, the smaller n is and the smaller the step amplitude Δu selected. According to Kuono/Oikawa and Kučera, it is advisable to correct the estimate of the standard deviation calculated using eqn. 3.28 with a factor K_1 (Fig. 3.40)

$$s = K_1 s^{(*)}. \tag{3.31}$$

Estimation of the 50% breakdown voltage using eqn. 3.27 exhibits only

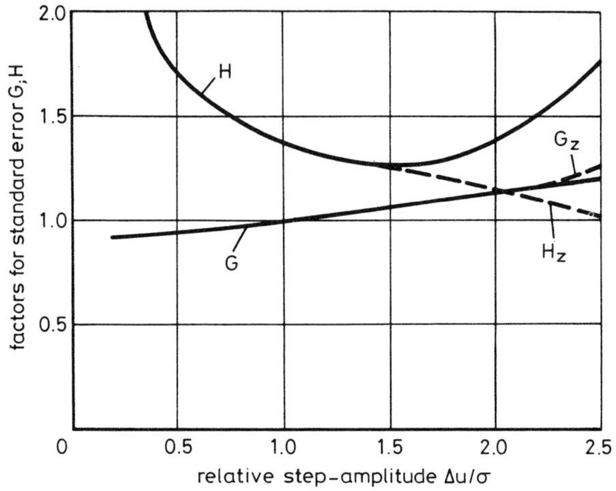

Fig. 3.39 Factors for standard error of mean value (G) and standard deviation (H), after Dixon and Mood [3.53]

G_z and H_z should be used instead of G and H if the mean value lies midway between two steps; if it is on the actual steps, one should retain G and H.

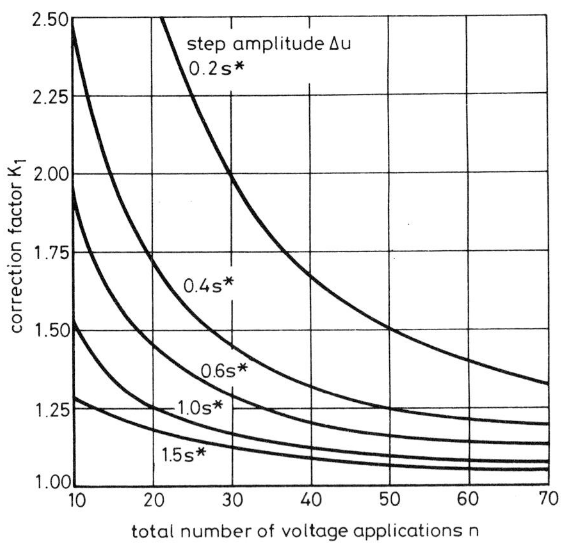

Fig. 3.40 Correction factor (K_1) for calculating the standard deviation in the up-and-down method of Kuono and Oikawa [3.28]

small errors, especially if Δu is not chosen too small. A correction can be made using

$$u_{d50} = u_{d50}^{(*)} + sK_2 \tag{3.32}$$

where K_2 is obtained from Fig. 3.41. Using the factor K_3 reproduced in the same diagram, confidence limits can, according to Kučera be estimated for u_{d50}:

$$u_{d50}^{(0,u)} = u_{d50} \pm s\lambda_{(1+\varepsilon)/2} K_3 . \tag{3.33}$$

When performing the up-and-down method, it proves expedient to choose $\Delta u \approx \sigma$ and to keep it constant (use s^* or s for estimates of σ). If only a rough estimate is required, $\Delta u \approx (0.01 \ldots 0.10) u_{d50}$ can be selected and variations even be permitted in this range. The estimate for u_{d50} should

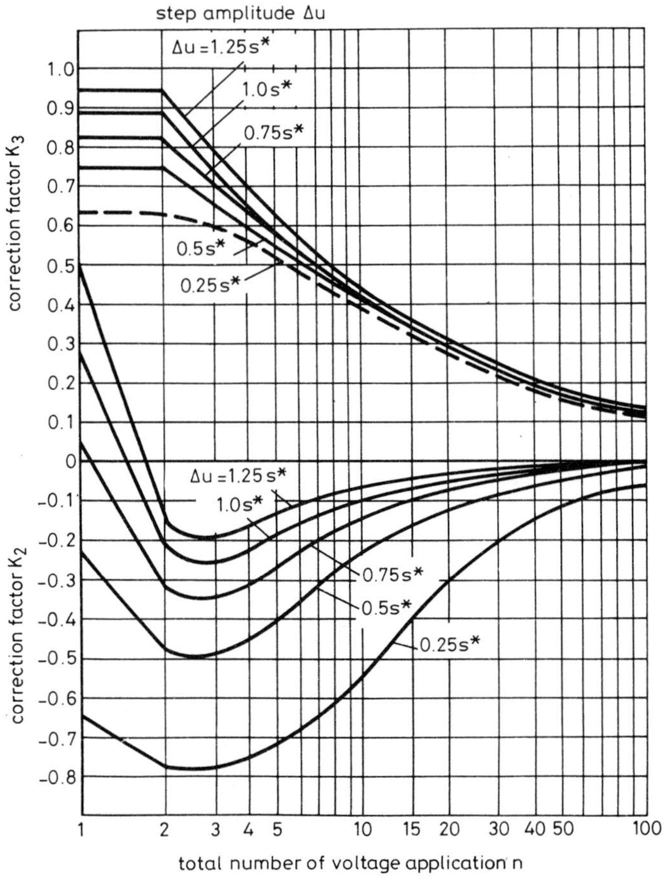

Fig. 3.41 Correction factor for calculating the mean value (K_2) and the confidence region of the mean value (K_3) in the up-and-down method of Kučera [3.54] [3.55]

then be calculated as the arithmetic mean of voltage values $u_{l1}, u_{l2} \dots u_{ln}$. The initial voltage u_{00} hardly affects the result. In the up-and-down method, the total number that can be evaluated of voltage applications n should generally not be less than 20; only in rare cases will more than $n = 60$ voltage applications be opted for. Within these limits, the method has proved highly satisfactory for estimating the 50% breakdown voltage of self-restoring insulation. Constant-voltage or rising-voltage tests (see Sections 3.2 and 3.3) are preferable for estimating the standardisation (i.e. the complete parameters of the standard deviation).

Example 3.4.1.1 The test procedure reproduced in Fig. 3.37 with $u_{00} = 120$ kV, $\Delta u = 5$ kV and $n = 20$ is to be evaluated. 9 breakdowns and 11 non-breakdowns are obtained: $k = 9$ and $q = 11$ are therefore determined (k is the number of rarer events). The lowest breakdown occurs at 140 kV; this voltage step is given the notation u_0 ($i = 0$). The breakdowns per voltage step k_i are thereby obtained (Fig. 3.37*c*) and from them the quantities $A = 1 \cdot 3 + 2 \cdot 4 + 3 \cdot 1 = 14$, as well as $B = 1^2 \cdot 3 + 2^2 \cdot 4 + 3^2 \cdot 1 = 28$. According to eqn. 3.27, $u_{d50}^{(*)} = 145.3$ kV, and according to eqn. 3.28 $s^* = 5.8$ kV. These two quantities exhibit standard deviations $s_m = 1.89$ kV and $s_s = 2.82$ kV (eqns. 3.29 and 3.30). It is expedient, however, to improve the estimates by correction. Using eqn. 3.31 and Fig. 3.40, an estimate of the standard deviation $s = 6.7$ kV is obtained. On the other hand, the correction of the mean value (eqn. 3.32, Fig. 3.41) is slight: $u_{d50} = 144.8$ kV. The confidence limits for u_{d50} are calculated using eqn 3.33 (K_3 from Fig. 3.41). $u_{d50}^{(u)} = 140.4$ kV and $u_{d50}^{(0)} = 149.3$ kV are obtained.

An 'extended' up-and-down method has been put forward by Carrara and Dellera. It is based on the principle that the method can also be employed using $m > 1$ voltage applications per voltage step (Fig. 3.37: $m = 1$), moving on to the next step when either m voltage applications have taken place without breakdown (Δu rise) or when one breakdown has occurred (Δu drop; Fig. 3.42). If the performance function for $m = 1$ is assumed to be a normal distribution $N(u_{d50}^{(1)}; \sigma_1^2)$, the performance functions for $m > 1$ can be calculated using the 'enlargement law' (see Section 6.3) (Fig. 3.43). According to Fig. 3.43 with $m = 1$, the value $p^{(1)} = 0.13$ corresponds to the value $V^{(5)}(u_{d50}^{(5)}) = p^{(5)} = 0.5$ of the performance function for $m = 5$ voltage applications per step. Determination of the 50% quantile for $m = 5$ thus simultaneously supplies the 13% quantile for $m = 1$. Fig. 3.44 again shows this relationship for various m, wherein the breakdown probabilities $p(1)$ for $m = 1$, corresponding to the 50% breakdown voltages $u_{d50}^{(m)}$, are reproduced as a function of m. In practice, one can therefore proceed to determine the number of voltage applications m per stage required (e.g. $m = 7$) from Fig. 3.44, corresponding to the quantile required (e.g. $u_{d10}^{(1)}$; i.e. $p^{(1)} = 0.10$). The extended up-and-down method is then performed (as in Fig. 3.42, but $m = 7$), and the empirical quantile $u_{d50}^{(m)*}$ determined, as described above for $m = 1$. The voltage value obtained is simultaneously an estimate for the quantile $u_{d10}^{(1)*}$ being sought. If $u_{d50}^{(1)*}$ is, moreover, determined in a known manner with $m = 1$, then two empirical quantiles of the normal distribution

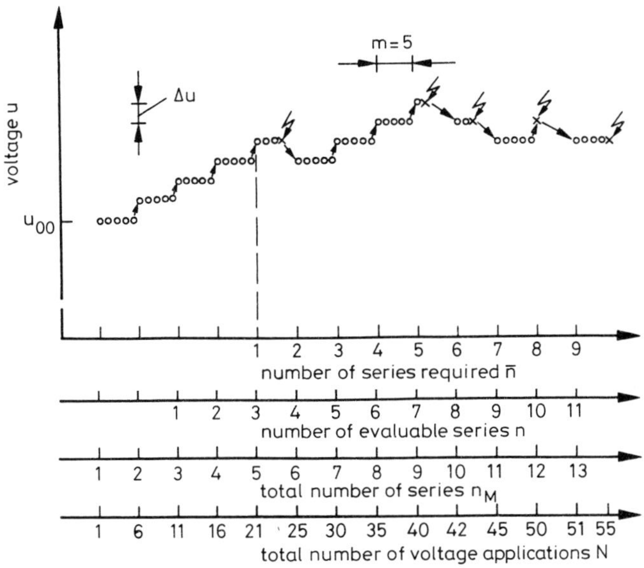

Fig. 3.42 Extended up-and-down method (diagrammatic)
$m=5$ voltage applications per step

$N(u_{d50}^{(1)}; \sigma_1^2)$ are obtained, which can be used to determine σ_1 simply. The extended up-and-down method consequently provides better estimates for the dispersion than the 'classical' method.

It should be borne in mind, of course, that for $m>1$ the performance functions are no longer normal distributions (Fig. 3.43). The up-and-down method presupposes normally distributed populations, however. In the region around $u_{d50}^{(m)}$, the performance functions $V(m)(ud)$ must therefore be approximated by normal distributions $N(u_{d50}^{(m)}; \sigma_m^2)$, where the relationship between σ_m and σ_1 is reproduced in Fig. 3.45 as a function of m. Since the prerequisite of a normal distribution for $V^{(m)}(ud)$ is only roughly applicable, it is sufficient to calculate $u_{d50}^{(m)}$ as an arithmetic mean, thereby avoiding the more time-consuming calculations using eqns. 3.27 and 3.32.

By simulation of the extended up-and-down method, Carrara and Dellera have obtained a good deal of information which is significant in selecting the test parameters. If the accuracy of the estimation is expressed by the standard deviation $\sigma_{u_d}^{(m)}$ of the mean value $u_{d50}^{(m)}$ (related to the standard deviation σ_m of the normal distribution with which $V^{(m)}(u_d)$ is approximated), the accuracy increases with the number of acceptable series n (Fig. 3.46). The standard deviation $\sigma_{u1}^{(m)}$ of the first breakdown voltage (cf. Fig. 3.38), of course, affects the accuracy of the estimate for a small number of acceptable series n. $\sigma_{u1}^{(m)}$ has no further influence when $n>30$.

The up-and-down method counts the events n with the first breakdown (Fig. 3.37). According to Fig. 3.42, with the extended method this is the

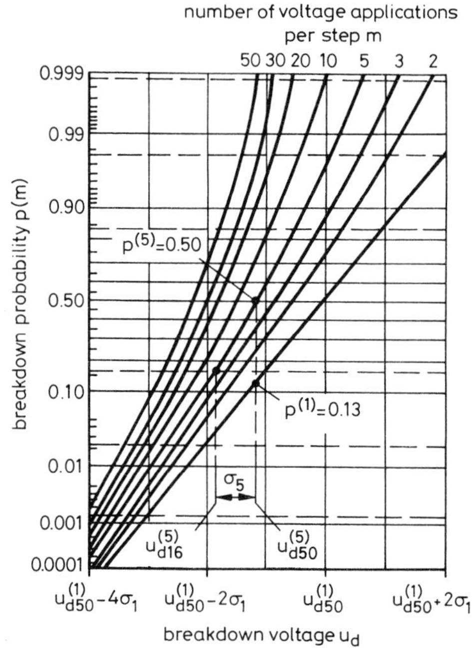

Fig. 3.43 Breakdown probability $p^{(m)}$ of at least one out of m voltage applications producing breakdown [3.58]

$p_1^{(1)} = N(u_{d50}^{(1)}; \sigma_1^2)$ is assumed to be known; associated with each voltage u_d is a probability $p^{(m)}(u_d)$ and a corresponding value $p^{(1)}(U_d)$ for $m = 1$ voltage application

number of series required $\bar{n}$. Since this first breakdown is randomly dispersed, the number of acceptable series n is a function of its standard deviation $\sigma_{u1}^{(m)}$. The number of acceptable series can be greater or less than the number of series required and is essentially a function of the number of series n (Fig. 3.47). n and $\bar{n}$ are more or less identical for $n > 20$, but with a small n the influence of $\sigma_{u1}^{(m)}$ is considerable.

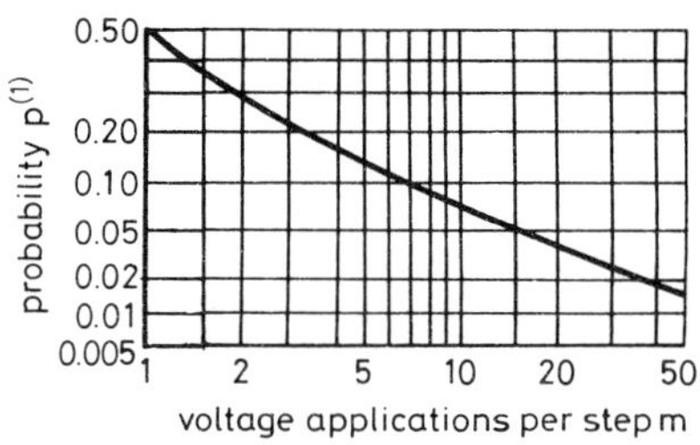

Fig. 3.44 Probability $p^{(1)}$ at a voltage $u_{d50}^{(m)}$, as a function of the number of voltage applications m [3.58]

At least one out of m voltage applications produces breakdown, with a probability $p^{(m)}$

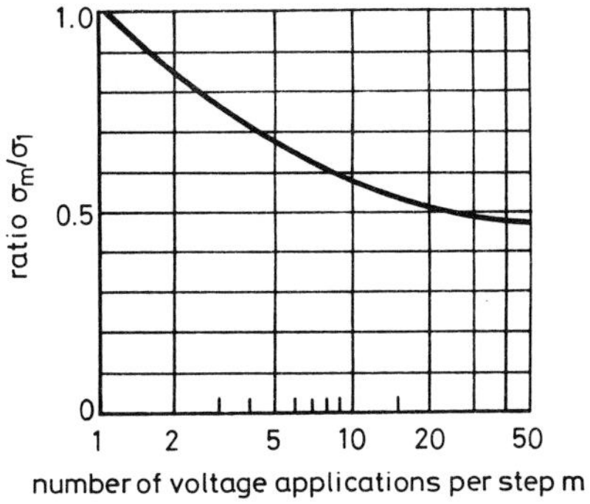

Fig. 3.45 Ratio of standard deviations of approximated normal distributions for *m* voltage applications and for one voltage application per step [3.58]

The work involved in employing the extended up and down method is governed by the maximum total number of series n_M, which is naturally greater than the number of acceptable series (Fig. 3.48). The work required can be estimated from Figs. 3.47 and 3.48.

The mean total number of voltage applications

$$\bar{N} \approx nm \tag{3.34}$$

determines the accuracy of the estimates. From simulations, the standard

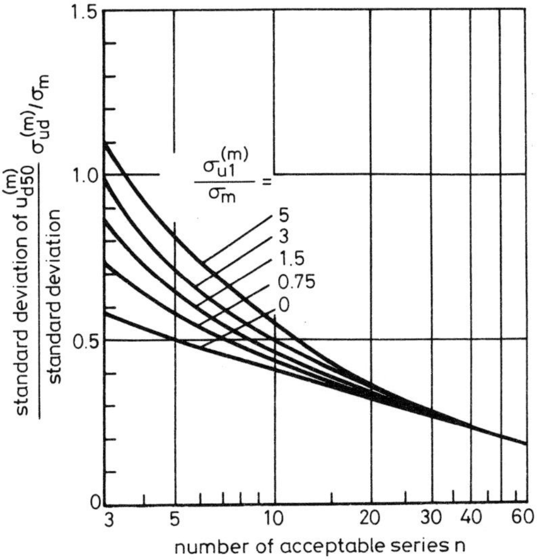

Fig. 3.46 Accuracy of estimate of mean value $u_{d50}^{(m)}$, expressed by its relative standard deviation, as a function of the number of acceptable series *n* [3.58]

Parameters: relative dispersion of first breakdown $\sigma_{u1}^{(m)}/\sigma_m$; step amplitude $\Delta u = \sigma_m$

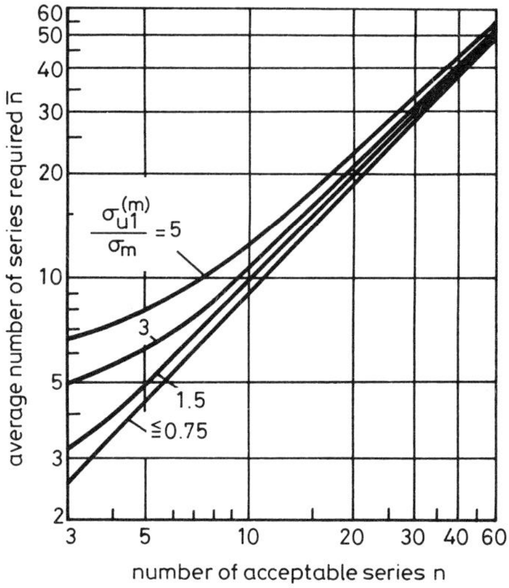

Fig. 3.47 Average number of series required $\bar{n}$, as a function of the number of acceptable series n [3.58]

Parameters: relative dispersion of first breakdown $\sigma_{u1}^{(m)}/\sigma_m$; step amplitude $\Delta u = \sigma_m$

deviations of the standard deviation $\sigma_1 - \sigma_\sigma^{(1)}$ and the estimated values have been determined for relevant quantiles $u_{dq} - \sigma_u^{(m)}$ (Fig. 3.49). These can be used to estimate confidence limits, as is demonstrated in the following example.

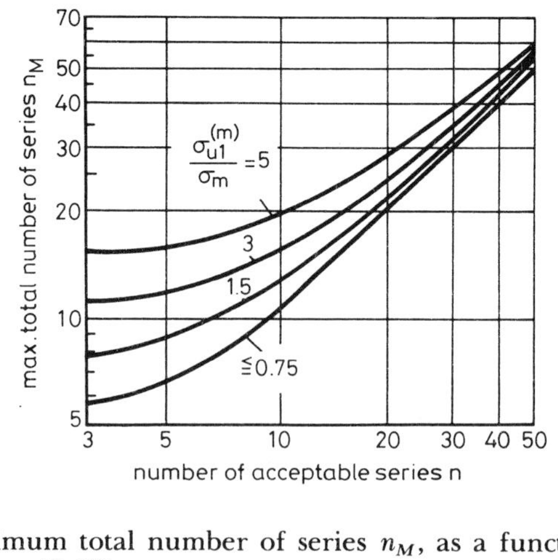

Fig. 3.48 Maximum total number of series n_M, as a function of the number of acceptable series n [3.58]

Parameters: relative dispersion of first breakdown $\sigma_{u1}^{(m)}/\sigma_m$; step amplitude $\Delta u \approx \sigma$

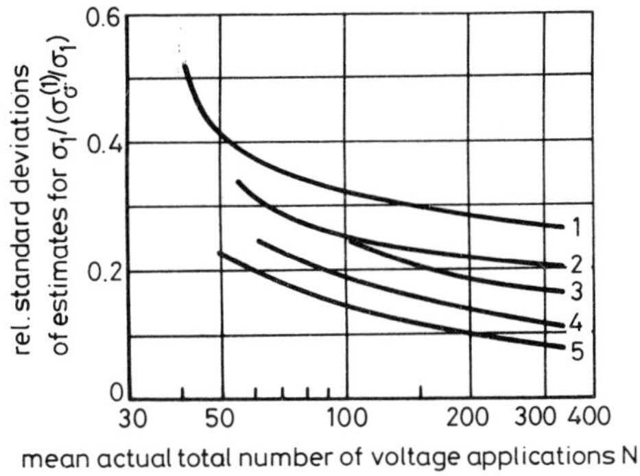

a

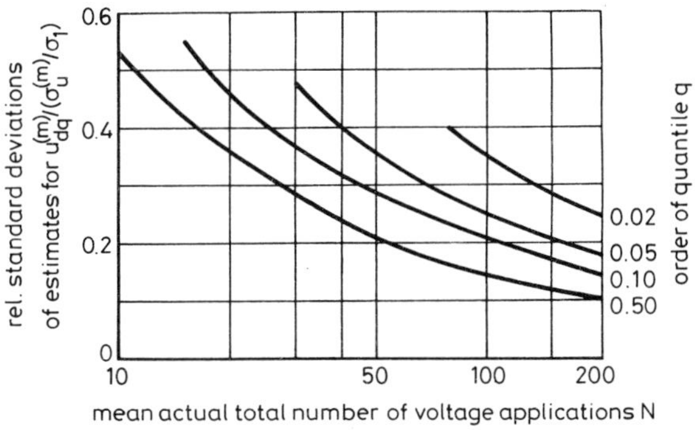

b

Fig. 3.49 Simulated accuracy for up-and-down method [3.58]

a Standard deviation $\sigma_\sigma^{(1)}$ of estimates for $\sigma^{(1)}$
1 u_{d50} with $n=25$; $m=1$ and u_{d11} with $m=6$ (any n)
2 u_{d50} with $n=25$; $m=1$ and u_{d05} with $m=13$ (any n)
3 u_{d50} with $n=25$; $m=1$ and u_{d02} with $m=34$ (any n)
4 u_{d50} (any n; $m=1$)
5 u_{d94} and u_{d06} with $m=11$ (any n)
b Standard deviation $\sigma_u^{(m)}$ of estimates for $u_{dq}^{(m)}$

Example 3.4.1.2 The 50% breakdown voltage, the 2% breakdown voltage and the standard deviation of an insulating clearance are to be estimated using the up-and-down method. First u_{d50}^* is determined with $m=1$ voltage application per step. To preclude the influence of the dispersion of the first breakdown, $n=30$ is chosen on the basis of Fig. 3.46. The initial voltage u_{00} has hardly any influence on the result: empirical values for the arrangement being investigated show that breakdown has still not occurred at $u_{00}=1000$ kV. We also know from earlier investigations that the variation coefficient is roughly 0.06. Since the step amplitude Δu should be set at roughly σ_1, we obtain approximately $\Delta u \approx 0{,}06 u_{00} = 60$ kV. The test procedure is given in Table 3.7*a*. By simple averaging we obtain $u_{d50}^* = 1194$ kV.

Table 3.7 Extended up-and-down method

a) $m = 1$ for estimation of u_{d50} ($n = 30$); *b*) $m = 34$ for estimation of u_{d02} ($n = 5$)

Events: $n = 30$; $m = 1$

Voltage u_i kV	1	2	3	4	5	6	7	8	9	10	11	12	13	14	15	16	17	18	19	20	21	22	23	24	25	26	27	28	29	30	31	32	Events per voltage step
1300														×										×								×	3
1240											×		○		×		×						○		×		×		×		○		9
1180				×		×				○		○				○		×		×		○				○		○		○			11
1120			○		○		×		○										○		○												6
1060		○						○																									1
1000	○																																0

a)

Events: $n = 5$; $m = 34$

Voltage u_i kV	1	2	3	4	5	6	Events per voltage step
1119			×				1
1083		○		×		×	3
1047	○				○		1

b)

For $n=30$, Fig. 3.46 produces a standard deviation of the mean value of $\sigma_{u_d}^{(1)}=0.25\sigma_1$. And with the estimate made above, $s_1=0.06\times1194\,\text{kV}=72\,\text{kV}$, we obtain $\sigma_{u_d}^{(1)}=0.25s_1=18\,\text{kV}$. The range $[u_{d50}^*\pm2s_{u_d}^{(1)}]=[1158\,\text{kV}; 1230\,\text{kV}]$ can be taken as the 95% confidence region for u_{d50}.

The next step is to determine u_{d02}^*. The number of voltage applications per step needed for $p^{(1)}=0.02$ is obtained from Fig. 3.44 as $m=34$. The step amplitude must be set at $\Delta u\approx\sigma_{34}$. From Fig. 3.45 it follows that $\sigma_{34}\approx0.5\sigma_1\approx36\,\text{kV}$ and therefore $\Delta u=36\,\text{kV}$. The total number of voltage applications $\bar{N}$ should be between 100 and 200. With $\bar{N}=mn$ and $m=34$, $n=5$ is chosen. If we assume that $\sigma_{u1}=\sigma_{u_d}^{(1)}$ (dispersion of first breakdown equal to that of the mean value), then it follows that $\sigma_{u1}=\sigma_{u_d}^{(1)}=0.25\sigma_1\approx18\,\text{kV}$ and $\sigma_{u1}/\sigma_{34}=0.5$. We consequently also obtain from Fig. 3.47 for $n=5$ the average number of series required $\bar{n}=4$, and from Fig. 3.48 the maximum total number of series as $n_M=6$. The 2% breakdown voltage is already specified as the initial voltage in order to manage with the smallest possible number of series. The initial voltage is

$$u_{00}\approx u_{d02}=u_{d50}^*-2.05\sigma_1=1047\,\text{kV}.$$

All the test parameters are thus specified, and the up-and-down test can be conducted with $m=34$ (result in Table 3.7*b*). By averaging we obtain $u_{d50}^{(34)*}=1083\,\text{kV}=u_{d02}^{(1)*}$, with which σ_1 can now be calculated as the difference between two quantiles

$$\sigma_1^*=s_1=\frac{u_{d50}^*-u_{d02}^*}{2.05}=54\,\text{kV}.$$

According to Fig. 3.49*a*, using curve 3, which corresponds roughly to the test parameters used here (u_{d50} with $n=30$, $m=1$ and u_{d02} with $n=5$, $m=34$), the standard deviation $\sigma_\sigma^{(1)}$ of the standard deviation σ_1 can be estimated from the mean total number of voltage applications

$$\bar{N}=(\bar{n}-0.25(n-1))m=102$$

as $s_s^{(1)}=0.25s_1=13.5\,\text{kV}$. The range $[s_1\pm2s_s^{(1)}]=[54\pm27\,\text{kV}]=[27\,\text{kV}; 81\,\text{kV}]$ is a 95% confidence region for the standard deviation being sought. According to Fig. 3.49*b*, the dispersion $\sigma_u^{(34)}$ of the 2% breakdown voltage ($q=0.02$) with $\bar{N}=102$ can be determined as $s_u^{(34)}=0.35s_1=19\,\text{kV}$. The 95% confidence region for the 2% breakdown voltage is then $[u_{d50}^{(34)}\pm2s_u^{(34)}]=[u_{d02}^{(1)*}\pm2s_u^{(34)}]=[1083\,\text{kV}\pm38\,\text{kV}]=[1045\,\text{kV}; 1121\,\text{kV}]$.

Finally, the above confidence region can be calculated for u_{d50} using the more accurate value $s_1=54\,\text{kV}$. From Fig. 3.46 $n=30\sigma_{u_d}^{(1)}/\sigma_1=0.25$ is obtained for $n=30$ and consequently $s_{u_d}^{(1)}=0.25s_1=13.5\,\text{kV}$. The 95% confidence region for the 50% breakdown voltage is then $[u_{d50}^*\pm2s_{u_d}^{(1)}]=[1194\,\text{kV}\pm27\,\text{kV}]=[1167\,\text{kV}; 1221\,\text{kV}]$.

Final result of confidence estimations for

50% breakdown voltage: $u_{d50}^{(0,u)}=(1194\pm27)\,\text{kV}$,

2% breakdown voltage: $u_{d02}^{(0,u)}=(1083\pm38)\,\text{kV}$,

standard deviation: $\sigma_1^{(0,u)}=(54\pm27)\,\text{kV}$.

3.4.2 Methods for determining lower-order quantiles

'Small' quantiles of the breakdown voltage can, as has already been treated in general terms in Section 3.3.4, be determined by rising-voltage tests with m voltage applications per voltage step. If the test is interrupted on each occasion after the first breakdown (Fig. 3.50), then, with n repetitions, one will obtain the empirical cumulative-frequency function of the breakdown voltage U_{dq} at which at least one voltage application out of m produces breakdown. If n is sufficiently large and if the initial voltage u_0 has been varied in fine stages (see Section 3.3.1), then the expectation $E(U_{dq}) = \bar{u}_{dq}$ will, for example, be such a lower-order quantile.

Methods have now been investigated by many authors that, with an extremely small number of breakdowns and with a tolerable total number of voltage applications, provide information about quantiles $q \leqq 0.1$, i.e. for statistical withstand voltages. Similar methods can then also be applied to 'large' quantiles $q \geqq 0.9$, i.e. for statistical operating voltages of surge diverters or spark-gaps.

In the simplest case, such a rising-voltage test is performed once only from an initial voltage

$$u_{00} = u_{d50} - ks \tag{3.35}$$

with $m > 1$ voltage applications ($n = 1$; estimates are used for u_{d50} and s; $k > 0$). The 'statistical withstand voltage' will then be the highest voltage step

$$u_{dq} = u_{d50} - \beta s, \tag{3.36}$$

at which no breakdown occurs in all m voltage applications. The number of the step at which the first breakdown occurs is a variate A. We then

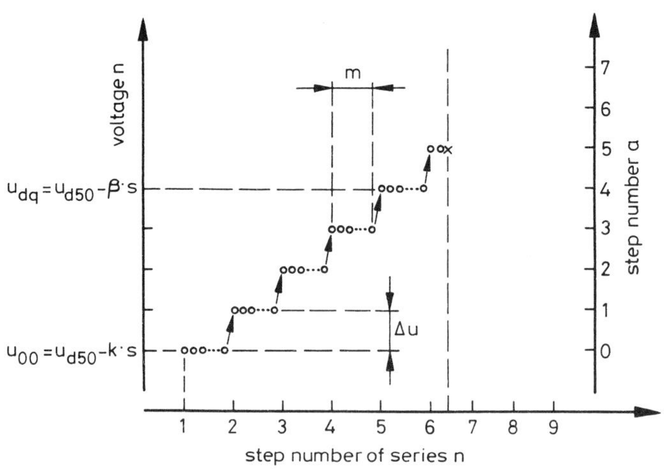

Fig. 3.50 Determination of a small quantile of the breakdown voltage in a rising-voltage test ($m > 1$)

obtain for the variate withstand voltage:

$$U_{dq} = u_{00} + (A-1)\Delta u \tag{3.37}$$

and for its expectation:

$$E(U_{dq}) = u_{00} - \Delta u + \Delta u\, E(A) = \bar{u}_{dq}. \tag{3.38}$$

The probability of the first breakdown occurring at step a is calculated, by analogy to the reasoning producing eqns. 3.1 and 3.25, as

$$P(A=a) = [1-(1-V(a))^m \prod_{i=0}^{a-1} (1-V(i))^m, \tag{3.39}$$

where $V(a)$ and $V(i)$ are the breakdown probabilities (performance-function values) associated with the steps. With the expectation

$$E(A) = \sum_{a=1}^{\infty} a[1-(1-V(a)^m] \prod_{i=0}^{a-1} (1-V(i))^m \tag{3.40}$$

and eqn. 3.38, the expectation of the statistical withstand voltage, is obtained as:

$$E(U_{dq}) = \bar{u}_{dq} = u_{d50} - s\left\{k + \frac{\Delta u}{s} - \frac{\Delta u}{s} \sum_{a=1}^{\infty} a[1-(1-V(a))^m] \times \prod_{i=0}^{a-1} (1-V(i))^m\right\} = u_{d50} - \bar{\beta}s, \tag{3.41}$$

as a function of the initial value u_{00} (in accordance with eqn. 3.35, expressed by k), of the number of voltage applications per step m, and of the step amplitude Δu. The dispersion of the step number at which the first breakdown occurs is

$$D^2(A) = E(A^2) - [E(A)]^2$$

and the dispersion of the withstand voltage consequently

$$D^2(U_{dq}) = (\Delta u)^2 \left\{ \sum_{a=1}^{\infty} a^2[1-(1-V(a))^m] \prod_{i=0}^{a-1} (1-V(i))^m - \left[\sum_{a=1}^{\infty} a[1-(1-V(a))^m] \prod_{i=0}^{a-1} (1-V(i))^m \right]^2 \right\}. \tag{3.42}$$

Using eqn. 3.36, the dispersion of the withstand voltage can also be represented as the dispersion s_β of the factor β.

Investigations exist into the influence of the initial voltage u_{00} (expressed by k; Fig. 3.51), of the number of voltage applications m (Fig. 3.52) and of the step amplitude Δu (Fig. 3.53), based on the assumption that $V(u_d)$ — and by the same token $V(a)$ — satisfy a normal distribution. If the initial voltage is selected as $u_{00} < u_{d50} - 3s$, it affects neither the expectation nor the dispersion of the statistical withstand voltage. The number m of voltage applications per step, on the other hand, has a considerable influence on the expectation, but only a slight influence on the dispersion. The larger m is, the lower will be the order of the anticipated quantile (Fig. 3.52).

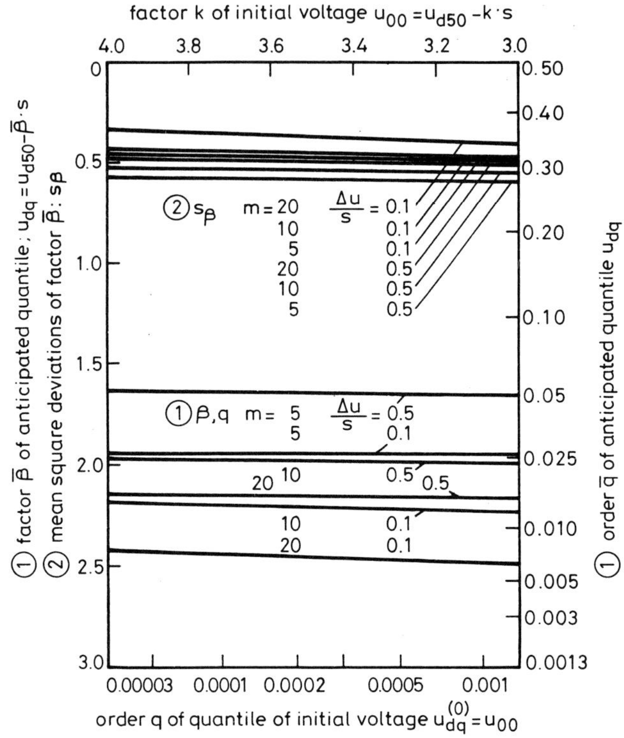

Fig. 3.51 Dependence of the anticipated quantile u_{dq} and of the mean square deviation s_β on the initial voltage u_{00}

Number of voltage applications per step m; step amplitude $\Delta u/s$

Similarly, a small step amplitude produces small quantiles; the effect on the dispersion increases with the step amplitude (Fig. 3.53). Let us clarify the relationships with an example.

Example 3.4.2.1 The 10% impulse breakdown voltage of an insulation arrangement is to be determined by the single-shot rising-voltage method (Fig. 3.50). The parameters u_{d50} and s must be estimated from empirical values; preliminary tests may be instituted if required. The parameters must be chosen so that incorrect estimates do not seriously affect the result. The initial voltage is specified as $u_{00} = u^*_{d50} - 4s^*$, so that $u_{00} < u_{d50} - 3s$ could also be ensured with a fairly uncertain estimate of s. If the step amplitude $\Delta u \geqq 0.25s^*$ is selected, the test result will not depend greatly on Δu (Fig. 3.53). With a view to minimising the dispersion s_β of factor β, $\Delta u = 0.25s^*$ is used. According to Fig. 3.52, the expectation of the highest step without breakdown will correspond to the 10% breakdown voltage ($\beta = 1.28$) when $m = 2$ is selected. The result with $m = 2$ is of course rather uncertain since the variations are considerable ($s_\beta = 0.6$). With a 90% certainty, the statistical

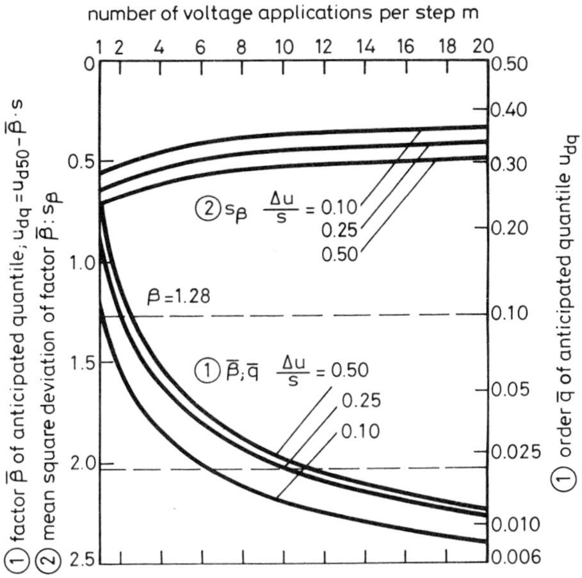

Fig. 3.52 Dependence of the anticipated quantile u_{dq} and of the mean square deviation s_β on the number of voltage applications per step m.

Initial voltage $u_{00} = u_{d50} - 3s = u_{d0013}$; step amplitude $\Delta u/s$

withstand voltage determined in this way will have a breakdown probability just less than 30%.

$$u_{d10} : u_{d50} - \beta s;\ \beta = 1.28;\ s_\beta = 0.6$$

$$u_{d50} - (\beta - 1.28 s_\beta)s = u_{d50} - 0.512s = u_{d30}.$$

(The numerical value 1.28, and also β, is the 90% quantile of the standardised normal distribution: Table 2.6*a*.) Should the statistical withstand voltage determined be less than the 10% breakdown voltage with 90% certainty, then $\bar{\beta} - 1.28 = 1.28 s_\beta$ must be chosen (i.e. $\bar{\beta} \approx 1.9$). The number of voltage applications for this is $m = 8$. The test should therefore be conducted with $u_{00} = u^*_{d50} - 4s^*$; $\Delta u = 0.25 s^*$ and $m = 8$. One can expect breakdown to occur roughly at the eighth step. Thereafter, 70 impulses are required, only one of which should result in breakdown, however.

The method can, of course, be modified by repeating the voltage rise n times. Vibholm, Pedersen and Thyregod have investigated this method on the assumption that $V(u_d)$ can be described by a double exponential distribution with the parameters u_{d63} and γ. According to their findings, the expectation of the step at which the first breakdown occurs is approximately

$$E(A) = \bar{a} = \frac{\gamma}{\Delta u}\left(-0.5772 - \frac{u_{00} - u_{d63}}{\gamma} - \ln m - \ln \frac{\exp(\Delta u/\gamma)}{\exp(\Delta u/\gamma) - 1}\right) \quad (3.43)$$

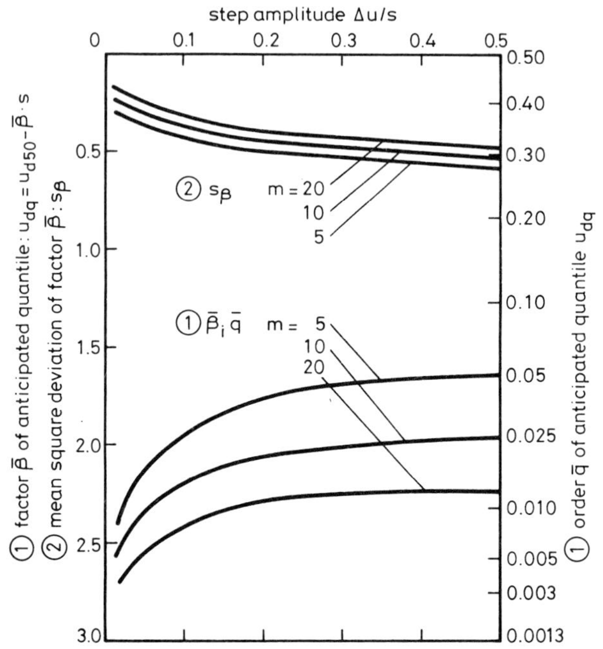

Fig. 3.53 Dependence of anticipated quantile u_{dq} and of mean square deviation s_β on step amplitude $\Delta u/s$

Initial voltage $u_{00} = u_{d50} - 3s = u_{d0013}$

and the dispersion

$$D^2(A) = s_a^2 = \frac{\pi^2}{6}\left(\frac{\gamma}{\Delta u}\right)^2. \tag{3.44}$$

We consequently obtain, for the expectation of the voltage at which the first breakdown occurs:

$$E(U_{dq}^{(1)}) = \bar{u}_{dq}^{(1)} = \gamma\left(-0.5772 + \frac{u_{d63}}{\gamma} - \ln m - \ln\frac{\exp(\Delta u/\gamma)}{\exp(\Delta u/\gamma) - 1}\right). \tag{3.45}$$

The breakdown probability associated with voltage $\bar{u}_{dq}^{(1)}$ is

$$P(\bar{u}_{dq}^{(1)}) = q = \frac{0.72}{ms_a}. \tag{3.46}$$

The mean square deviation of the step number s_a is of course not derived from eqn. 3.44, in which the measure of dispersion γ of the double-exponential distribution taken as the basis is unknown, but is calculated in a known manner (eqn. 2.48) from the realisations of step number $a_j (j = 1, \ldots, n)$ of the first breakdown obtained with n voltage rises:

$$s_a = \sqrt{\frac{1}{n-1}\sum_{j=1}^{n}(a_j - \bar{a})^2}, \tag{3.47}$$

with

$$\bar{a} = \frac{1}{n} \sum_{j=1}^{n} a_j. \tag{3.48}$$

The number of voltage rises n only affects the accuracy of the estimate of q (eqn. 3.46). A confidence region can be obtained for $n = 10$ with a confidence coefficient $\varepsilon = 0.95$ as $[q \pm 0.5q]$, and for $n = 25$ as $[q \pm 0.3q]$.

In a practical case one would prescribe the order q of the required quantile $\bar{u}_{dq}^{(1)}$, e.g. $q = 0.001$. To estimate the test expenditure, Fig. 3.54 can be used to determine the total number of voltage applications required, for which $N \approx \bar{a}mn$. It should be pointed out that the expenditure can be considerably reduced by applying the test voltage to parallel insulating clearances (see Chapter 6). The extended up-and-down method (see Section 3.4.1) can also be modified so that the test result obtained is directly a desired quantile as the mean value of the step occurring.

A method (Fig. 3.55) has been developed by Powell and Ryan, which provides the technically significant 10% and 90% breakdown voltages u_{d10} and u_{d90} directly as expectations. The method uses $m = 7$ voltage applications per series and $n = 16$ series with a step amplitude of $\Delta u \approx s^*$. For u_{d10}, one starts in the vicinity of the 50% breakdown voltage and reduces the voltage in steps of Δu until no further breakdown occurs in a series of $m = 7$ voltage applications (Fig. 3.55*a*). This series is the first acceptable series. The extended up-and-down method (Fig. 3.42) is then adopted until the 16 series are obtained, from whose voltages u_{d10} is calculated as an arithmetic

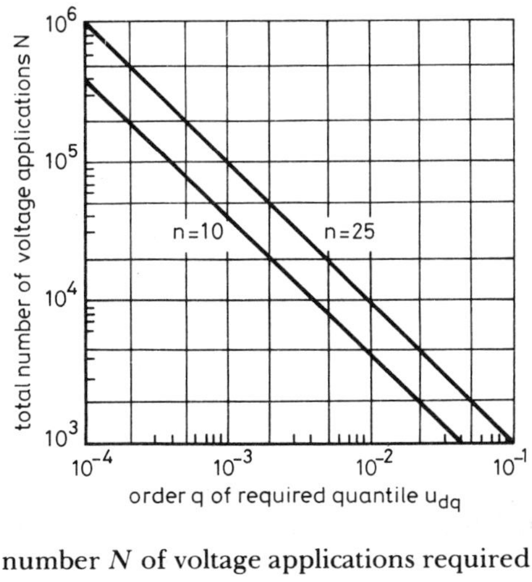

Fig. 3.54 Total number N of voltage applications required to determine a quantile u_{dq} [2.68]

n Number of voltage rises (sample size)
m Number of voltage applications per step
$\bar{a}$ Expectation of step number at which first breakdown occurs
$N \approx \bar{a}mn$

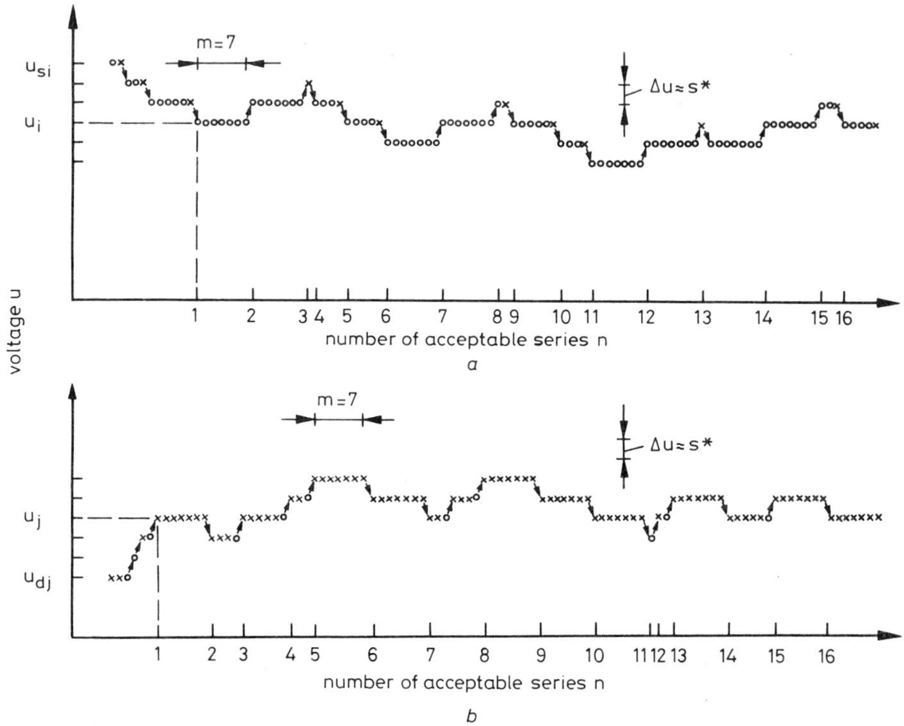

Fig. 3.55 Determination of 10% and 90% breakdown voltage, after Powell and Ryan [3.63]

a 10% breakdown voltage $u_{d10} = \sum_{n=1}^{16} u_{in}$
b 90% breakdown voltage $u_{d90} = \sum_{n=1}^{16} u_{jn}$

mean. A similar procedure is adopted to determine u_{d90}, but breakdowns now occur within the series, in place of non-breakdowns (Fig. 3.55*b*).

A method which is basically quite similar (Fig. 3.56), proposed by Fryxell, is more widely used. With $m = 3$ voltage applications per step, the method (a test) begins at a voltage u_{00} at which no breakdown is anticipated. The voltage is then raised by $\Delta u \approx (0.025 \ldots 0.05) u^*_{d50}$ until breakdown is recorded at a step. After reducing the voltage by $2\Delta u$, the process is continued with $m = 25$ voltage applications (β test). If a breakdown occurs, the voltage is again reduced by $2\Delta u$, otherwise raised by Δu. If the first breakdown again occurs with a voltage rise in the β test, the test is interrupted. The highest voltage step at which breakdown occurs neither in the α test nor the β test, is the statistical withstand voltage u_{st}. With a normally distributed breakdown voltage, the expectation of this withstand voltage is equivalent to the 0.6% quantile of the breakdown voltage (u_{d006}); its 95% confidence region is formed by the interval $[u_{d50} - 3.6s;\ u_{d50} - 1.4s]$ (corresponding roughly to $[u_{d0002};\ u_{d08}]$).

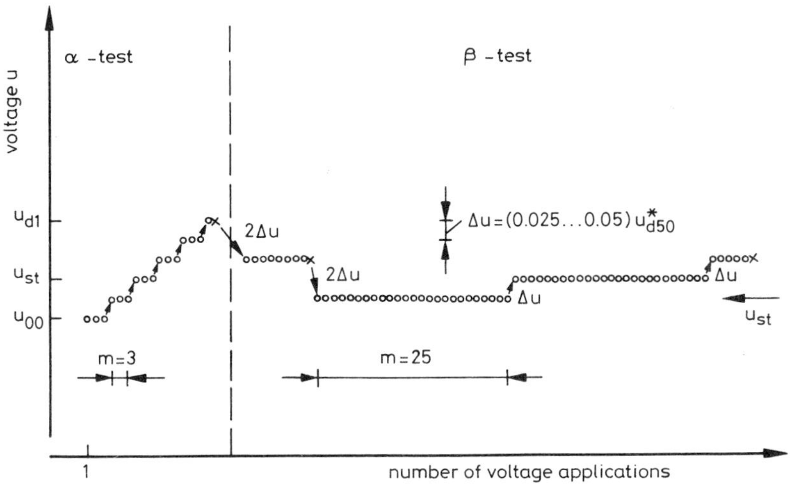

Fig. 3.56 Determination of a withstand voltage, according to Fryxell [3.65]
Expectation $u_{st} \approx u_{d006}$

95% confidence region $u_{d0002} \leqq u_{st} \leqq u_{d08}$

Fryxell's method has proved to be less sensitive in relation to the unknown type of distribution function of the breakdown voltage and manages with a very small number of breakdowns (in Fig. 3.56 with three). The efficacy of the method can be further improved by slight modifications, e.g. as suggested by Paderta. If $\Delta u \geqq 0.04u^*_{d50}$ is chosen for the method (Fig. 3.56), Δu should always be the same when raising or reducing the voltage. Only when $\Delta u < 0.04u^*_{d50}$ should one proceed as proposed by Fryxell (always reducing by $2\Delta u$). The number of voltage applications can of course be varied in the α test and β test. Paderta recommends working with $m=8$ in the α test and with $m=52$ voltage applications in the β test, and interrupting the test without breakdown after the first series in the β test ($\Delta u = 0.02u_{00}$). Generally speaking, only two breakdowns will then occur during the whole test, whereas, in all, a good 100 voltage applications are needed. When a normal distribution is assumed, u_{d50} and s can be estimated, so that any quantile can be calculated approximately.

A method of Bakken (Fig. 3.57) only prescribes a voltage rise with $m=1$ application in the α test, until a breakdown occurs at u_{d1}. The author works with $m=20$ in the subsequent β test. The voltage is always reduced by $\Delta u = (0.01–0.04)u_{d1}$ until no breakdown occurs at three consecutive steps (Fig. 3.57*a*). The highest of these three steps is the statistical withstand voltage u_{st}. In the case of a normally distributed breakdown voltage, it is below the 8% breakdown voltage u_{d08} with a confidence coefficient $\varepsilon = 0.977$. Paderta recommends that the voltage be first of all reduced from u_{d1} at $\Delta u = 0.05u_{d1}$, but after a breakdown has occurred at two steps then reducing it by only $\Delta u = 0.025u_{d1}$. The method can be used accordingly to determine a statistical operating ('response') voltage u_{dd} (Fig. 3.57*b*).

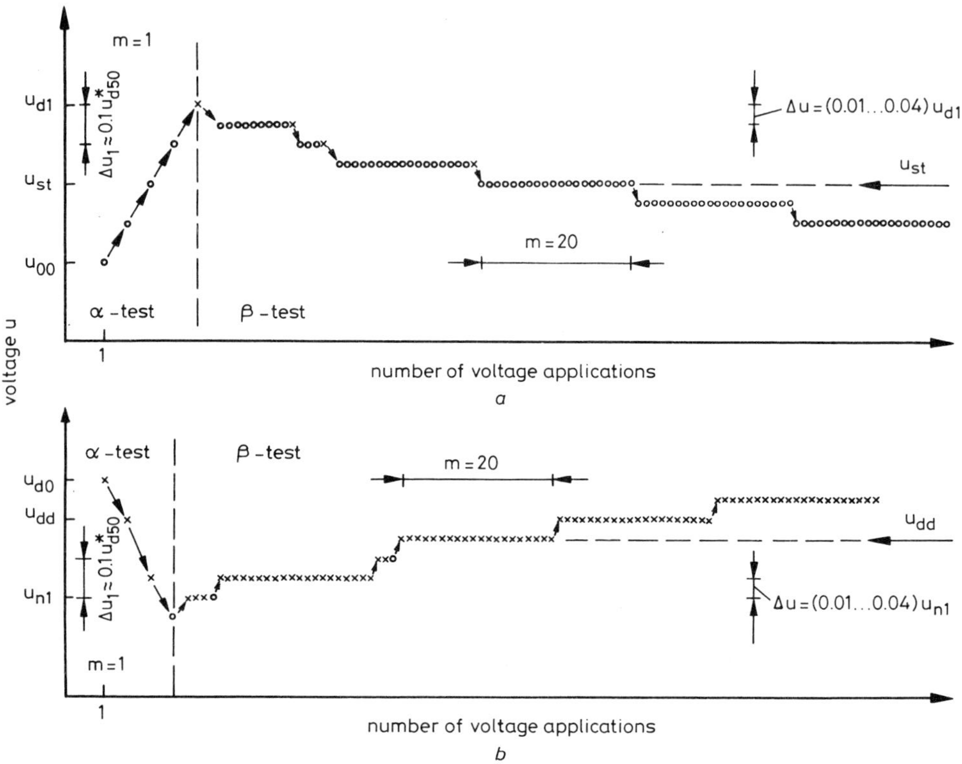

Fig. 3.57 Determination of a withstand voltage and operating voltage, after Bakken [3.9]

a Withstand voltage: 97.7% confidence region; $u_{st} \leqq u_{08}$
b Operating voltage: 97.7% confidence region; $u_{dd} \geqq u_{92}$

Again, only voltage reductions figure in a method recommended by Hylten–Cavallius and Fonseca. In this instance the test starts at a peak voltage[1] $\hat{u}_{d0}$ (Fig. 3.58*a*) at which a breakdown will occur with certainty, and the instantaneous value u_{d0m} of the breakdown voltage is measured. The peak value is then reduced to u_{d0m}, but at least by $\Delta u = 0.02\hat{u}_{d0}$. Voltage applications are repeated at this step until a breakdown occurs, for which the instantaneous value is again measured. The voltage is again reduced in the manner described. The process continues until a predetermined number N of voltage applications have taken place. Preselection of N depends on the desired order q of the anticipated quantile u_{dq} and its dispersion, as well as on the type of distribution (Fig. 3.58*b*). The formula required for

[1] Up to now, all voltages have represented peak voltages, unless specifically stated otherwise. Here, peak values are specifically indicated by a circumflex ($\hat{u}_{d0}$) and instantaneous values by the index m (u_{d0m}).

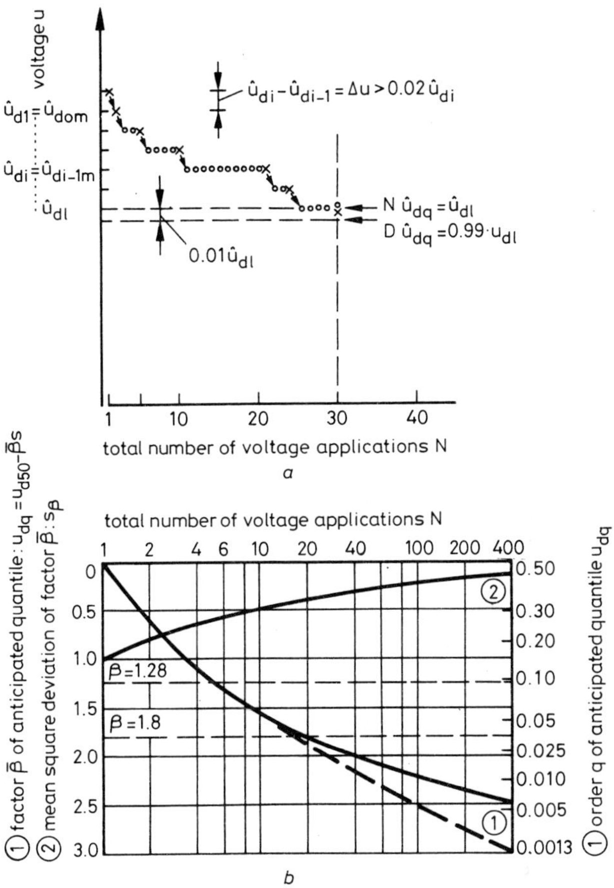

Fig. 3.58 Method for determining a withstand voltage, after Hylten–Cavallius [3.68]

a Procedure;
b Anticipated quantile u_{dq} and mean square deviation s_β

D Breakdown; *N* Non-breakdown
—— calculated for Weibull distribution ($x_0 = -3$; $EX = 0$; $D^2X = 1$)
- - - calculated for normal distribution $N(0; 1)$

the purpose is identical to that described for the rising-voltage method with $m > 1$ voltage applications per series (Figs. 3.50, 3.52; eqn. 3.36).

Example 3.4.2.2 The 10% breakdown voltage is to be determined. The number of voltage applications is obtained as $N = 6$ from Fig. 3.58. One must of course anticipate a mean square deviation of factor β of $s_\beta = 0.6$. Consequently, the quantile $\hat{u}_{dq}$ obtained has, with 90% certainty, a breakdown probability which is only $q \leqq 0.3$ ($\hat{u}_{d10} = \hat{u}_{d50} - \bar{\beta}s$ with $\bar{\beta} = 1.28$; because $s_\beta = 0.6$, then $\hat{u}_{d50} - (\bar{\beta} - 1.28 s_\beta)s = \hat{u}_{d30}$; see Example 3.4.2.1). Should the limit of the 90% confidence region be at u_{d10}, then $\bar{\beta}_1 = 1.28 + 1.28 s_\beta^*$ must be chosen, it being borne in mind that s_β^* is a value corresponding to the freshly selected N (Fig. 3.58*b*). If $s_\beta^* \approx 0.4$ is assumed, one obtains $\bar{\beta} = 1.28 + 1.28 \times 0.4 \approx 1.8$. This $\bar{\beta}_1 = 1.8$ corresponds to a required total number of voltage applications of $N = 18$.

The test should be interrupted after $N = 18$ voltage applications. If the last application fails to produce a breakdown, the associated voltage $\hat{u}_{dl}$ is regarded as the quantile $\hat{u}_{dq}$ associated with $N = 18$. If, on the other hand, a breakdown does occur, then $\hat{u}_{dq}$ should be equated to 0.99 $\hat{u}_{dl}$. The method produces good agreement with that presented by Fryxell (Fig. 3.56).

Which of the methods presented here for determining a small quantile of the breakdown voltage should be used will depend on the test piece concerned, the test expenditure possible and on the accuracy required. In the case of completely self-restoring insulation, the number of breakdowns occurring will not be important, and the extended up-and-down method (Figs. 3.42, 3.55), as well as the methods of Powell/Ryan (Fig. 3.55) and Hylten-Cavallius (Fig. 3.58), can be directly used. If, on the other hand, a fresh test piece has to be used after each breakdown, the extended voltage-rise test (Fig. 3.50) and the methods of Fryxell (Fig. 3.56) and Bakken (Fig. 3.57) are preferable. It should be borne in mind, of course, that non-breakdowns can also produce changes. In this instance only preliminary tests can produce clarity. If appropriate, a new test piece should be used for each individual test. In this case, methods involving a small number of individual voltage applications (e.g. Fig. 3.58) will be particularly economical.

Chapter 4

Statistical evaluation of standardised test methods

The insulating capacity of an insulation system is guaranteed provided no breakdown-producing discharge processes occur in it. When considering actual technical situations, the insulating capacity has to be described by appropriately selected voltages. The characteristic voltage for the insulating capacity of an insulation arrangement — the so-called withstand voltage — is assigned to the event 'non-breakdown' (highest voltage at which 'non-breakdown' is still observed), but it has to be derived from the complementary event 'breakdown' since only this event is capable of being measured. This is carried out using the statistical withstand voltage. This is a quantile of order p (in most cases $p \leqq 0.10$) of the breakdown voltage. Insulation arrangements employed at the same insulation level on a system need to have co-ordinated rated withstand voltages (see Section 4.1). The rated withstand voltages are agreed (standardised) voltage values which the insulation is capable of withstanding in a standardised test method. Such test methods are used worldwide as type tests and routine tests on insulation. They are inherently only meaningful when considered from a statistical standpoint. We shall now try to establish concisely a relationship between what has been stated in relation to the variate 'breakdown voltage' (see Section 3) and the standardised test methods, as well as in relation to insulation co-ordination.

4.1 Aims and problems of insulation co-ordination

In service, the insulation of equipment and plant is subjected to the operating voltage (alternating voltage; occasionally direct voltage), overvoltages and surges (switching and lightning surges). Insulation co-ordination is used

- to match the insulating capacity of the insulation to the anticipated stresses
- by overvoltage protection measures to limit overvoltages to maximum values
- to ensure co-ordination between different types of insulation.

Types of insulation are categorised into insulation groups, according to: the significance of the insulation in the power system (e.g. phase-to-earth, phase-to-phase and isolating-gap insulation); the capacity of an insulation arrangement to re-establish its insulating capacity once it has lost it (regenerating capacity: self-restoring, e.g. outdoor air insulation; partially self-restoring, e.g. compressed-gas or oil insulation; non-self-restoring, e.g. solid insulation); and according to the value of the insulation (e.g. comparison between air insulation and transformer insulation):

- Insulation group 1: isolating clearances, fuse holders, between networks
- Insulation group 2: insulators, bushings, switching devices (apart from isolating gaps), power transformers, instrument transformers, cables, etc.
- Insulation group 3: prefabricated switchgear with individual items of gear assigned to insulation group 2 (i.e. consideration of enlargement effects; see Section 6)
- Insulation group 4: neutral-point insulation
- Insulation group 5: insulation of rotating machines

A rated withstand voltage is established for each insulation group for each insulation voltage by a method which we do not intend to go into here, but group 1 has the highest rated withstand voltage and group 5 the lowest. The internationally specified rated withstand voltages for phase-to-earth insulation, for example, are given in Table 4.1 for medium-voltage systems, in Table 4.2 for high-voltage systems and in Table 4.3 for extra high-voltage systems. From this selection of rated withstand voltages, the most suitable are chosen for insulation groups and devices, taking the above criteria and the insulation class into account. The insulation class depends on the treatment of the neutral point and on the risk of overvoltages (NE, effectively earthed; N, not effectively earthed; SE, effectively earthed and not prone to overvoltages).

Table 4.1 Rated withstand voltages for medium-voltage systems (IEC 71 [4.1])

2N) and 3N) represent insulation groups and classes according to [4.2]

Insulation voltage U_m kV	Rated lightning withstand voltage $\hat{U}_{nsts}$ List 1^{3N}) kV	List 2^{2N}) kV	Rated alternating withstand voltage U_{nstw} 2N) 3N) kV
3.6	20	40	10
7.2	40	60	20
12	60	75	28
17.5	75	95	38
24	95	125	50
36	145	170	70

Table 4.2 Rated withstand voltages for high-voltage systems (IEC 71 [4.1])

2N) and 3N), etc., represent insulation groups and classes according to [4.2]

Insulation voltage U_m kV	Reference voltage $\frac{\sqrt{2}}{\sqrt{3}} U_m$ kV	Rated lightning withstand voltage $\hat{U}_{n\,st\,s}$ kV		Rated alternating withstand voltage $U_{n\,st\,w}$ kV	
52	42.5	250		95	
72.5	59	325		140	
123	100	450	3N) 4N) 2NE)	185	4N) 2NE) 3NE)
145	118	550	2N) 1NE)	230	2N) 3N)
170	139	650	4N)	275	4N)
245	200	750	3NE) 2SE) 3SE)	325	2SE) 3SE)
		850	2NE)	360	2NE) 3NE)
		950	3N) 1SE)	395	
		1050	2N) 1NE)	460	2N) 3N)

Example 4.1.1 Using Table 4.2, the rated withstand voltages are to be selected for insulation voltage 245 kV and insulation class NE. The following values are obtained:

	Rated lightning withstand voltage $\hat{U}_{nsts}$	Rated alternating withstand voltage U_{nstw}
Insulation group 1	1050 kV	460 kV
Insulation group 2	850 kV	360 kV
Insulation group 3	750 kV	325 kV

(Insulation groups 4 and 5 are omitted for 245 NE.)

By analogy to the rated withstand voltages of insulation, rated operating voltages are specified for voltage limiting devices (diverters/arresters; level spark-gaps).

Table 4.3 Rated withstand voltages for extra-high-voltage systems (IEC 71 [4.1])

(2SE) and 3SE), etc., represent insulation groups and classes according to [4.2]

Insulation voltage U_m kV	Reference voltage $\frac{\sqrt{2}}{\sqrt{3}}U_m$ kV	Rated switching withstand voltage $\hat{U}_{n\,st\,sch}$ relative	kV		Ratio $\frac{\hat{U}_{n\,st\,s}}{\hat{U}_{n\,st\,sch}}$ —	Rated lightning withstand voltage $\hat{U}_{n\,st\,s}$ kV	
					1.13	850	
		3.06	750				
300	245				1.27	950	
					1.12		
		3.47	850				
		2.86					
362	296				1.24	1050	
					1.11		
		3.21	950	2SE) 3SE)			
		2.76					
420	343				1.24	1175	2SE) 3SE)
					1.12		
		3.06	1050	2NE) 3NE)	1.24	1300	3NE)
		2.45			1.11		
525	429						
					1.36		
		2.74	1175		1.21	1425	2NE) 1SE)
					1.10		
					1.32		
		2.08	1300	2SE) 3SE)	1.19	1550	3SE)
					1.09		
					1.38		
765	625	2.28	1425	$2NE_2$) $3NE_2$)	1.26	1800	2SE) $3NE_2$)
					1.16		
		2.48	1550	$2NE_1$) $3NE_1$)	1.26	1950	$3NE_1$)
					1.47	2100	$2NE_2$) 1SE)
					1.55	2400	$2NE_1$) $1NE_2$)

The classical method of insulation co-ordination consists of testing the insulation in a new condition according to standard procedures dealt with in the following (see Sections 4.2 and 4.3). If only insulation arrangements for which the (correctly selected) rated withstand voltages have been demonstrated are used in a system, then the insulation co-ordination is regarded as assured. It is obvious that this relatively simple method is incapable of taking adequate account of either the stochastics of the insulating capacity (i.e. of breakdown) and the stochastics of the overvoltage, or of reductions in the insulating capacity due to ageing, wetting or pollution. From the economic's standpoint, however, the classical method can in no way be dispensed with: one just needs to be aware of its limitations.

The statistical method of insulation co-ordination assumes a knowledge of the density functions of the overvoltages $f(u_u)$ in the system concerned, and of the performance functions $V(u)$ of the insulation arrangements. These performance functions can only be determined experimentally for completely self-restoring insulation, whereas it is impossible, for example, to determine the performance functions empirically for the insulation of a large transformer or even to estimate it with sufficient reliability. The statistical method briefly described below can consequently only be applied to outdoor insulation. In most cases, only estimates are available for the density function $f(u_u)$, or the results of model investigations (network analysers; digital computer programs). Recordings of overvoltages on systems have indeed been made, but the scope of such measurements is small, they are expensive, and are not suitable for direct generalisation. In consequence, due to uncertain assumptions for insulation co-ordination, there are naturally uncertainties in the result and limits in the applications. Nevertheless, only this method does justice to the random nature of the stresses on the system and of the insulating capacity. It is therefore certain that the method will be further extended in the future.

The mathematical derivation of the method involves consideration of the variates overvoltage U_u and breakdown voltage U_d of the insulation. Failure of the insulation will occur precisely when $U_u \geqq U_a$. If a variate

$$D = U_d - U_{ü} \tag{4.1}$$

is introduced, $D \leqq 0$ becomes the failure condition. The failure probability (risk R) is then:

$$R = F_D(d^* = 0) = \int_{-\infty}^{0} f_D(d^*)\, dd^*, \tag{4.2}$$

whereby the density function $f_D(d^*)$ for the difference D (eqn. 4.1) can easily be calculated using the convolution integral (Table 2.1). For voltages $u = u_{ü}$ with density functions of the overvoltages $f_{ü}(u)$ or breakdown voltage $f_d(u)$:

$$f_D(d^*) = \int_{-\infty}^{+\infty} f_{ü}(u) f_d(u + d^*)\, du. \tag{4.3}$$

From eqns. 4.2 and 4.3 it follows that:

$$R = \int_{-\infty}^{0} \int_{-\infty}^{+\infty} f_{\ddot{u}}(u) f_d(u + d^*)\, du\, dd^*$$

and finally

$$R = \int_{-\infty}^{+\infty} f_{\ddot{u}}(u)\, V_d(u)\, du, \tag{4.4}$$

where $V_d(u)$ is the performance function of the breakdown voltage. Fig.

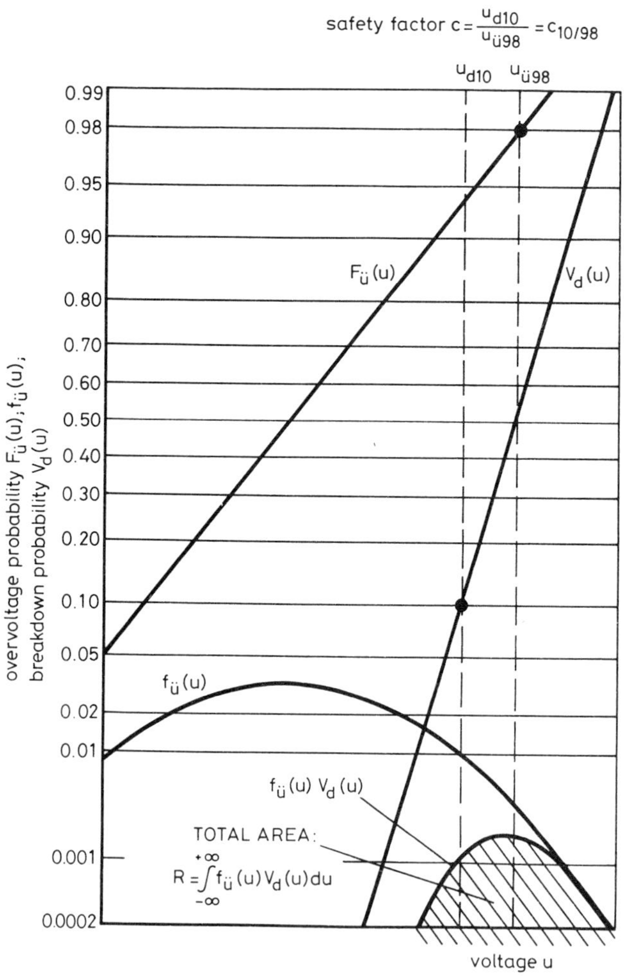

Fig. 4.1 Determination of the failure probability R from the density function $f_{\ddot{u}}(u)$ of the overvoltages and the performance function of the breakdown voltage $V_d(u)$ (diagrammatic for representation on normal-probability paper)

4.1 illustrates the relationships; the failure probability R is identical to the hatched area.

If a normally-distributed density function of the overvoltages $\phi(u;\, \bar{u}_{ü};\, \sigma^2_{ü})$ and a normal distribution of the breakdown voltages $\Phi(u;\, \bar{u}_d;\, \sigma^2_d)$ is assumed, then

$$R = \Phi\left(\frac{\bar{u}_{ü} - \bar{u}_d}{\sqrt{\sigma^2_{ü} + \sigma^2_d}}\right), \tag{4.5}$$

so that the failure probability R can easily be obtained from a normal distribution table (Table 2.6). Eqn. 4.4 should be used as the basis for other distribution functions.

In accordance with its derivation, the failure probability is related to the number of overvoltages. With a known annual overvoltage rate, the annual anticipated failure number can then be calculated, and thereby an approximate value for the cost of damage. Statistical insulation co-ordination merely consists in specifying a permissible failure probability R_z (in most cases $R_z = 10^{-4}$ to 10^{-6}) and dimensioning the insulation in such a way (i.e. by displacing $V_d(u)$) that, for the calculated failure probability, R is just equal to R_z or R is in a preselected range of R_z. It becomes clear at the same time how helpful a reduction of the overvoltages is in designing the insulation more economically.

To sum up, one may observe that classical (deterministic) insulation co-ordination is based on the most representative fixed values of the variates overvoltage and insulating capacity (breakdown voltage), and establishes a certain gap (safety factor) between these values, with which the rated withstand voltages can be calculated. Different safety factors (rated withstand voltages) then ensure the co-ordination of differently graded insulation clearances. In the statistical method, one works with the variates themselves: the entire distribution function of the insulating capacity must be suitably determined by prescribing a failure probability (i.e. the insulation has to be suitably dimensioned). Co-ordination is achieved by prescribing appropriately graded failure probabilities for insulation clearances of differing significance.

Both methods require that statistical relationships be taken into consideration. Before the procedure for dimensioning insulation arrangements is explained by reference to two examples (Section 4.4), to make it easier to understand, we shall consider the procedure for demonstrating the insulating capacity.

4.2 Proving rated alternating withstand voltage and rated DC withstand voltage

The alternating (or direct) voltage test is intended to demonstrate that the insulation can withstand the service voltage, occasional overvoltages and (only for lower insulation voltages) switching surges. To demonstrate the rated withstand voltages U_{nst}, the test piece undergoes a procedure that is practically identical for alternating and direct voltage.

The voltage is raised from zero (or from a value at which switching surges are certain not to affect the test result), in such a way that the voltage measuring instruments can be read off reliably, but the practical test duration is not extended. This is generally ensured when the voltage is raised above $0.75\,U_{nst}$ at $0.02\,U_{nst}$/s. After the rated withstand voltage U_{nst} is reached, it is kept constant for a prescribed test duration t_n, which is governed by the aim of the test. $t_n = 1$ minute is usually chosen. Considerably longer test durations have to be adopted in the case of pollution problems or for tests on synthetic insulation, because of the discharge processes, which take place slowly. For routine tests on components insulated with epoxy resin, test durations of $t_n = 10$ hours have been considered necessary, and earlier even $t_n = 100$ hours. After the test duration t_n has elapsed, the alternating voltage must be quickly reduced (but not switched off suddenly, because of surges). In the case of alternating voltage, the voltage is reduced by isolating the test piece and feeding capacitances from the system and discharging via a suitable resistance. The test is successful if no breakdown occurs during the entire procedure. Further details of the test are determined by the equipment standards.

The rated operating voltages U_{na}, at which the voltage limiting devices should operate 'with certainty' (e.g. spark gaps break down), are demonstrated by the following procedure. The voltage is applied and raised as described above, but this time until breakdown occurs. The voltage U_{da} at which breakdown occurs is recorded. The rated operating voltage is regarded as demonstrated when, for a prescribed number n of such cycles, the values U_{dai} $(i = 1, \ldots, n)$ are always less than U_{na}. The number n is specified in the equipment standards.

An overall statistical evaluation of these procedures is not possible. In the case of gas and liquid insulation (without heavy-current partial discharges), the influence of test duration or rate of rise of the voltage on the test result is relatively small, and the dispersion of the breakdown voltage when determining the rated operating voltage is scarcely greater than the error of voltage measurement. The demonstration will therefore take place quite reliably. What is more, the performance of the insulation arrangements with alternating or direct voltage is of subordinate importance in co-ordination, compared with that with impulse voltages.

The situation is quite different when demonstrating the rated withstand voltage of solid insulation, for which the breakdown voltage is greatly dependent on time (see Section 5.5), and very great dispersion of the breakdown voltage can be expected, especially with short voltage-application times. The performance with a service voltage applied for a year may be of greater relevance to insulation co-ordination than that with overvoltages. This fact has yet to be comprehensively incorporated in insulation co-ordination specifications. Finally, there is still a danger that, while demonstrating the rated alternating withstand voltage on solid insulation, partial discharges may cause irreversible damage reducing the reliability of the insulation in service. The tendency now is increasingly to couple proof of the rated withstand voltage with partial-discharge measurement. The partial-discharge test provides far more information about the state of the insulation than the Yes/No result with the rated alternating withstand voltage.

4.3 Demonstrating rated impulse withstand voltages

In contrast to demonstration of the rated continuous withstand voltages, the current standards for impulse voltages prescribe methods which are different and whose information content is not the same (see Section 4.3.2).

4.3.1 Procedures

Identical procedures are used to demonstrate the rated switching withstand voltage and rated lightning withstand voltage $\hat{U}_{nst}$. The procedures are subdivided according to the regenerating capacity of the insulation.

In the case of self-restoring insulation (air insulation; conditionally, gaseous insulation), two methods are permitted with equal justification:

(i) When the performance function of the breakdown voltage of the insulation is known, the test is successful if $\hat{u}_{d10} \geqq \hat{U}_{nst}$ is ensured for the 10% breakdown voltage. Since the performance function is also the prerequisite for statistical insulation co-ordination, this method takes closest account of statistical considerations.

(ii) The test is successful if not more than two breakdowns occur ($k \leqq 2$) in $n = 15$ applied impulses at $\hat{U}_{nst}$.

Example 4.3.1.1 The probability p_P of the test being successful is calculated using the binomial distribution (see Example 2.3.1.3.2). With the individual probabilities it is

$$p_P = p_0(k=0;\ n=15) + p_1(k=1;\ n=15) + p_2(k=2;\ n=15) \qquad (4.6)$$

and with the cumulative probability

$$p_P = p_\Sigma(k \leqq 2;\ n=15). \qquad (4.7)$$

If the breakdown probability of the insulation is $p = 0.01$ at $\hat{U}_{nst}$, then, according to eqn. 4.7 and a table of the cumulative probability of the binomial distribution, it is immediately possible to state that the test is successful with a probability $p_P = 0.9996$.

In the case of non-self-restoring insulation (insulation arrangements consisting of solid materials, combinations of solid and liquid materials or solid and gaseous materials), the test is successful if no breakdown occurs in $n = 3$ applications of $\hat{U}_{nst}$ (method *c*): $k = 0$).

In the case of combinations of self-restoring and non-self-restoring insulation (e.g. SF_6 insulation of switchgear, in which the gas breakdown clearances are self-regenerating but the flashover distances on spacers are not self-regenerating), the test is successful if, out of 15 applied impulses, no breakdowns occur in the non-self-restoring part ($k = 0$), and no more than two breakdowns in the self-restoring part ($k \leqq 2$). The method assumes that the breakdown location can be clearly determined. If this is not the case, then no breakdowns may be permitted (method (iii)).

The rated impulse operating voltage $\hat{U}_{na}$ is demonstrated if, out of $n_1=5$ applied impulses with $\hat{U}_{na}$, either all five produce breakdown ($k_1=5$), or if $k_1=4$ breakdowns occur and ten further voltage applications ($n_2=10$) result in breakdown ($k_2=10$).

Example 4.3.1.2 With the individual probabilities of the binomial distribution, and with a known breakdown probability p, the probability of the test being successful is

$$p_P = p_5(n_1=5;\ k_1=5) + p_4(n_1=5;\ k_1=4)p_{10}(n_2=10;\ k_2=10). \qquad (4.8)$$

With $p=0.99$, using eqn. 4.8, we find that the test is successful with a probability $p_P=0.994$.

All the procedures must be executed for both polarities, but it is quite sufficient to test with one polarity when it is sufficiently certain that the other polarity will produce a higher breakdown voltage (rated withstand voltage) or lower breakdown voltage (rated operating voltage). Details are specified in equipment standards.

4.3.2 Evaluation

The three procedures described

(i) $\hat{u}_{d10} \geqq \hat{U}_{nst}$
(ii) $n=15;\ k\leqq 2$ at $\hat{U}_{nst}$
(iii) $n=3;\ k=0$ at $\hat{U}_{nst}$

are formally treated as equivalent in the result of a test. The differences in the procedures, however, suggest that their information contents are not the same. To test this hypothesis, the probabilities of the test being successful p_P, as a function of the breakdown probability of the insulation p, have been calculated and reproduced (Fig. 4.2) for all three methods:

According to method (i), all insulation arrangements for which $p(\hat{U}_{nst}) \leqq 0.10$ applies to their breakdown probability at the rated withstand voltage are fault-free, and the probability of the test being successful is $p_P=1$. When $p(\hat{U}_{nst}) > 0.10$, the test is not successful ($p_P=0$). Methods (ii) and (iii) do not have such a sharp demarcation. In this instance, it is a distinct possibility for insulation arrangements which have been classed as fault-free by method (i) ($p \leqq 0.10$) not to pass the tests (manufacturer's risk) and for defective insulation ($p > 0.10$) to pass the test (operator's risk). Because of the smaller sample size, method (iii) is even less precise than method (ii). Yet it is precisely method (iii) that is important in the context of non-self-restoring technical insulation arrangements, since the performance function cannot be determined for them. Consequently, the use of methods (ii) and (iii), which are inherently practicable, involves the manufacturer and operator in certain risks, which, as Fig. 4.2 shows, are greater for the operator/user than they are for the manufacturer. The risks must not be exaggerated, however, since in classical insulation coordination there is in fact a consider-

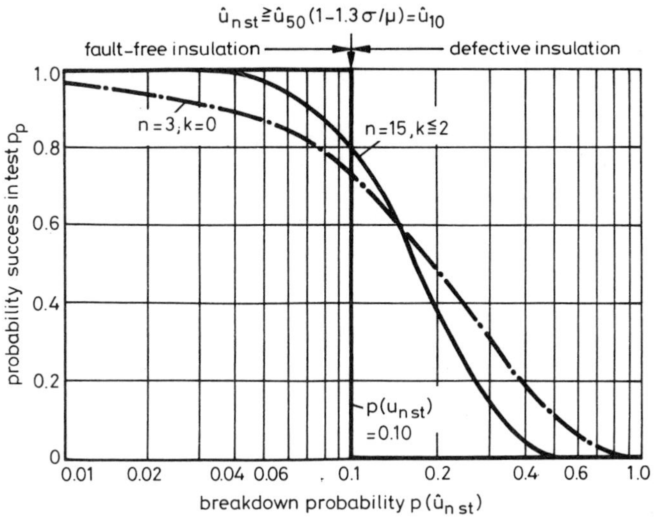

Fig. 4.2 Probability of success of a test to demonstrate the rated impulse withstand voltage

able safety margin between the maximum permissible overvoltage and the rated withstand voltage. If one nevertheless wishes to reduce the risks, then the insulation arrangements should be dimensioned not on the basis of $p(\hat{U}_{nst})=0.10$, but $p(\hat{U}_{nst})=0.02$, for example.

Uncertainties also exist with method (i), since the performance function has to be determined empirically and it represents only an estimate of the true performance function. It is also possible here to calculate the lower tolerance limit for the 10% quantile $\hat{u}_{d10}$ for a normal distribution (see Section 2.3.2.2), and to work with this value instead of a point estimate for $\hat{u}_{d10}$.

4.4 Advice on the dimensioning of insulation

When designing/dimensioning insulation arrangements, a completely empirical approach is often still adopted. First, attempts are made to estimate the requisite insulation dimensions using empirical values, and sometimes with semi-empirical methods of calculation (e.g. on the basis of the Schwaiger relationship); then a laboratory sample is made and tested by method (ii) or (iii). One needs to be aware of the fact that an optimum can hardly be achieved this way. There is an urgent need at least to estimate the performance function, then check this experimentally, and evaluate the insulating capacity by method (i). The various insulation clearances have to be co-ordinated according to their significance and regenerating capacity.

Such co-ordination must both satisfy the rules laid down by insulation co-ordination, and also undertake additional optimisation of equivalent (in the sense of present-day insulation co-ordination) insulation clearances.

Within the confines of this book, it has only been possible to make a brief mention of the problem area of insulation co-ordination. Extensive specialist literature has been devoted to the subject, a selection of which appears in the Bibliography.

There is naturally a close interaction between physical knowledge of the breakdown process and the specifications of insulation co-ordination: for example, the problems of insulation stress are studied with impulse voltages whose waveform differs from the standard lightning and switching voltages. There is a marked move towards co-ordination of conductor/conductor insulation arrangements. It is anticipated that, in the future, not only will co-ordination of insulation arrangements take place when new, as in the past, but that greater consideration will be given to reduction of insulating capacity (e.g. as a result of pollution or ageing).

In the following we intend to explain how the concepts of insulation co-ordination, but with prescribed rated withstand voltages and also with permissible fault-probabilities, can be taken into consideration when designing/dimensioning insulation.

Example 4.4.1 Let us consider a single-phase metal-clad SF_6-insulated tubular gas cable (Fig. 4.3). In the sense of conventional insulation co-ordination, the tubular gas cable exhibits exclusively equivalent conductor/earth insulating clearances, which can be advantageously subdivided as follows:

G gas breakdown distances
Ü flashover distances on spacer
F solid breakdown distances in spacer.

These distances/clearances possess very different insulation characteristics: in the gas (*G*), the insulating capacity is completely regenerated after a breakdown (e.g. in a test); at the interface (*Ü*), it is partly restored; and in

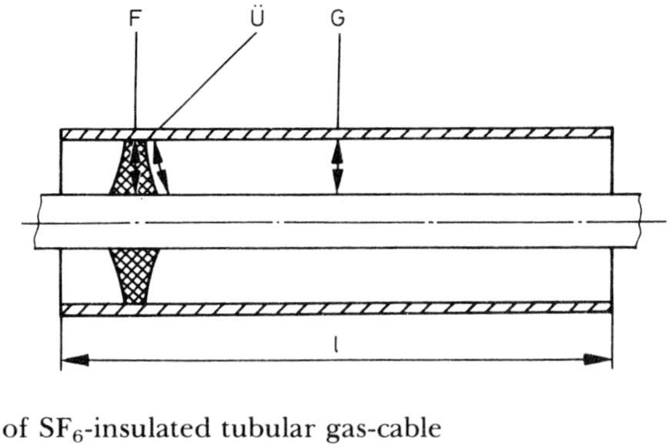

Fig. 4.3 Unit of SF_6-insulated tubular gas-cable

F solid breakdown-distance in spacer
Ü flashover distance over spacer
G gas breakdown distance
l length of unit

the solid material (F) it does not recover at all. In addition, the solid breakdown voltage is greatly dependent on time. In none of the clearances must partial discharges be permitted, since the inception voltages and breakdown voltages coincide ($u_i = u_d$). In order from the outset to relieve points in the insulation that are electrically critical and are at the same time difficult to cope with technically, and to keep any discharges occurring in the test away from such points, and to minimise maintenance, it is advisable to grade the insulating capacity of these distances/clearances in such a way that, for the 2% breakdown voltages, for example:

$$\begin{matrix} (G) & (\ddot{U}) & (F) \\ u_{d02} < & u_{d02} < & u_{d02} \end{matrix} \tag{4.9}$$

Using the method for precalculation of the performance function of the breakdown voltage in SF_6 [2.9] (see Section 5.3), it is possible to design the geometry of the insulation clearances in such a way that the co-ordination required by eqn. 4.9 is achieved. To do so, one must start with the rated withstand voltage U_{nst} of a test section consisting of m units (Fig. 4.3 shows one of these units). If the type of distribution is known, the rated withstand voltage of the unit U_{Bst} can be calculated from the rated withstand voltage of the test section U_{nst} using the enlargement law (see Section 6.3). Assuming a double-exponential distribution, which is expedient for SF_6 insulation for example:

$$U_{Bst} = U_{nst} + \gamma^* \ln m, \tag{4.10}$$

where γ^* is an estimated value for the dispersion of the double-exponential distribution.

In the procedure thereafter, the possible impreciseness of the precalculation has to be taken into consideration. On the one hand, this can be done by proceeding with the unit in accordance with method (i), but demanding $U_{Bst} \leqq u_{d02}^{(B)}$ ($u_{d02}^{(B)}$ is the 2% breakdown voltage of the unit). In addition, an accuracy limit $\delta = (0.02 \ldots 0.05) U_{Bst}$ can be adopted for the calculated 2% breakdown voltages of the individual insulation clearances. The desired values for the 2% breakdown voltages can be calculated using eqn. 4.9 and the specifications for U_{Bst} (eqn. 4.10), δ and ε. For the gas clearance we obtain:

$$U_{Bst} + \delta \leqq u_{d02}^{(G)} \leqq U_{Bst} + 2\delta,$$

for the flashover distance

$$U_{Bst} + 2\delta + \varepsilon \leqq u_{d02}^{(\ddot{U})} \leqq U_{Bst} + 3\delta + \varepsilon$$

and for the solid distance

$$U_{Bst} + 3\delta + 2\varepsilon \leqq u_{d02}^{(F)} \leqq U_{Bst} + 4\delta + 2\varepsilon.$$

If the geometry is finally optimised so that the calculated 2% breakdown voltages lie in the required ranges, then test samples of the unit should be constructed and their breakdown-voltage performance function measured. It should be more or less identical to that calculated for the gas clearance.

For SF_6 insulation, it is possible to approximate this with a double-exponential distribution (parameters $u_{d63}^{(B)}$, $\gamma^{(B)}$). A check of the required rated withstand voltage of the test section in accordance with method (i) must follow without fail:

$$U_{nst} \leqq u_{d10} = u_{d63}^{(B)} - \gamma^{(B)}(2.25 + \ln m). \tag{4.11}$$

With a greater outlay, the method can of course also be performed without calculation, with exclusively experimental fitting of the individual insulation clearances to the desired values.

Example 4.4.2 The dimensioning of a UHV isolator described in the following is based on a publication of Carrara *et al.*

For an isolator (Fig. 4.4), insulation co-ordination evaluates the isolating-clearance insulation ('longitudinal insulation') more highly than the earth insulation, since the longitudinal insulation (with the isolator open) either isolates networks from each other (there are then different voltages at the two isolator poles), or isolates parts of a network (one pole energised, the other earthed), when one is working on a part of a network. We intend to consider the latter case in greater detail in the following using the method of statistical insulation co-ordination (see Section 4.1), on the basis of experimental investigations.

The empirical performance functions of the switching breakdown voltage (regardless of whether it was D or H_S that broke down) were determined for various isolating clearances D, earth insulating clearances H_S and heights H, H_S and H_G (Table 4.4). This relative frequency of overall breakdowns

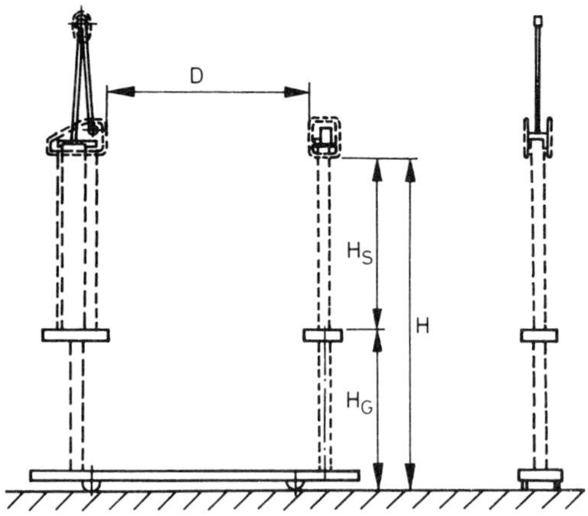

Fig. 4.4 Determination of principal electrical dimensions of an isolator (Example 4.4.2 [4.3])

D isolating clearance ('longitudinal insulation')
H_S earth insulating clearance ('earth insulation')
H_G basic height
H overall height

Table 4.4 Experimental results of measurements on an isolator [4.30]

Basic height H_G m	Earth insulating clearance H_S m	Overall height H m	Isolating clearance D m	Total 50% breakdown voltage u_{d50} kV	Variation coefficient v %	Ratio of longitudinal breakdown to breakdown to earth $\frac{\sum N_L}{\sum N_T} = \varphi$
			4	1475	5.1	0.48
3.8	4.2	8	5	1530	5.9	0.12
			6	1530	7.5	0.05
			4	1580	6.3	0.72
3.8	5.7	9.5	5	1650	5.2	0.57
			6	1730	5.9	0.11
			4	1530	6.5	0.51
5.3	4.2	9.5	5	1560	9.3	0.24
			6	1580	7.0	0.014

can be approximated by a normal distribution (Fig. 4.5: h_T; Table 4.4: parameters u_{d50} and variation coefficient $v = s/u_{d50}$). The proportion of breakdowns of the insulation clearance were additionally determined for each set switching voltage (Fig. 4.5: h_L; ϕ in Table 4.4). These results are to be generalised for the dimensioning of the isolator.

The ratio ϕ of longitudinal to overall breakdown is of crucial importance in insulation co-ordination. With a constant isolating clearance D, the ratio ϕ increases with the height $H = 3.8\ \text{m} + H_S$ of the isolator (with the earth

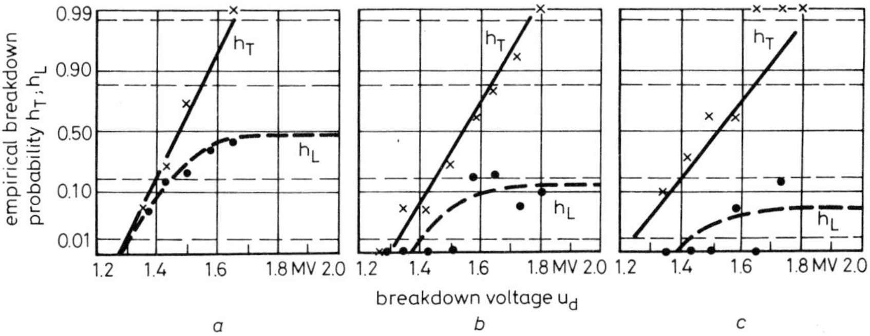

Fig. 4.5 Relative breakdown frequencies of complete arrangements (h_T) and longitudinal insulation (h_L) of isolator in Fig. 4.4 [4.30]

a) $H = 8$ m; $H_S = 4.2$ m; $D = 4$ m
b) $H = 8$ m; $H_S = 4.2$ m; $D = 5$ m
c) $H = 8$ m; $H_S = 4.2$ m; $D = 6$ m

insulating clearance H_S naturally becoming longer: Fig. 4.6), and is reduced accordingly as the isolating clearance D increases, if H is kept constant (Fig. 4.7). For a prescribed value of ϕ, these diagrams can in themselves be a useful tool in dimensioning, if one also knows the 50% breakdown voltage of the arrangement as a whole (Fig. 4.8) and at the same time makes the variation coefficient $v=0.06$ (constant) (Table 4.4; $H_G=3.8$ m). The procedure can be further simplified by plotting lines of equal 50% breakdown voltage u_{d50} in a D/H diagram (Fig. 4.9, derived from Fig. 4.8).

Similarly, lines of equal ratio ϕ of longitudinal to overall breakdown can be plotted in a D/H diagram (Fig. 4.10). In both diagrams the permissible parameter values form areas: for example, a maximum value of $\phi=0.10$ can be prescribed (i.e. a breakdown of the isolating clearance D is only expected after ten breakdowns of earth insulating clearance H_S); then, according to Fig. 4.10, the parameter area of D and H to the right of the lines $\phi=0.10$ is permissible.

Assuming $v=s/u_{d50}=0.6$, then, using u_{d50} from Fig. 4.9, the total failure probability R_T in the sense of statistical insulation coordination can be

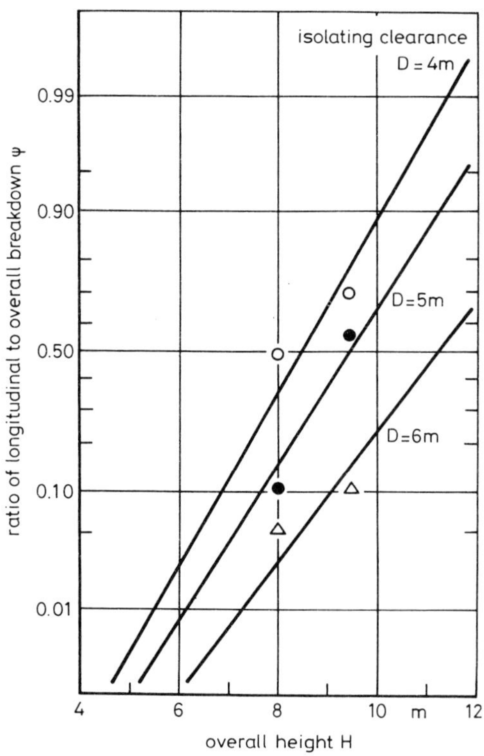

Fig. 4.6 Ratio ϕ of longitudinal to overall breakdowns, as a function of the overall height $H=H_S+3.8$ m of an isolator [4.30]

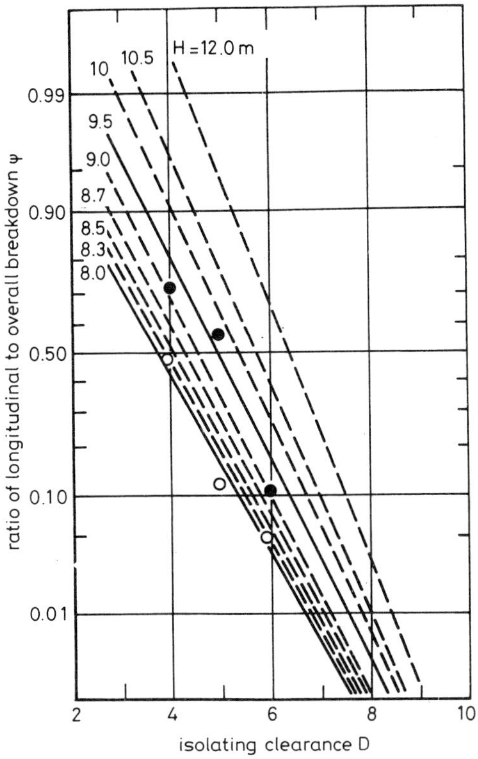

Fig. 4.7 Ratio ϕ of longitudinal to overall breakdown, as a function of isolating clearance D for $H=H_S+3.8$ m for an isolator [4.30]

calculated in accordance with eqn. 4.4 with the performance function $V(u)=\Phi(u;\, u_{d50};\, \sigma^2)$[1] and the assumed normally distributed switching overvoltages $f_{\ddot{u}}(u)=\varphi^*(u;\, 925\text{ kV};\, (139\text{ kV})^2)$:[2]

$$R_T=\int_{-\infty}^{+\infty}\varphi^*(u;\, 925\text{ kV};\, (139\text{ kV})^2)\Phi(u;\, u_{d50};\, \sigma^2)\, du. \tag{4.12}$$

If the ratio ϕ of longitudinal to overall breakdown is taken into consideration (Fig. 4.10), then the failure probability of longitudinal breakdowns is

$$R_L=\varphi R_T. \tag{4.13}$$

The failure probabilities for the two insulation clearances (R_T, R_L) have

[1] $\Phi(u;\, u_{d50};\, \sigma^2)$ characterises the distribution functions of the normal distribution with the parameter $\mu=\mu_{a50}$ and σ^2.

[2] $\phi^*(u;\, 925\text{ kV}(139)^2)$ characterises a density function of the normal distribution with the parameters $\mu=925$ kV and $\sigma^2=(139\text{ kV})^2$. ϕ^* is not the same as the ratio ϕ of longitudinal and overall breakdowns!

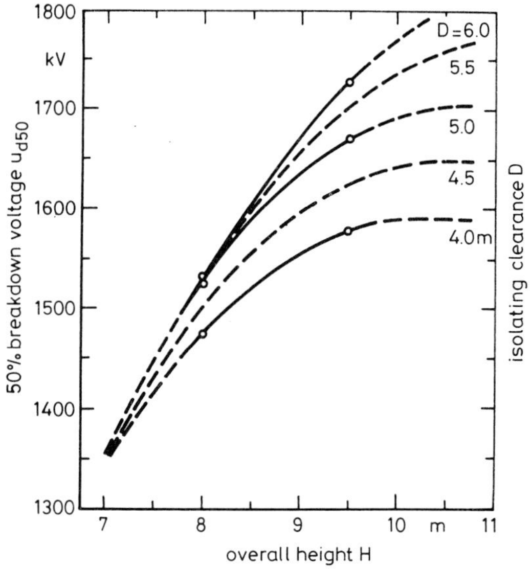

Fig. 4.8 50% breakdown voltage, as a function of overall height $H = H_S + 3.8$ m and isolating clearance D for an isolator [4.30]

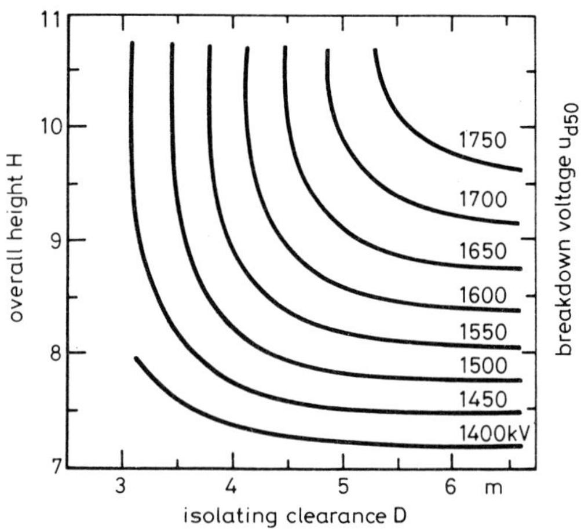

Fig. 4.9 Lines of constant breakdown voltage u_{d50}, as a function of isolating clearance D and overall height $H = H_S + 3.8$ m of an isolator [4.30]

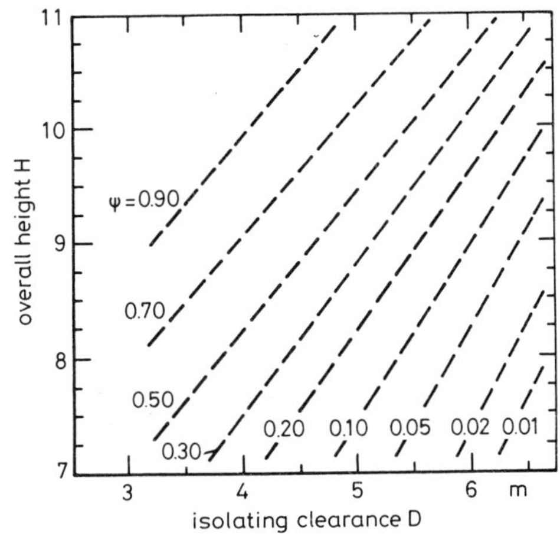

Fig. 4.10 Lines of constant ratio ϕ of longitudinal and overall breakdown, as a function of isolating clearance D and overall height $H = H_S + 3.8$ m of an isolator [4.30]

again been reproduced in the D/H diagram (Fig. 4.11*a*) for further consideration. The permissible parameter range of D and H will be limited by specifying the permissible failure probabilities. A value of $R_T = 10^{-4}$ is to be permitted for the total failure probability, and $R_L = 10^{-5}$ for the longitudinal failure probability. This gives a parameter range (Fig. 4.11*b*) bounded by the corresponding lines $R_L = B = 10^{-5}$ (up to point S) and $R_T = A = 10^{-4}$ (from S onwards). All combinations of the values of D and H which lie to the left of this boundary (range I in Fig. 4.11*b*) are unsuitable for isolator dimensioning. For the values to the right of the boundary, with a prescribed permissible ratio of longitudinal and overall breakdown of $\phi = 0.1$ (according to Fig. 10), yet another difference should be drawn: in range II $\phi \leqq 0.1$, in range III $\phi > 0.1$ (Fig. 4.11*b*). Co-ordination between earth insulation and longitudinal insulation only exists for parameter combinations of D and H in range II. Carrara raises the question, however, as to whether we ought not to highlight the extremely low absolute failure probability $R_L = 10^{-5}$, dispense with co-ordination, and permit both ranges (II and III). This can be very economical, as is shown in the following:

All conditions are satisfied if one chooses $H = H_S + 3.8$ m $= 8.15$ m (i.e. $H_S = 4.35$ m) and $D = 5.45$ m (point S). If one is forced, on pollution grounds for example, to increase H to 9 m ($H_S = 5.20$ m), then D can be reduced to 4.90 m (point T) for a constant failure probability. If one wishes to maintain co-ordination, however, D must be increased to 6.10 m (point V). The longitudinal failure probability is reduced to a value of $R_T < 10^{-6}$ (which is not demanded at all); the isolator becomes larger and more expensive.

A consideration of the probability with which the standard test method (ii) ($n = 15$; $k \leqq 2$; see Section 4.3.1) is successful is to be found in Carrara *et al.*, based on Fig. 4.11.

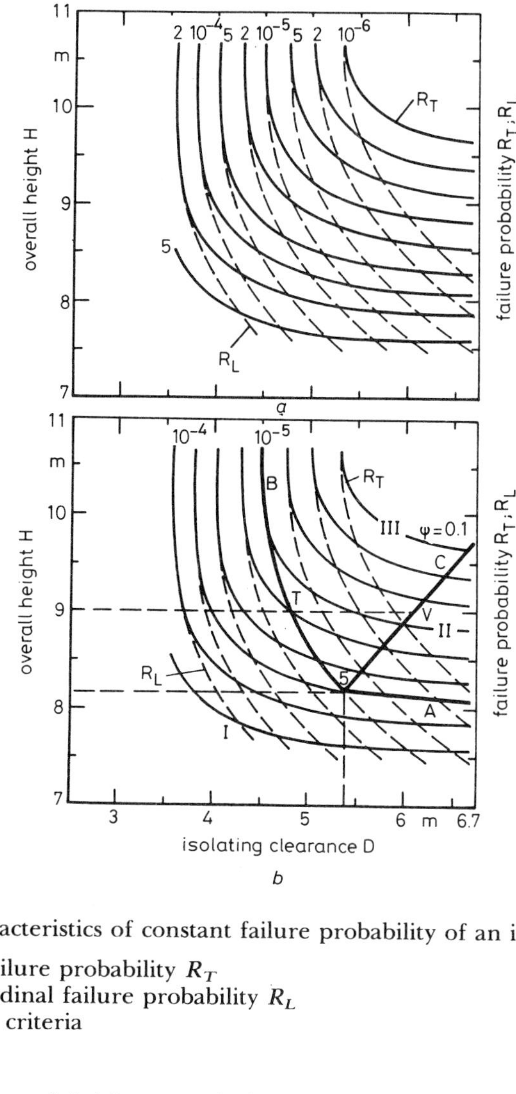

Fig. 4.11 Characteristics of constant failure probability of an isolator [4.30]

a —— total failure probability R_T

--- longitudinal failure probability R_L

b dimensioning criteria

Examples 4.4.1 and 4.4.2 are typical of how the dimensioning of insulation can be approached. Both methods naturally have to be modified for special applications: they are merely intended to encourage the reader to adopt a similar approach. It is absolutely essential to derive requisite generalisations from a limited number of test results. While these generalisations in Example 4.4.1 are involved in the calculation method for the SF_6 breakdown voltage, this is performed directly in Example 4.4.2: a whole family of curves is derived from a relatively small number of test results (e.g. Figs. 4.6 to 4.11).

Chapter 5

Statistical description of insulating capacity

The basis when planning experiments in high-voltage engineering must always be the information available concerning the physical processes involved and the random variables describing them. When determining the test parameters and sample size, assumptions have to be made in relation to the expectation and dispersion (or coefficient of variation) of the variate under investigation (see Sections 3.2.1, 3.3.1 and 3.3.4, for example). The better the agreement between these assumptions and the subsequent test result, the greater will be the accuracy of the test conducted. Advice is therefore given in the following Chapters as to the type of distribution which can be expected, the distribution parameters and functional parameters, for basic electrode arrangements in gaseous, liquid and solid insulating materials. The values given should be taken only as a rough guide: anyone conducting an experiment should of course also make use of special information published in relation to the specific problem, and especially his or her own personal experience. Recommendations are also given for the individual insulation arrangements as to how randomness and independence of realisations can be ensured in the sample.

5.1 Choosing the variate

Consideration so far has been restricted to that of the random process 'electrical breakdown', and in particular to its description using the variate 'breakdown voltage'. In the first place, there are other significant random processes in high-voltage insulation engineering (e.g. inception and propagation of partial discharge, and ageing processes described by characteristic dielectric properties); secondly, the breakdown process can also be described using different variates.

We shall be concentrating on the randomly occurring breakdown process, since experimental and mathematical methods are to the fore. In this context, the breakdown process concerned serves as an example: the methods presented for it can also be applied accordingly to other random processes, bearing physical relationships in mind, of course. A few special comments need to be made about the statistics of partial discharges (see Section 5.6).

The choice of variate to describe the breakdown process will depend on the physical or technical problem posed. The variate breakdown voltage U_d

is directly investigated in most experiments: the measuring instruments provide direct readings of the realizations u_d. If the experiments relate to measurements on insulation arrangements to be installed on a system having a particular service voltage, then it will be advisable to stay with the variate 'breakdown voltage'. The performance function of the breakdown voltage will then be comparable with the insulation level, the test voltages or the distribution functions of the overvoltages. One must ensure that the voltage used in the experiment is characterised by clear, unambiguous parameters. Far too little use is made, for example, of endurance tests (particularly on solid high-polymer insulation) to determine breakdown-voltage performance functions (what is usually determined is breakdown time distribution functions for a particular breakdown voltage). Protracted application of an alternating voltage to a test piece can indeed also be used to determine breakdown voltage performance functions for different stress durations (Fig. 5.1). Breakdown voltage performance functions for different stress durations are needed in the statistical consideration of the effect of time (see Section 6.7).

In the case of slightly non-uniform insulation arrangements, we are often interested in breakdown voltage generalisations. The Schwaiger relationship can be used for simple calculation of the variate maximum breakdown field-strength E_{dh} from the variate breakdown voltage U_d for an insulation arrangement with an electrode gap d and degree of uniformity η:

$$E_{dh} = \frac{U_d}{d\eta}. \tag{5.1}$$

If the curvature factor e_h, which describes the influence of electrode curvature on the breakdown process, is also known for the insulation arrangement concerned, then the variate dielectric strength E_d can be introduced as a material property:

$$E_d = \frac{E_{dh}}{e_h} = \frac{U_d}{d\eta\, e_h}. \tag{5.2}$$

This strength is of course also a function of the duration of the applied stress (cf. Fig. 5.1*c*). It is sometimes technically expedient and physically meaningful to use a random time factor to describe the random nature of the breakdown processes

$$K_t = \frac{E_d(t)}{E_{d0}}, \tag{5.3}$$

where E_{d0} is the strength at a given time (e.g. $t_0 = 1$ minute). Such an approach proves satisfactory when considering impulse-voltage breakdown, for example, where the random nature is caused by the random provision of initiatory electrons with respect to time and the random discharge development with respect to time. By analogy to time factor K_t, it is always advisable to introduce selected (relative) variates if they can be used to describe the random breakdown process concerned and permit generalisation of the test results.

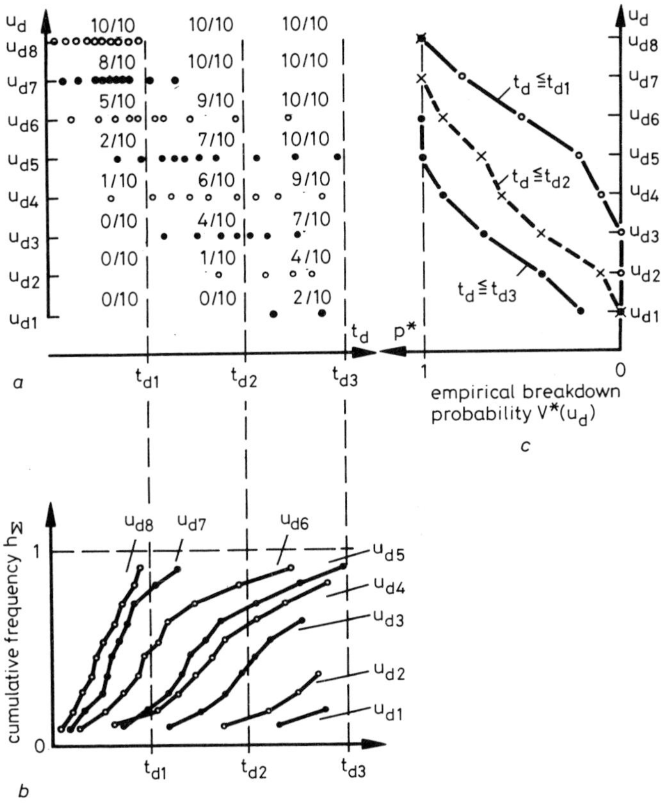

Fig. 5.1 Determination of breakdown-voltage distribution functions for endurance tests (diagrammatic for eight voltage levels, each with $n = 10$ test pieces)

a Test results; pairs of values $(u_d\,;\,t_d)$
b Empirical distribution functions of breakdown time

$$\left(h_{\Sigma} = \frac{k(t_d < t_{di})}{n+1}\right)$$

c Empirical performance functions of breakdown voltage for three voltage application durations $(t_{d1}\,;\,t_{d2}\,;\,t_{d3})$

Finally, it is advantageous in many problem situations to consider the variate breakdown time T_d. In tests at a given voltage, realisations t_d of this continuous variate are obtained; as in the case of rising-voltage tests (see Section 3.3.3), empirical distribution functions $F^*(t_d)$ can be determined and be approximated by theoretical distribution functions. The performance functions are of particular interest for the breakdown voltage and the variates derived from it (see Section 3.1.2), but it is exclusively distribution functions that are given for the breakdown time. In special cases it also seems quite reasonable to determine breakdown time performance functions (e.g. impulses of the same amplitude but different duration t_d).

Choosing the variate is therefore not a mathematical problem: it should be made purely according to criteria of physical and/or technical expediency, and possibly take account of the time and expense involved in evaluating the tests.

5.2 Air insulation

Atmospheric air is, for insulation arrangements at low voltages up to ultra-high transmission voltages, the most important and consequently the most thoroughly investigated insulating material. The almost overwhelming number of publications will force us in the following to present only selected aspects. For detailed problems, we would refer readers to the literature: the monograph of Meek/Craggs would be an excellent starting point.

5.2.1 Test problems

Atmospheric air is the only insulating material that can be regarded as completely self-restoring. Atmospheric air is, however, subject to climatic variations (temperature; pressure; humidity) that affect its insulating capacity. The rated withstand voltages for a reference atmosphere (20°C; 101.3 kPa; 11 $g \cdot m^{-3}$ water) can be converted to different test conditions using correction factors. Conversely, it is customary to correct breakdown voltages to a reference atmosphere. The standard correction factors are of course empirical values from comparative international measurements and are consequently averages. Numerous recent measurements have highlighted the problems of climatic correction, which can be expected to improve. Different climatic conditions and their (still) unsatisfactory correction are a major cause of dispersion and poor reproducibility.

The inception of discharge depends on the provision of initiatory electrons, which are made available in the atmospheric air and at electrode surfaces, in particular by cosmic radiation and by UV radiation. Consequently, both types of radiation have a clear effect on the breakdown voltage: as the intensity of the irradiation is increased, the breakdown voltage is reduced — especially in the case of slightly nonuniform arrangements without stable partial discharges and in disturbed slightly nonuniform arrangements, the effect of cosmic radiation (Fig. 5.2) being less in the region of its natural fluctuations than that of artificial UV irradiation (Fig. 5.3). UV radiation is also produced in the laboratory, for example, by the igniting sparks of an impulse generator. The effect of this radiation depends to a great extent on the material and geometry of the electrodes.

In atmospheric air, there is always a certain amount of dust, the composition of which is usually unknown. Thin coverings of dust on the electrodes do not greatly affect the breakdown voltage, but moving dust can result in considerably reduced breakdown voltages on slightly nonuniform arrangements. Moving dust produces charging of particles at the electrodes, thereby affecting the ignition processes. The result of the relationships outlined is,

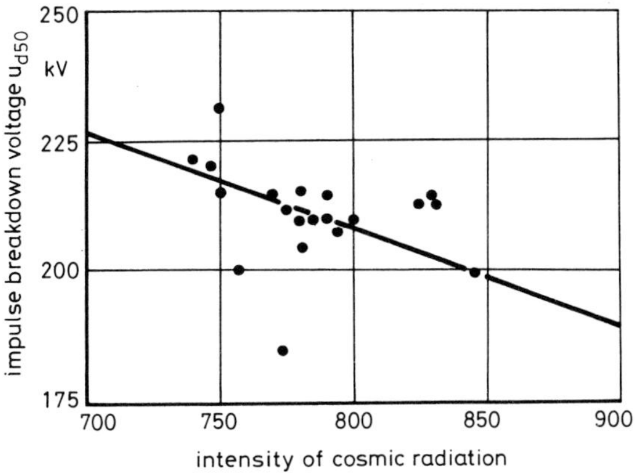

Fig. 5.2 Influence of intensity of cosmic radiation on breakdown voltage u_{d50} of a sphere spark-gap

Sphere radius 12.5 cm; electrode gap $d = 6$ cm in air; impulse voltage with rate of rise of 0.25 kV/μs

for example, that far longer breakdown times are measured on a sphere spark-gap in filtered air than in normal ambient air (Fig. 5.4). The greater dispersion observed in filtered air is due to the fact that, in contrast to normal ambient air, there are only very few dust particles present (effect of 'enlargement law'; see Chapter 6).

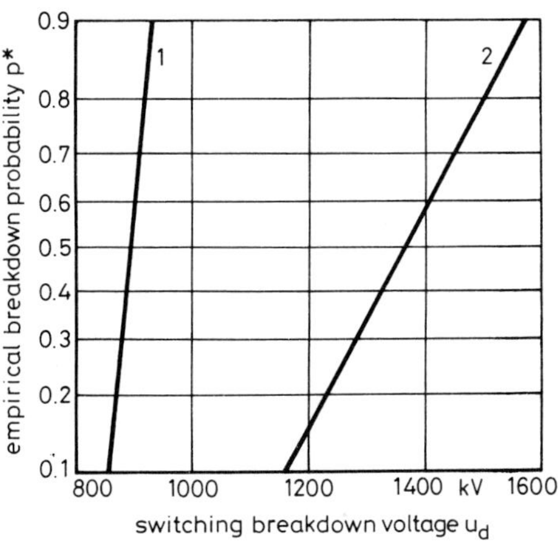

Fig. 5.3 Influence of UV radiation on the performance function of switching breakdown voltage

1.5 m sphere with interference-electrode plate; $d = 2.5$ m in air
1 with UV irradiation
2 without UV irradiation

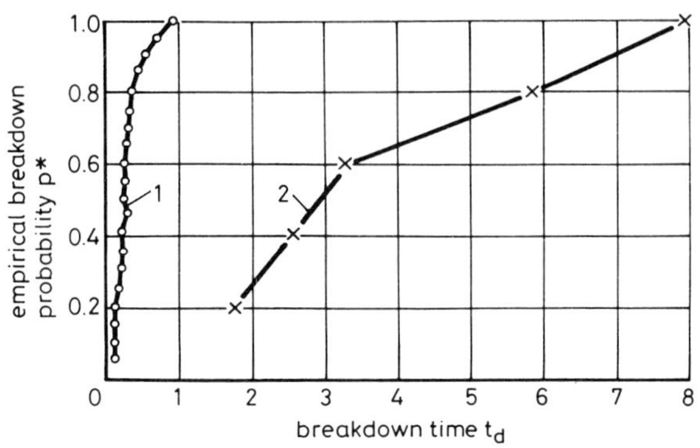

Fig. 5.4 Influence of dust on the breakdown time of a sphere spark-gap in air

Sphere radius $r = 6$ cm; gap $d = 2.5$ cm; $\bar{u}_{\sim} = 69.5$ kV
1 in normal ambient air
2 in filtered air

In a high-voltage laboratory, the test result is also governed by the way the test piece is set up, by the test procedure (see Chapter 3) and the parameters of the test piece. The test result is also influenced by the energy dispensed by the test plant, particularly in the case of heavy-current discharge processes prior to breakdown (leader). Series resistances consequently need to be chosen with great care, particularly if a limitation of breakdown energy appears desirable in relation to erosion or modification of the electrodes.

On slightly nonuniform air insulation, the rate of rise of a 'continuous' voltage does not affect the test result. On arrangements exhibiting partial discharges, the breakdown voltage rises as the rate of rise is increased.

With impulse voltages, the time gap between successive voltage applications has practically no effect, based on the sequences customary with impulse generators ($t \geqq 10$ s).

The phenomena being dealt with affect the test result in a manner which is difficult to determine. Breakdown-voltage distribution functions measured under seemingly identical conditions are not identically reproducible (Fig. 5.5), a situation exacerbated by the uncertain estimates resulting from the limited sample-size. Analysis of the reproducibility of measured breakdown voltages (Table 5.1), which is greatly dependent on electrode arrangement and gap, reveals that relative deviations in measurements (e.g. u_{d50}) of up to $v = 0.055$ are possible in the same laboratory. Since the result can vary over a range of at least $\pm 2v$, excessively high demands must not be placed on the reproducibility of tests in atmospheric air.

5.2.2 Slightly nonuniform air insulation

In slightly nonuniform fields, it is advantageous if the insulating clearance is relatively uniformly stressed. The inception of an independent discharge (streamer) produces breakdown. It should, of course, be borne in mind that

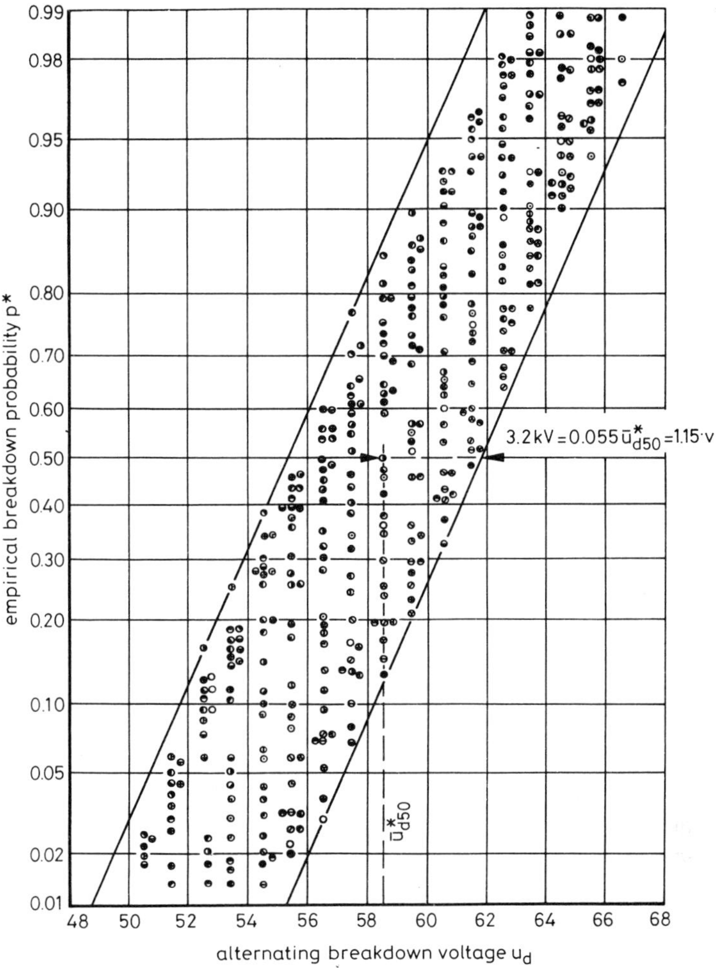

Fig. 5.5 Reproducibility of breakdown-voltage distribution functions (normal-distribution paper)

Measurements of 30 empirical cumulative frequency functions, determined on a point/plane arrangement ($d = 100$ mm) in air under identical conditions (mean values: $\bar{u}^*_{d50} = 58.6$ kV; $\bar{s}^* = 2.8$ kV)

the influence of dirt, rain, dew etc. on the electrode surface in free air can easily produce disturbed fields, making them highly nonuniform (see Section 5.2.3). The 'classical' measuring instrument, the sphere spark gap, represents an arrangement with a slightly nonuniform field. Similarly, the electrodes of high-voltage test plants, and in more recent times also insulating arrangements in the medium-voltage range (metal-clad switchgear), are designed so that the fields are slightly nonuniform. In laboratories, appropri-

Table 5.1 Uncertainty in the reproducibility of breakdown-voltage measurements on air spark-gaps

Expressed by the relative deviations of the mean values of repeated measurements under identical conditions: first figure in each instance is for 50-Hz alternating voltage; second figure for lightning voltage; third for switching voltage

Electrode arrangement		Relative deviation for test result (e.g. u_{d50}) for measurements in the same laboratory on the same test-piece over a short period v_1	as for v_1, but over a long period v_2	in different laboratories on practically the same test-pieces v_3	as for v_2, but similar test-pieces with the same principal dimensions v_4
Rod/rod ($d > 1.2$ m)	~	0.012	—	0.038	—
	Λ	0.018	0.023 . . . 0.035	0.040	—
	Ω	0.015	—	0.058	—
Rod/plate ($d > 1.2$ m)	~	0.107	—	0.025	—
	Λ	0.015	0.024 . . . 0.030	0.061	—
	Ω	0.030	—	0.049	—
Rod/plate ($d < 1.2$ m)	~0.029	—	0.045	—	—
	Λ	0.037	—	0.090	—
	Ω	0.055	—	0.058	—
Insulator strings (dry)	~	—	0.034	—	0.090
	Λ	—	0.035 . . . 0.054	(0.080)	0.06 . . . 0.085
	Ω	—	0.05	—	0.07 . . . 0.085
Insulator strings (wet)	~	—	0.041	0.098	0.130
	Λ	—	—		—
	Ω	—	0.05		0.095 . . . 0.130

ate precautions, e.g. cleaning, should be taken to keep such air insulation free from disturbance.

In such conditions, the dispersion of the breakdown voltage is relatively small and is generally approximated by normal distributions. Comparing this with the qualitatively similar ignition process in SF_6, which produces extreme-value-distribution breakdown voltages (see Section 5.3.2), it must be assumed that the boundary conditions (air flows, dust etc.), which in free air are of a highly random nature, are the cause of the normal distribution.

If the expectation of the breakdown voltage $\bar{u}_d^*$ is to be estimated prior to a series of tests, a suitable method of calculation is that using Schwaiger's expression (eqn. 5.1), and the values of the maximum breakdown field strength E_{dh} from Schumann (Fig. 5.6):

$$\bar{u}_d^* = E_{dh} d\eta; \qquad (5.4)$$

where d is the electrode gap and η the degree of uniformity.

Further details of pre-calculation of the expectation are to be found in the Bibliography. The expectation of the breakdown voltage is only slightly reduced with the duration of action of the applied voltage; a time effect (see Section 6.7) can practically be neglected in most cases (Fig. 5.7*a*). This is confirmed both by the well known classical nomograms for sphere spark gaps, and by recent measurements with large-area slightly curved electrodes. The surface quality of the electrodes, on the other hand, exerts a great influence: with rough surfaces, microscopic field strength rises can reduce

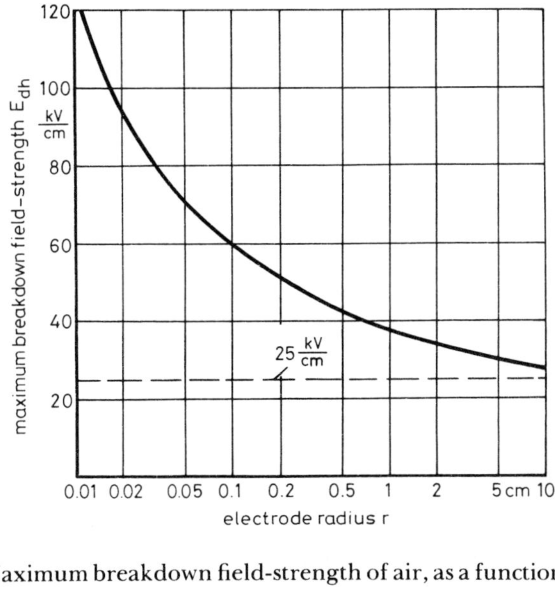

Fig. 5.6 Maximum breakdown field-strength of air, as a function of electrode radius

Electrode gap d assumed to be far greater than electrode radius; cylinders

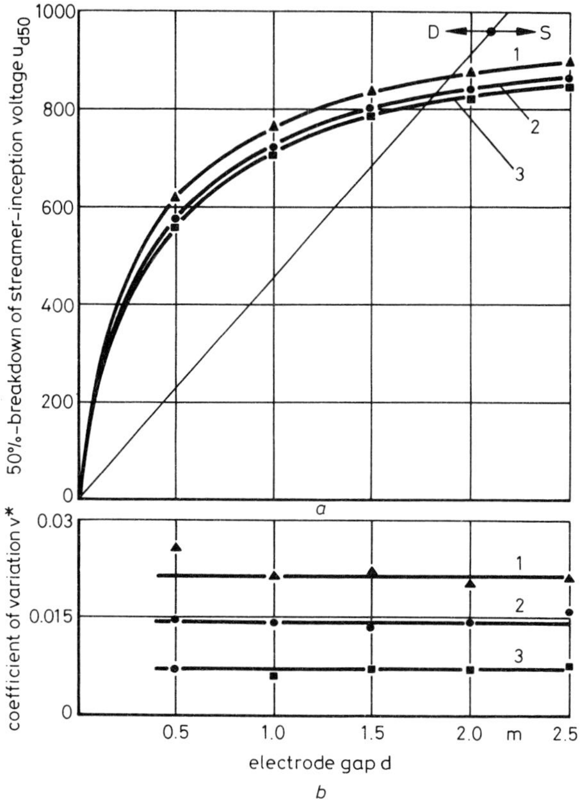

Fig. 5.7 Expectation (a) and coefficient of variation (b) of normally-distributed breakdown voltage of sphere/plane arrangement in air

Sphere radius $r = 0.25$ mm; *D* breakdown; *S* streamer inception
1 Lightning voltage +1.2/50
2 Switching voltage +180/2300
3 Switching voltage +800/7400

the macroscopic maximum breakdown field strength to considerably below 25 kV/cm (cf. Fig. 5.6). This effect, which occurs in particular with large-area electrodes, ought to be more satisfactorily clarified by preliminary tests when planning tests proper.

The anticipated coefficient of variation depends in the first instance on the duration of voltage application and thereby on the type of voltage (Fig. 5.7*b*): for lightning voltage $v^* \leqq 0.025$, for switching voltage $v^* \leqq 0.017$, and for alternating voltage $v^* \leqq 0.010$.

5.2.3 Highly nonuniform air insulation

In arrangements with a nonuniform field, the voltage requirement (*e*) of the stable partial discharge before breakdown determines the amplitude of

Table 5.2 Coefficients of variation for a rod/plate arrangement

Roughly identical to those for practical electrode arrangements (see Table 5.3)

Type of voltage	Voltage requirement e of stable partial discharge (specific breakdown voltage) kV/cm	Coefficient of variation v^*
50 Hz alternating voltage	Streamer: 4.5 Leader: ≈ 1	<0.05 (increasing with electrode gap)
Direct voltage		
positive	5	<0.01
negative	9 . . . 11	<0.03
Lightning voltage		
positive	5	0.01 . . . 0.02
negative	≈ 15	somewhat larger than positive
Switching voltage		
positive	Streamer: 4.5; Leader: ≈ 1	0.04 . . . 0.06
negative	Streamer: ≈ 10; Leader: ≈ 1	roughly the same as positive

the breakdown voltage and is identical to the 'specific' breakdown voltage (u_d/d). Considerations are based on the rod/plate arrangement (Table 5.2) and remain clear, provided only one kind of stable partial discharge existed before breakdown. This is the case with lightning voltage and direct voltage: the expectation of the breakdown voltage $\bar{u}_d^*$ is then linearly connected to the electrode gap d through the voltage requirement e of the partial discharge (Fig. 5.8, curve 1):

$$\bar{u}_d^* = ed. \tag{5.5}$$

Eqn. 5.5 can also be used for alternating and switching voltage if only streamer discharge has existed in stable form prior to breakdown ($d < d_0 \approx 1.5$ m). When $d > d_0$, the breakdown voltage only continues to increase slightly with the gap (Fig. 5.8, curve 2). According to the model presentations of Lemke, for $d \geqq d_0$ the expectation of the breakdown voltage is determined from the series connection of streamer discharge (e_S) and leader

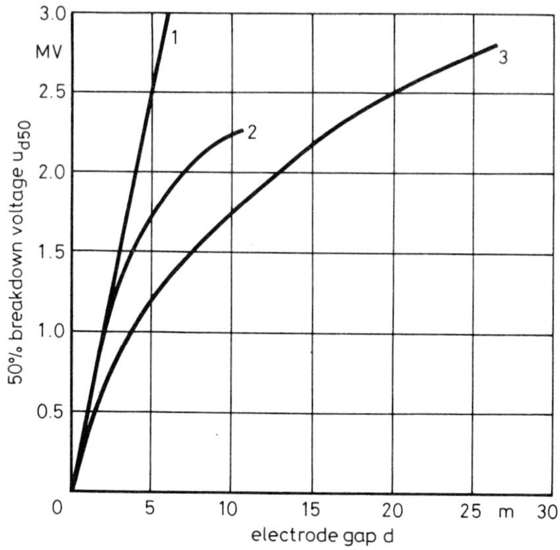

Fig. 5.8 50% breakdown voltage of rod/plate arrangement

1 Positive direct-voltage and +1.2/50 lightning voltage
2 50 Hz alternating voltage
3 Positive switching voltage with front times corresponding to the minimum breakdown voltage

discharge (e_L):

$$\bar{u}_d^* = e_L d + (e_S - e_L)\, d_0 \left(1 + \ln \frac{d}{d_0}\right),$$

$$\bar{u}_d^* = 1 \frac{\text{kV}}{\text{cm}} d + 3.5 \frac{\text{kV}}{\text{cm}} d_0 \left(1 + \ln \frac{d}{d_0}\right). \tag{5.6}$$

d_0 should be calculated from a measured breakdown value for $d > d_0$. The known minimum of the breakdown voltage exists for switching voltage and is dependent on the front time; the position of the minimum, in its turn, depends on the electrode gap. The characteristic for the minimum breakdown voltage (Fig. 5.8, curve 3) is still appreciably lower than that for alternating voltage, but the breakdown voltage can also be estimated for it in the way described (eqn. 5.6).

If one wishes to dispense with a preliminary test to determine d_0, the expectation of the breakdown voltage $\bar{u}_d^*$ can also be estimated for a highly nonuniform arrangement from the characteristic for a point/plane arrangement (Fig. 5.8) and from a spark-gap factor k applicable to the arrangement (Table 5.3):

$$\bar{u}_d^* = k\bar{u}^*_{d\text{rod/plate}}. \tag{5.7}$$

The wealth of measurements of the breakdown characteristic of arrangements with a highly nonuniform field similarly show that the breakdown

Table 5.3 Spark-gap factors for selected practical electrode arrangements in air

The first electrode named is at high-voltage potential; the second is earthed

No.	Spark gap	Gap factor k
1	Rod/plate	1.00
2	Rod/metal frame	1.05
3	Bundle conductor/plate	1.15
4	Bundle conductor/tower window	1.20
5	Bundle conductor/metal frame	1.30
6	Rod/rod (rod length 3 m)	1.30
7	Bundle conductor/metal frame (above or alongside bundle conductor)	1.35
8	Rod/rod (rod length 6 m)	1.40
9	Bundle conductor/stranded conductor	1.40
10	Bundle conductor/tower cross-arm	1.55
11	Bundle conductor/rod (rod length 3 m)	1.65
12	Bundle conductor/rod (rod length 6 m)	1.90
13	Bundle conductor/rod (above bundle conductor)	1.90

Table 5.4 Coefficients of variation of breakdown voltage, as a function of amplitude of 50% breakdown voltage

Range of 50% breakdown voltage kV	Coefficients of variation v_i^*				
	Rod/plate v_1^*	Rod/rod v_2^*	Post insulator v_3^*	Insulator string v_4^*	Overall v^*
<600	0.034	0.037	0.032	0.036	0.034
600 . . . 900	0.055	0.045	0.054	0.035	0.044
900 . . . 1200	0.045	0.055	0.056	0.038	0.042
1200 . . . 1500	0.051	0.063	0.063	0.043	0.051
>1500	0.057	0.052	0.070	0.049	0.053
Coefficient of variation v_v^* of coefficients of variation v_i^*	0.27 . . . 0.52	0.28 . . . 0.5	0.16 . . . 0.44	0.27 . . . 0.62	0.16 . . . 0.62
Number of measurements of v^* available per voltage range	6 . . . 45	6 . . . 48	8 . . . 33	11 . . . 133	—

voltages can be expediently described by a normal distribution for all kinds of voltage. Guide values for the coefficient of variation are given in Table 5.2. The variation coefficients are of course greatly dependent on the particular arrangement and on its dimensions, and therefore also on the voltage amplitude (Table 5.4). When planning breakdown measurements, however, it will in most cases be sufficient to work with the guide values.

Arrangements with large-area electrodes exhibiting points of disturbance should be classified between highly and slightly nonuniform arrangements. It may happen that the discharge producing breakdown is initiated either at the points of disturbance (as initially stable partial discharge) or at the undisturbed part of the electrode. In the result, one then obtains 'multiplicative' mixed distributions for the switching breakdown voltage (Fig. 5.9). If one attempts to indicate coefficients of variation for them, these will amount to as much as $v^* = 0.20$.

A similar tendency is observed with lightning voltages, but the breakdown probabilities can only be reproduced with great difficulty (Fig. 5.10), since,

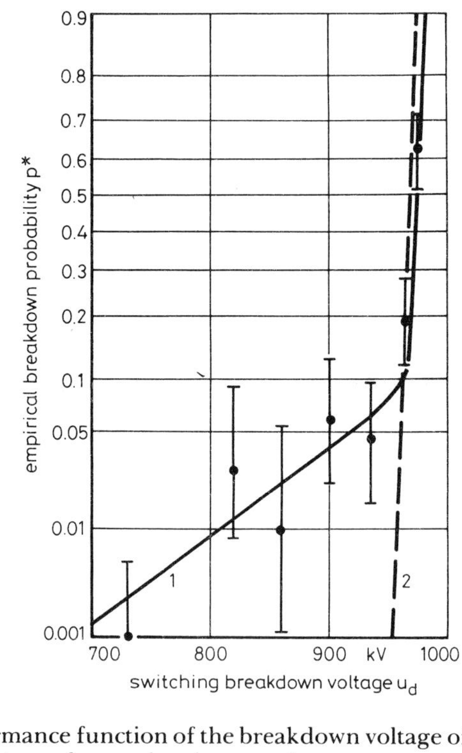

Fig. 5.9 Performance function of the breakdown voltage of a sphere/plane arrangement with an extremely rough sphere-surface

Sphere radius $r = 75$ cm; electrode gap $d = 50$ cm
1 Rough sphere surface
2 Polished metal sphere

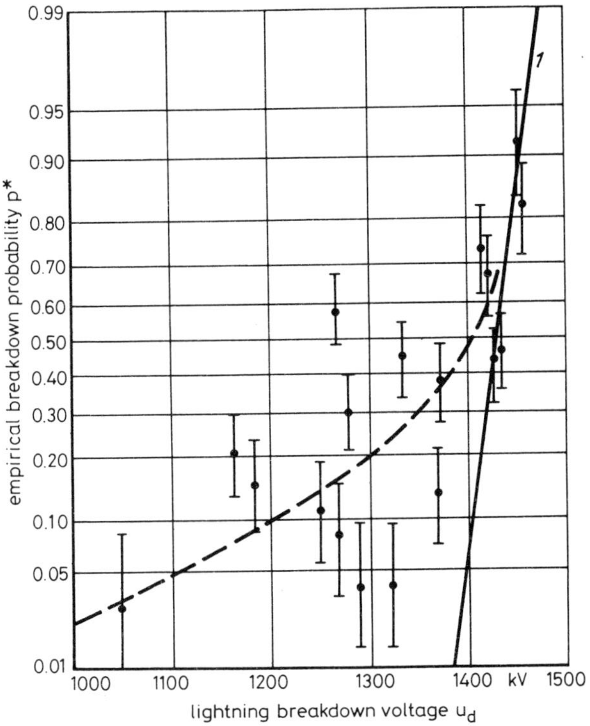

Fig. 5.10 Reproducibility of breakdown probabilities for lightning voltage

Sphere ($r = 75$ cm) with disturbing spike ($l = 0.35$ cm)/plane; $d = 1$ m
1 Performance function of sphere/plane without irregularity

during the short duration of action of the voltage, initiatory electrons are provided in the very small disturbance point area in a very random manner.

With alternating voltage, a partial discharge is always initiated at the defect; the breakdown-voltage distribution function can be approximated by a normal distribution (especially with technical defects or a large number of defects; coefficient of variation $v \leqq 0.10$), or even by a Weibull distribution. If streamer discharge occurs at the irregularity at low voltages and glow discharge at high voltages, then the breakdown voltage obeys an additive mixed distribution (Fig. 5.11). In the case of arrangements with disturbed slightly nonuniform fields in air, preliminary tests are essential for optimum determination of the test parameters.

Multi-electrode systems are currently being intensively experimentally investigated in relation to isolating clearances and conductor/conductor insulation. In such systems, the field strength distribution is dependent on geometry and electrode potentials, and the test results are difficult to systematise. Preliminary tests are required, for which hints are given, for example, by Carrara *et al.*, Thiane *et al.* and Bondarenko and Böhme.

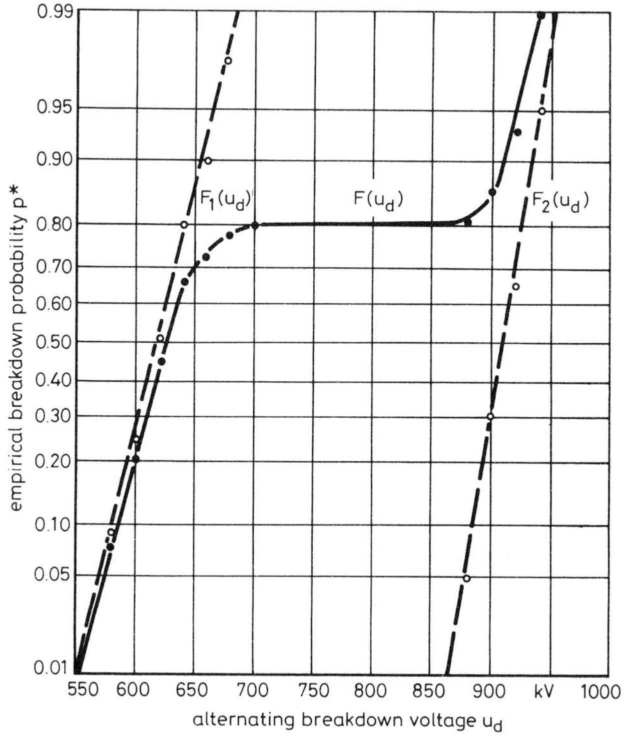

Fig. 5.11 Cumulative frequency function of alternating breakdown voltage of a disturbed sphere/plane arrangement in air

Sphere ($r = 75$ cm) with spike irregularity ($l = 0.35$ cm)/plane; $d = 1$ m; 25 kV/s

$F(u_d)$ Cumulative frequency function of arrangement as a whole
$F_1(u_d)$ Cumulative frequency function for breakdown from streamer discharge
$F_2(u_d)$ Cumulative frequency function for breakdown from glow discharge

5.2.4 Insulators

The flashover voltage of insulators made from inorganic insulating materials is essentially dependent on the condition of the solid material/air interface.

In a dry state (dry flashover), the flashover voltage corresponds to a highly nonuniform air spark gap derived from the electrode geometry. The influence of the insulator is thus negligible. (In a slightly nonuniform basic field, an air gap between electrode and insulator can behave like an irregularity in a single-material system: see Section 5.2.3.)

In a wet state (wet flashover), the flashover voltage is reduced, since the discharge can to a certain extent develop in a water layer. Roughly speaking, it is reduced to about 0.85 times the value of the dry alternating flashover voltage and 0.90 times the value of the dry impulse flashover voltage. A normal distribution can generally be used equally well to approximate dry and wet insulators. The difference between the coefficients of variation for

Table 5.5 Coefficients of variation of flashover voltages of insulator strings

Kind of voltage	Coefficient of variation v^* Strings of long-rod insulators	Strings of cap-and-pin insulators
50 Hz alternating voltage	0.016 ... 0.028	0.010 ... 0.030
Switching voltage		
+	0.052 ... 0.088	0.040 ... 0.080
−	0.034 ... 0.052	0.013 ... 0.026
Lightning voltage (both polarities)	0.020 ... 0.030	0.005 ... 0.010
Sources	*Kucera, J.; Fiklik, V.*: Bull. EGU 6 (1968) 5–6	*Hutzler, B.; Rin, J.-P.*: IEEE Trans. PAS 98 (1979) 3; *Young F.S.*, u.a.: IEEE Trans. PAS 99 (1980) 2; *Schneider, K.-H.*: Electra (1979) 57
	Unpublished measurements at Dresden TU	

the two cases is only slight; the coefficients can be obtained from Tables 5.4 and 5.5.

In a polluted state (tracking breakdown) there is an appreciable reduction of the flashover voltage, since the discharge producing breakdown (prearcing) is fed via an electrolytically conducting surface. The tracking flashover voltage is reduced as the layer conductivity rises. Measurement of tracking flashover voltages is quite expensive and is mainly carried out in the form of constant-voltage tests (see Section 3.2.1). The strength of the voltage source can materially affect the test result: it is therefore necessary for the voltage across the pollution layer not to be more than 5% below the zero-load voltage of the test plant. Natural pollution layers are extremely varied and are practically unreproducible. Artificial pollution layers produced in the laboratory are also only reproducible within certain limits, so that the dispersion of the test results is considerable from this prerequisite alone. Since, for reasons of expense, sample sizes are often very small, the performance functions of the tracking flashover voltage are generally approximated by normal distributions. Reference should be made to the literature in connection with the planning and execution of the rather complicated tests. On the basis of treatment of service-life characteristics of solid insulating materials (see Section 5.5), Böhme, Pilling and Streubel interpret pollution flashover characteristics with the assumptions that the performance function of the tracking flashover voltage is approximated by

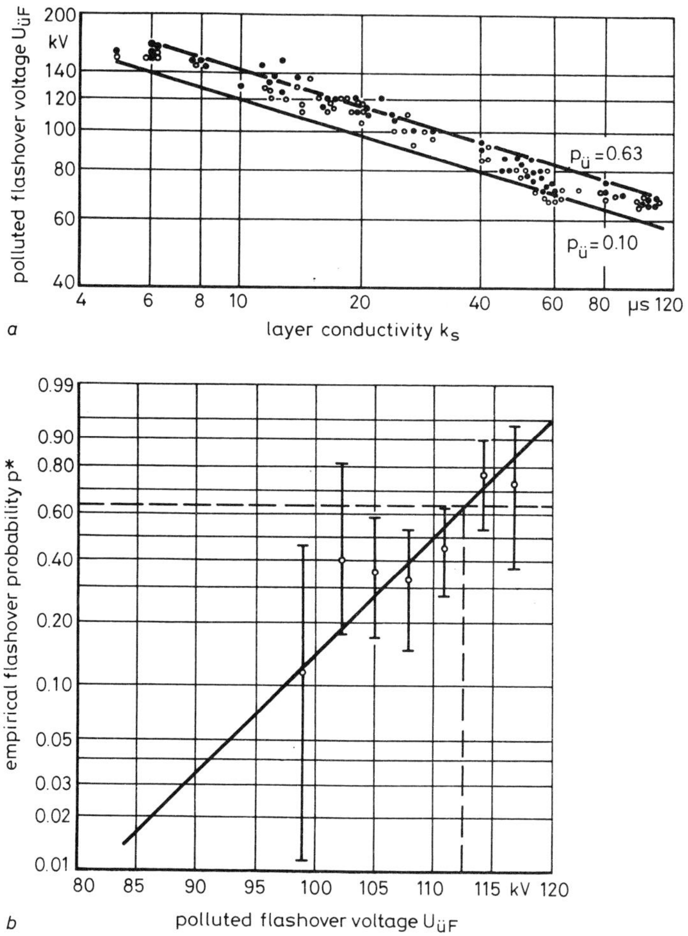

Fig. 5.12 Tracking flashover voltage of an LF 75/16 long-rod insulator

a Tracking flashover characteristic

$$u_{ü} = K_F \kappa_S - n_\kappa \,;\; n_\kappa = 0.3$$

Voltage applied

○ No breakdown

● Breakdown

b Weibull distribution of tracking flashover voltage calculated for $\kappa_{S0} = 20\ \mu$mho, each pair of values $(u_i\,;\,\kappa_{Si})$ being converted from (***a***)

$$u_0 = u_i \left(\frac{\kappa_{Si}}{\kappa_{S0}}\right)^{n_\kappa};$$

Voltages u_0 aggregated for intervals Δu wide

a Weibull distribution, and that with a logarithmic representation the tracking flashover characteristic follows a straight line (Fig. 5.12*a*). This presents the possibility of converting measured tracking flashover voltages for different layer conductivities to one conductivity value. For this layer conductivity, the tracking flashover voltage follows a Weibull distribution (Fig. 5.12*b*), but this still needs to be confirmed for other insulators.

In a clean, dry state, insulators made from organic materials behave like inorganic insulators. However, with wet and polluted surfaces partial discharges can cause surface erosion or tracking due to water diffusion, so that the independence of consecutive tests can hardly be ensured. The phenomena are similar to those of long-term breakdown in solid materials (see Section 5.5).

5.3 Compressed-gas insulation

Among the gases used for pressurised gaseous insulation, the electronegative gas sulphur hexafluoride (SF_6) is now of outstanding technical significance. The following brief treatment is therefore restricted to SF_6 and closely follows the monograph of Mosch and Hauschild. The details can be applied accordingly to other compressed gases and gas mixtures.

5.3.1 Testing problems

When electrical breakdown occurs in a sealed volume of gas, the insulating capacity is not regenerated as completely as it is in free air. The ability to recover and, by the same token, the independence of a series of tests, will be restricted according to the energy converted per unit volume. The extent of the plasma-chemical conversions taking place in the gas at breakdown is a function of energy. The energy is also responsible for the intensity of the space charge remaining in the gas volume after breakdown, which can affect a subsequent test. In the interests of non-dependence, and despite economic considerations, the test chamber should not be too small, and one should endeavour to minimise the energy converted at breakdown. Series resistors between test piece and voltage source are suitable for the purpose, as well as, in the case of continuous voltages, rapid disconnection. The most favourable solution is a spark gap in parallel with the test piece, triggered by breakdown and carrying the breakdown current.

A further cause of dependences in pressurised gaseous insulation is particles that can initiate breakdown but are then reduced in size in the breakdown spark. In subsequent tests they consequently reduce the breakdown voltage to a lesser extent, and the breakdown voltage rises with the number of voltage applications (Fig. 5.13). This phenomenon, which can only be put to practical use in single-material systems (e.g. compressed-gas capacitors), is referred to as the conditioning effect. It is advisable to condition pressurised gaseous insulation before the start of a statistically evaluable test series.

In constant-voltage tests with impulse voltages, the residual space charges after breakdown cause pronounced dependences, which cannot be elimi-

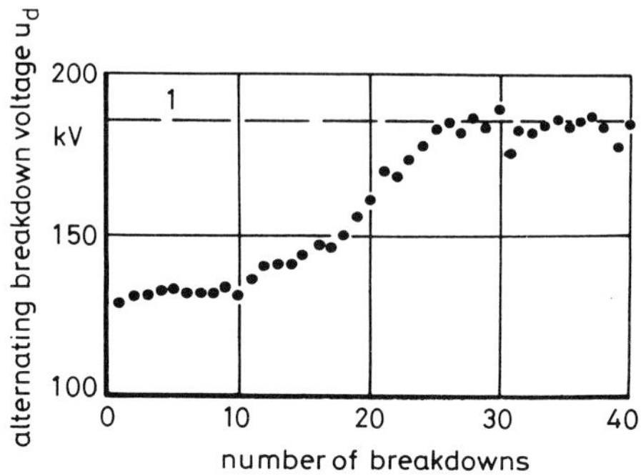

Fig. 5.13 Conditioning effect in SF6

Coaxial cylinders; $r_i = 0.6$ cm; $r_a = 2$ cm; $p_{20} = 0.4$ MPa; defined disturbing particle $d_p = 0.2$ mm (spherical)

1 Undisturbed field

nated, even by relatively long intervals (Fig. 5.14). One solution is to perform rising-voltage tests instead of constant-voltage tests and to convert the cumulative frequency functions to performance functions (see Section 3.3.4). In rising-voltage tests, the time intervals between consecutive breakdowns

										h_{10}	h_{50}
●	●	●	○	●	●	○	●	●	●	0.80	
●	○	○	●	●	○	●	○	○	○	0.40	
○	●	●	○	○	○	○	○	○	○	0.20	0.30
○	○	○	●	○	○	○	○	○	○	0.10	
○	○	○	○	○	○	○	○	○	○	0	

a

										h_{10}	h_{50}
●	●	●	●	●	○	●	●	●	●	0.90	
●	●	●	○	○	●	●	○	●	○	0.60	
○	○	○	○	●	●	○	●	○	○	0.30	0.44
●	●	○	○	○	○	○	○	○	○	0.20	
○	○	●	○	○	○	○	○	●	○	0.20	

b

Fig. 5.14 List of test values for constant-voltage tests with lightning voltage in SF_6

a Interval $t_p = 15$ s ($u_d = -150$ kV)
b Interval $t_p = 200$ s ($u_d = -160$ kV)

Coaxial cylinders; $r_i = 0.6$ cm; $r_a = 2.0$ cm; $p_{20} = 0.25$ MPa
● Breakdown; ○ No breakdown

are, of course, far longer than in constant-voltage tests. Space charges produced are consequently able to recombine or become diffused, the process being promoted by the relatively low stresses at the beginning of each individual test. On this basis, it is also possible to perform constant-voltage tests, applying a comparatively low alternating voltage in the intervals between two stresses (voltage applications) to foster recombination and diffusion of space charges (see Section 3.2.2). In most cases, the methods indicated produce samples which, according to the appropriate tests (see Section 2.5.3), can be regarded as independent.

In rising-voltage tests with direct or alternating voltage, the rate of rise of the voltage can affect the test result. It can be demonstrated for SF_6 insulation, on the other hand, that the influence of the rate of rise of the voltage can in fact be neglected in the range of practical interest (Fig. 5.15). Even if the actuator of a voltage source does not permit a linear voltage rise, there is no risk of the realisations obtained being affected. The rate of rise of the voltage can therefore be selected according to the criteria of the voltage source available and test economics.

When the test parameters that ensure non-dependence are specified, then, despite uniform testing techniques (cleaning electrodes, chamber

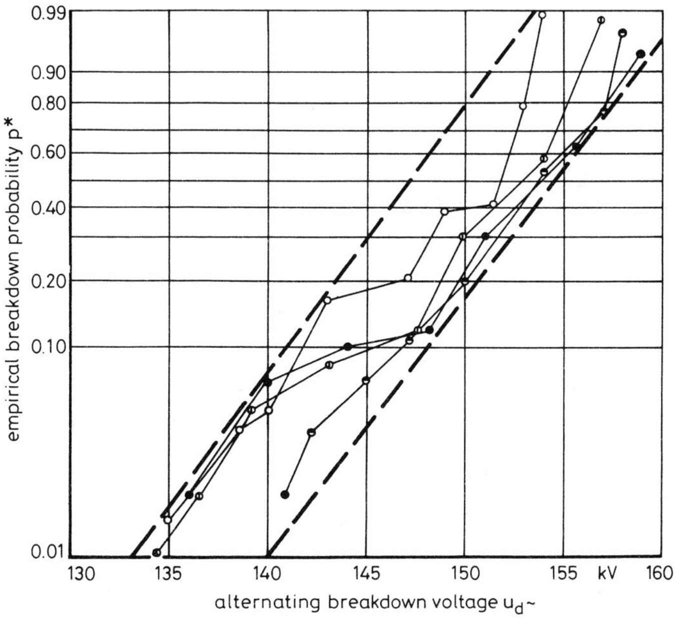

Fig. 5.15 Influence of rate of rise of voltage on distribution function of the breakdown voltage in SF_6; double-exponential paper

Coaxial cylinders, $r_i = 0.6$ cm; $r_a = 2.0$ cm; $p_{20} = 0.25$ MPa
Rate of rise of voltage:
● 0.03 kV/s; ◓ 0.3 kV/s; ⏀ 3.0 kV/s; ○ 13.0 kV/s

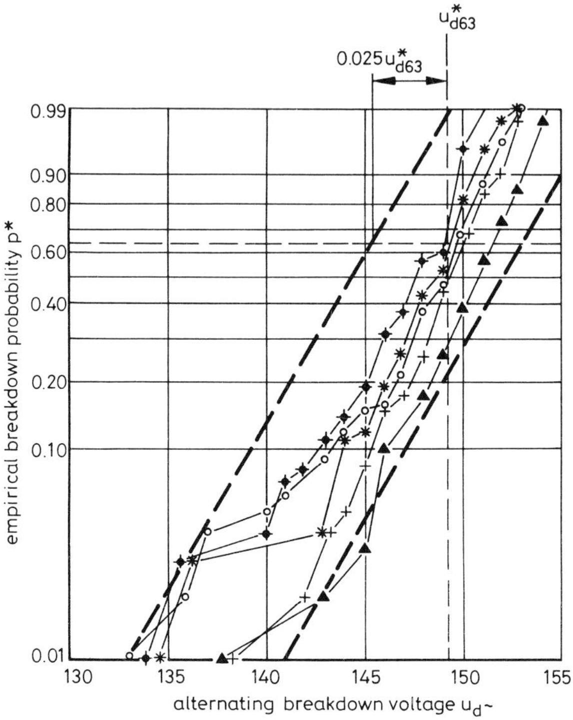

Fig. 5.16 Reproducibility of empirical distribution functions of breakdown voltage in SF_6

Five distribution functions determined empirically at four-week intervals. Coaxial cylinders: $r_a = 2.0$ cm; $p_{20} = 0.25$ MPa

mounting, evacuation, gas filling, etc.), one must, in repeated series of tests, expect empirical distribution functions in which differences are to be found not only in the random breakdown process, but also due to tolerances in assembly and gas-pressure setting, as well as in different conditioning and measuring error. It is nevertheless possible to achieve good reproducibility (Fig. 5.16) with a quite modest outlay on testing.

5.3.2 Slightly nonuniform SF_6 insulation

Pressurised gaseous insulation arrangements fundamentally exhibit slightly nonuniform electric fields. The random ignition of an independent discharge always produces breakdown. The randomness of the ignition consequently causes the statistical variations of the breakdown-voltage realisations obtained in large-scale tests. The stochastic modelling of this process produces extreme-value-distribution breakdown voltages. These can be expediently generalised as electric strengths (eqn. 5.2). An experimental check (Fig. 5.17) confirms that the empirical distribution function of the

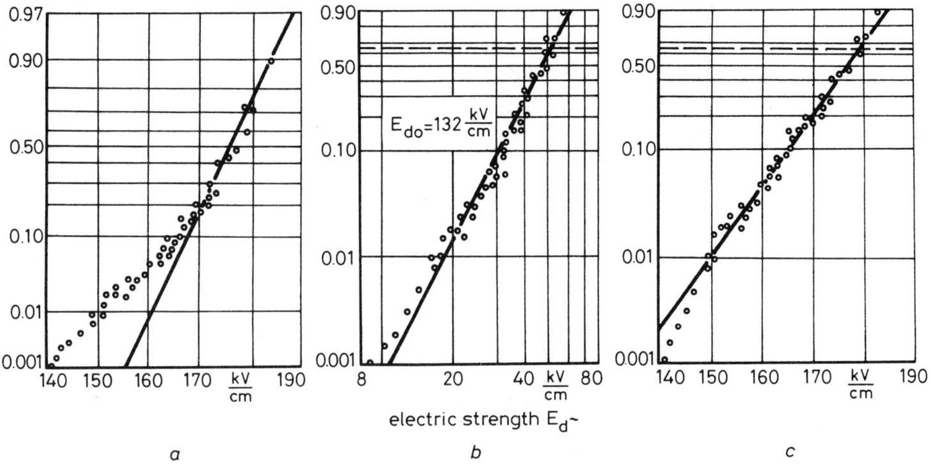

Fig. 5.17 Approximation of distribution function of electric strength of SF_6 at $p_{20} = 0.25$ MPa

a Normal distribution (Gauss paper)
b Weibull distribution (Weibull paper)
c Double exponential distribution (double exponential paper)

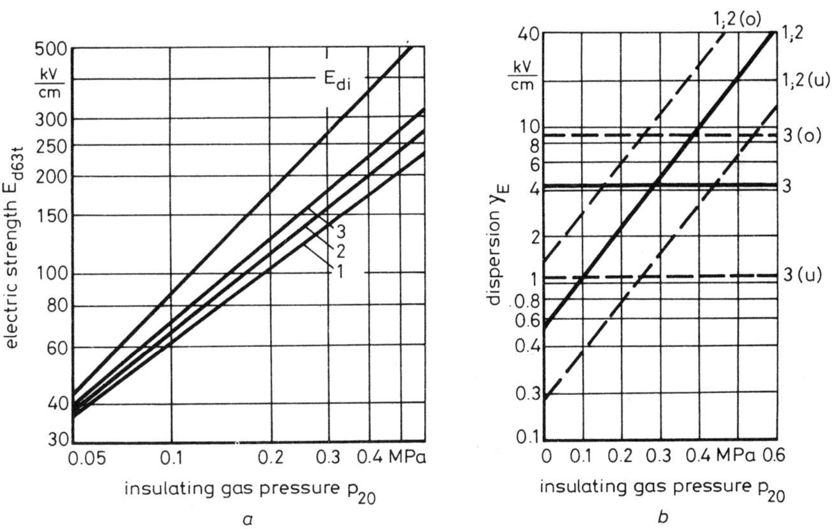

Fig. 5.18 Parameters E_{d63t} and γ_E of electric strength of SF_6 approximated by a double exponential distribution

1 50 Hz alternating voltage (identical to direct voltage)
2 Negative switching voltage 250/2500
3 Negative lightning voltage voltage 1.2/50

Mean values given for γ and 90% inter-decile ranges

electric strength cannot be described by normal distributions (Fig. 5.17*a*), but is far better described by extreme-value distributions (Fig. 5.17*b*, *c*).

One of the problems with an approximation by a Weibull distribution is that the initial value (u_{d0} or E_{d0}) reacts very sensitively to all the test parameters, especially the micro-roughness of the electrode surface, which can hardly be precisely adjusted, and to the micro-particles in the volume of gas. For practical reasons (quality of approximation, convenience, uniformity), it has proved advantageous to describe the electric strength by a double-exponential distribution. If the anticipated parameters are to be estimated prior to a series of tests in order to establish optimum test-parameters, then, by reference to Fig. 5.18:

$$u^*_{d63} = E_{d63t}\, d\eta \tag{5.8}$$

$$\gamma^*_u = \gamma_E\, d\eta; \tag{5.9}$$

where d is the electrode gap and η the degree of uniformity.

In physical respects, description by a double-exponential distribution is of course a compromise, since the fit is better at high pressures than it is at low pressures, and the range of values is unlimited, unlike the Weibull

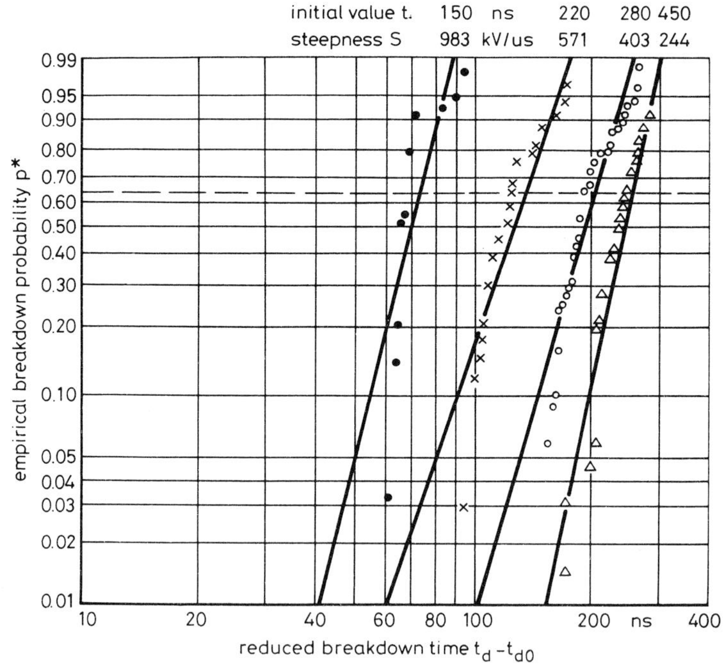

Fig. 5.19 Distribution function of breakdown time with lightning voltage in SF_6, approximated with three-parameter Weibull distribution; Weibull paper

Coaxial cylinders; $r_i = 0.75$ cm; $r_a = 2$ cm; $l = 12$ cm; $p_{20} = 0.25$ MPa; positive polarity

distribution (which is used by Nitta and Bortnik, for example, despite the uncertain determination of the initial value).

The three-parameter Weibull distribution (Fig. 5.19), on the other hand, can be used to handle breakdown-time problems in SF_6 with lightning voltage. In this instance the initial value t_{d0} of the breakdown time t_d can be obtained from physical considerations. With a known voltage/time characteristic (e.g. prescribed rate of rise of voltage S in kV/μs), the initial time t_{d0} is assigned to the voltage $u(t_{d0})$ at which breakdown would be anticipated according to eqn. 5.8 with continuous voltages. Such a calculation is easy to perform, and the parameter estimate is reduced to a problem of a two-parameter Weibull distribution (see Section 2.3.2.4).

With slightly nonuniform SF_6 arrangements with boundary layers, the electric strength can similarly be described by a double exponential distribution (Fig. 5.20*a*, *b*). The breakdown-voltage parameters for an arrangement being investigated can be calculated using eqns. 5.8 and 5.9, as well as with E_{d63tg} and γ_{Eg} from Fig. 5.21. If the transition from electrode to boundary layer is not free from field disturbances, of which there is a particular danger at high SF_6 pressures, then it is often more accurate to approximate strength and breakdown voltage with a normal distribution (Fig. 5.20*c*). Nevertheless, if one uses a double exponential distribution as an approximation, the error will be on the 'safe' side.

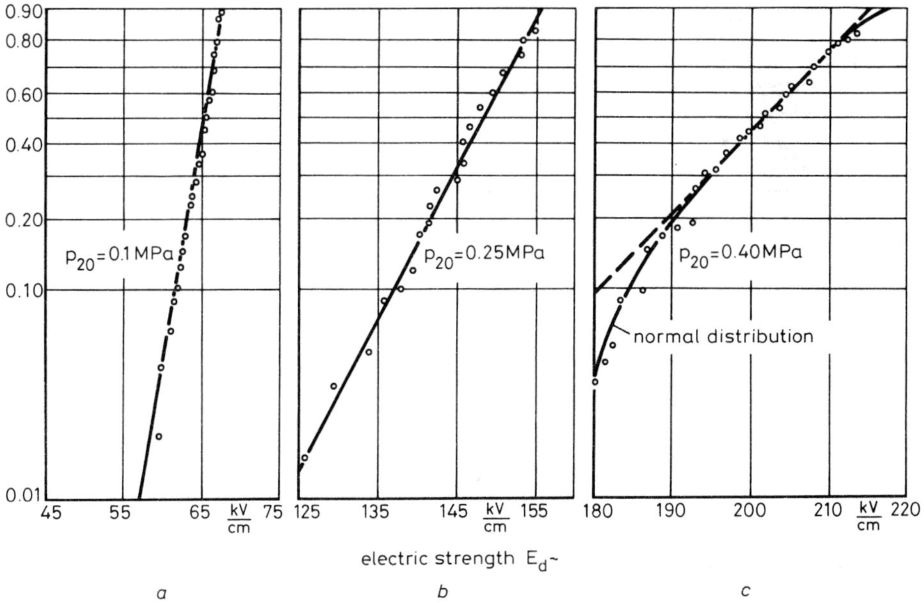

Fig. 5.20 Electric strength of boundary-layer arrangements in SF_6; longitudinal boundary-layer in each case; double exponential paper

a Coaxial cylinders; $r_i = 5$ cm; $r_a = 15$ cm
b and ***c*** Coaxial cylinders; $r_i = 0.675$ cm; $r_a = 2.0$ cm

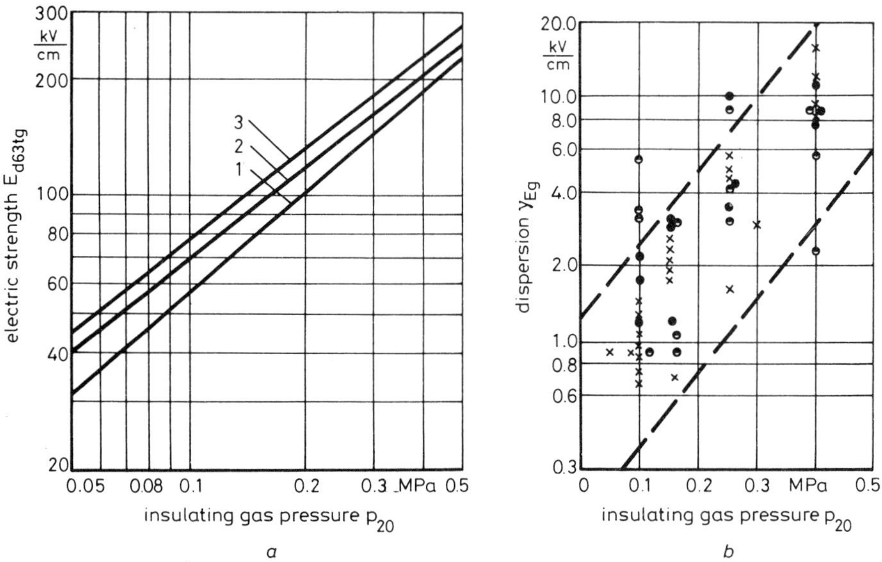

Fig. 5.21 Parameters E_{d63tg} and γ_{Eg} of the electric strength of boundary-layer arrangements in SF_6, approximated with a double exponential distribution

Alternating voltage: line 1; symbol ×
Negative switching voltage: line 2; symbol ●
Negative lightning voltage: line 3; symbol ◓
Broken lines: range of γ_{Eg} for alternating voltage

5.3.3 Disturbed, slightly nonuniform SF_6 insulation arrangements

Because of manufacturing errors, it is impossible to prevent disturbances to the slightly nonuniform designed field, causing partial discharges in the compressed-gas insulation. Such field disturbances may be stationary (sharp edges, burrs, adhering chips, fissures), or be free to move under the forces acting in the electrostatic field (particles). Both groups of field disturbances are often simulated experimentally in the laboratory to study the effects. There is basically a (sometimes extremely pronounced) reduction of the breakdown voltage compared with compressed-gas insulation, with extreme-value distributions no longer applicable.

With stationary defects (Fig. 5.22), the forms of partial discharge which are observed are principally streamer or eruption discharge and glow discharge. If breakdown always develops from a single kind of discharge, then the breakdown voltage can be advantageously approximated by a normal distribution. If, on the other hand, breakdown can develop either from streamer or from glow discharge, then additive mixed distributions of the breakdown will arise (Fig. 5.22: alternating voltage; $p_{20} = 0.25$ MPa).

With a freely moving point of disturbance (particle), normally-distributed breakdown voltages also occur for the standard types of voltage (alternating voltage: Fig. 5.23, 1 particle; direct and switching voltage: Fig. 2.25). If,

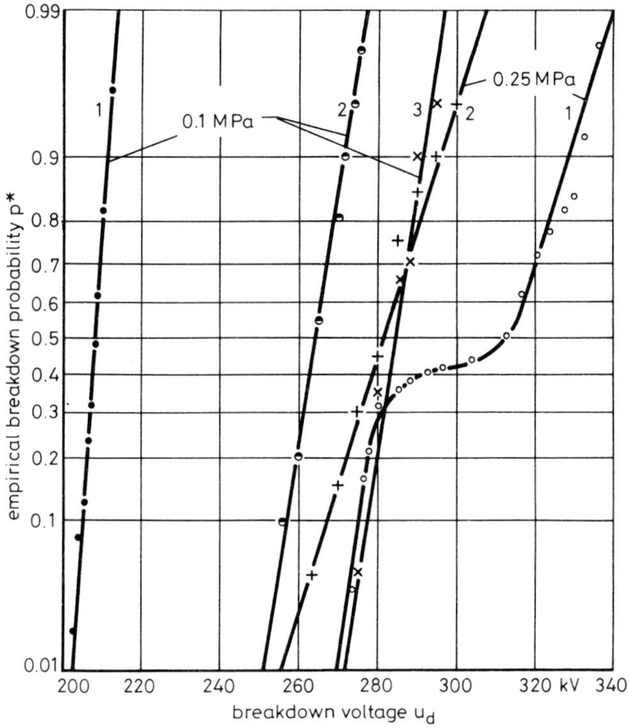

Fig. 5.22 Distribution function of breakdown voltage of SF_6 insulation with a stationary irregularity (Gauss paper)

Coaxial cylinders: $r_i = 5$ cm; $r_a = 15$ cm; disturbing spike on inner electrode, $l =$ 0.5 cm

1 50 Hz alternating voltage
2 Positive lightning voltage 1.2/50
3 Negative lightning voltage 1.2/50

however, there is more than one particle present in the insulation (Fig. 5.23), it follows from the 'enlargement law' (see Section 6.3) for identical particles that, as the number of particles increases, the type of distribution will become closer and closer to a double-exponential distribution. When there are a large number of particles, one should expect an extreme-value distribution, as in a slightly nonuniform field without particles. This situation implies that, in a technical sense, compressed-gas insulation arrangements that are free from points of disturbance exhibit far more micro-disturbances, which tend to govern the random process. If particle-affected SF_6 insulation is stressed with mixed voltages (e.g. combination of direct and switching voltages), then additive mixed distributions of the breakdown voltage will be obtained (Fig. 2.25).

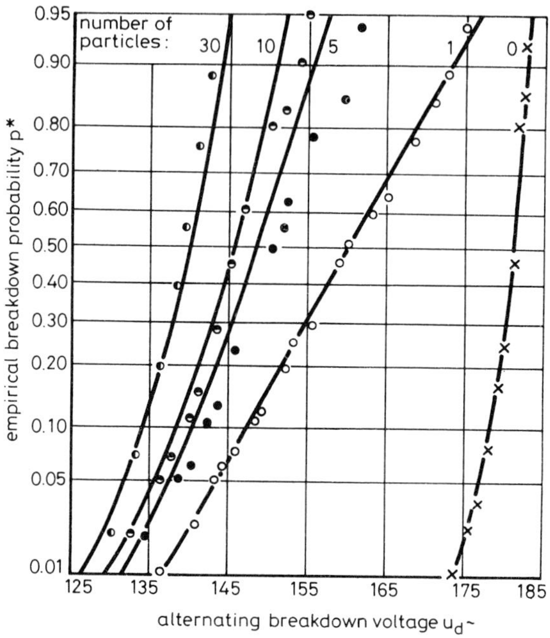

Fig. 5.23 Distribution function of alternating breakdown voltage of SF_6 insulation with free-moving particles (Gauss paper)

Spheres one inside the other; $r_i = 5$ cm; $r_a = 10$ cm; $d = 2$ cm; $p_{20} = 0.1$ MPa; spherical particles $d_p = 1$ mm

5.4 Liquid insulation

Insulating liquids have dielectric strengths between 100 and 1000 kV/cm, which are extremely dependent on their condition. Since mineral-oil based insulating oils continue to be of the most practical importance, we intend to explain a few basic relationships in the execution of statistical tests and the evaluation of an example. While this information can be applied accordingly to liquid dielectrics under indoor conditions, reference should be made to the literature with regard to liquid gases and associated problems. In combined solid/liquid insulating arrangements, it is largely the solid component which determines statistical behaviour. Section 5.5 should be taken into account accordingly.

5.4.1 Testing problems

The insulating capacity of liquids is only partly regenerated after electrical discharges: plasma-chemical reactions, for example, break the insulating oil down into low-molecular gaseous hydrocarbons and elementary carbon (soot). The gases are partially dissolved in the oil, but they can also remain in the oil as bubbles, which reduce its insulating capacity considerably. The soot produces an even more pronounced drop in the breakdown voltage, which will of course depend on the insulation arrangement concerned (soot bridge formation; Fig. 5.24). As has already been mentioned in connection with compressed gases (see Section 5.3.1), the aim should be to minimise the discharge energy converted in the space concerned. It is also possible to circulate the oil and to filter or even treat it. The oil flow hardly affects the breakdown voltage. The dependences due to decomposition products can consequently be kept within bounds, even with large sample sizes.

The residual space-charges in the oil after breakdown exhibit extremely low mobilities, thereby causing pronounced dependences in consecutive voltage applications. As in the case of compressed gases, the use of rising-voltage tests instead of constant-voltage tests presents an excellent opportunity for avoiding dependences. 'Interval stresses' (see Section 3.2.2) will presumably also be suitable for reducing or eliminating dependences in liquids.

The condition of the electrodes may also change under the influence of the insulating liquid, particularly of discharges, thereby causing dependences. For this reason, too, it is desirable to limit the energy converted at breakdown. If there are a few particles or gas bubbles present in the oil,

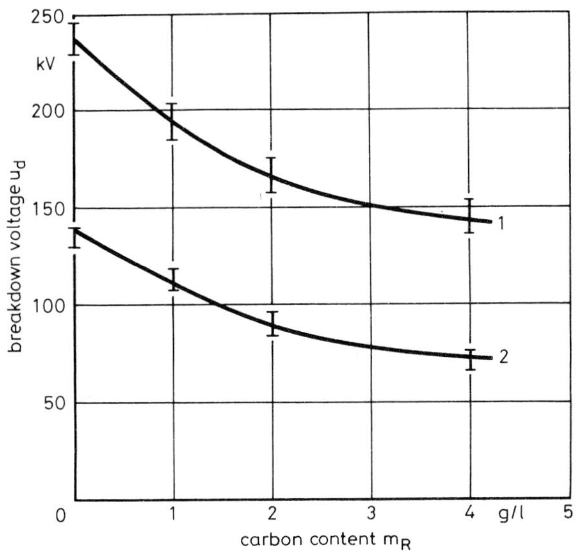

Fig. 5.24 Influence of carbon content on breakdown voltage in insulating oil (after Günther)

Contact gap of a small-oil volume circuit-breaker for 12 kV

1 Switching voltage
2 Alternating voltage

95% confidence region

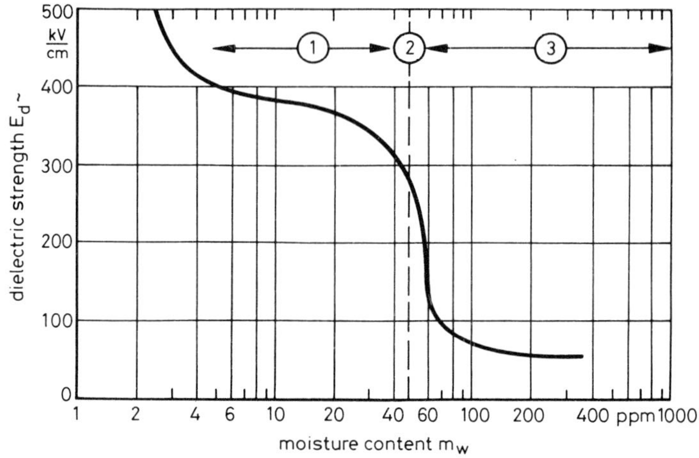

Fig. 5.25 Dielectric strength of insulating oil, together with moisture content

1 Water dissolved
2 Solubility limit
3 Water emulsified

they can be removed by a few conditioning breakdowns or purely by applying a voltage.

In rising-voltage tests with direct or alternating voltage, variations in the rate of rise of the voltage by a factor of 3 to 5 do not affect the test result. Variations in the rate of rise of the voltage caused by the actuator will consequently not affect the test result.

Testing technique is important, as it was in the case of compressed gases. It is the great dependence of dielectric strength on moisture content (Fig. 5.25) that necessitates strict measures for oil and test chamber drying as well as the prevention of humidification during the tests. Provided one is working with a well dried oil, even open test chambers can be used in tests lasting less than a day. Sealed chambers, permitting oil circulation and drying, are recommended for longer test series. Clear specifications should be provided for cleaning of electrodes and test chamber, drying process (plus any evacuation), filling process and oil circulation rate.

5.4.2 Distribution functions which can be used

Estimation of the distribution function of an insulating arrangement in oil or in another liquid cannot be presented in such a simplified manner as it was for gases (see Sections 5.2 and 5.3), since, in the final analysis, the condition of the liquid will govern the expectation and dispersion of the breakdown voltage. There has not been much recent work on breakdown in oil, so that we intend simply to indicate the distribution functions which can be used. A few details on the pre-calculation of expectations can be found in Mosch *et al.*, otherwise the specialist literature should be consulted.

In numerous, usually somewhat older, works, the breakdown voltage is described by a normal distribution. It now seems certain that such an approximation is eminently suitable for highly nonuniform insulation arrangements and for arrangements in which the highly stressed electrode area is small. With slightly nonuniform and uniform arrangements, its use is usually associated with relatively small sample sizes, in which case it is only suitable for describing an average trend. What is more, unfavourable test procedures and/or the aggregation of a large number of test series can produce normal distributions, since the random boundary conditions then presumably determine the type of distribution. A normal distribution should therefore only be used to describe average trends and for highly nonuniform electrode arrangements. A few rough values for the coefficients of variation — which should be treated with caution — are given in Table 5.6.

In American literature, extreme-value distributions were being applied to the problem of oil breakdown as early as the 1950s and were treated as a typical application in the classical statistics reference by Gumbel. More recent publications refer to these works and use both a double exponential distribution and a Weibull distribution. By virtue of the breakdown voltage in oil insulation arrangements, the two-parameter Weibull distribution is eminently suitable for approximation of the breakdown voltage. The step to the three-parameter Weibull distribution (see Section 5.3.2), which is rather tricky in practice, does not seem to be absolutely essential. It is interesting to observe that the duration of voltage application does not affect

Table 5.6 Approximate values for coefficients of variation of breakdown voltage of oil arrangements for approximation by a normal distribution

Electrode arrangement	Type of voltage	Coefficient of variation v v	References
Plate/plate ($d=0.5\ldots12$ mm)	Alternating voltage	$H\leqq0.10$	[5.67]
Rod/rod ($d=50$ mm; $r=10$ mm)	Lightning voltage	0.05	[5.65]
Point/plane ($d=20\ldots100$ mm)	Switching voltage	0.08	[3.50] [2.23] [3.36]
Plate/plate with pressboard longitudinal boundary layer ($d=12$ mm)	Lightning voltage	0.08	[5.67]
Plate/plate, surrounded (plus barrier) ($d=6\ldots12$ mm)	Alternating voltage	0.10 . . . 0.18	[5.67]

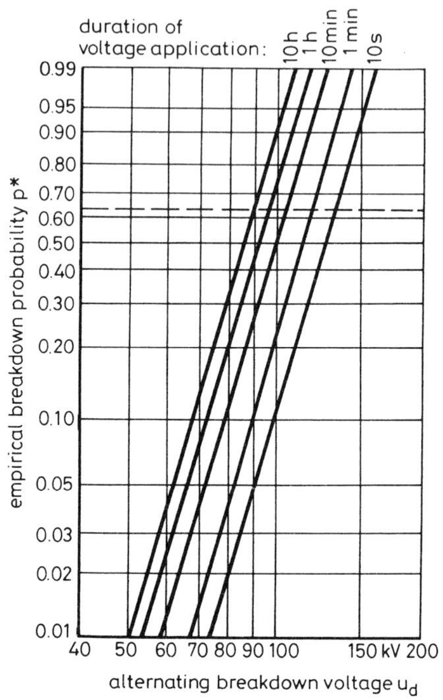

Fig. 5.26 Weibull distribution of alternating breakdown voltage of a plate/plate arrangement in transformer oil for various voltage-application times

the Weibull exponent (i.e. the rise of the distribution function) (Fig. 5.26). This trend has also been confirmed with enclosed electrodes. A Weibull distribution can be used successfully both for oil-insulating clearances and for oil/paper insulation, thereby presenting the possibility of uniform treatment of this group of insulating arrangements. Since oil insulation can often only be investigated experimentally as small models, conversion to larger practical insulating arrangements (e.g. transformer insulation), using the enlargement law, is important (see Chapter 6). Extreme-value distributions are also a good prerequisite. It is therefore advisable to describe the breakdown voltage of slightly nonuniform and uniform oil insulation with a Weibull distribution (or double-exponential distribution).

5.5 Solid insulation arrangements

The electric strength of solid insulation can reach values in excess of 1000 kV/cm, but it is dependent on the state of the insulation to such a degree that values above 100 kV/cm are rarely used in practice. As well as this, the insulating capacity of inorganic and organic insulating materials differs considerably with respect to time dependence: whereas there is a relatively small time dependence in the case of inorganic insulating materials (e.g. porcelain, glass), it is so crucial for organic materials, which are becoming increasingly popular (e.g. epoxy resin, polythene, etc.), that it is described by a life characteristic. The breakdown time thereby becomes a vital variate for technical evaluation, as well as the breakdown voltage. Our comments in the following will be restricted to high-polymer insulating materials, which, because of their favourable processability, are now widely used and will be even more widely used in the future. The treatment applies accordingly to other organic insulating materials (e.g. paper) and to discharge processes at organic boundary layers, in which thermal or plasma-chemical processes take place at the surface of the solid body.

In contrast-voltage tests (see Sections 3.2.3 and 5.1), breakdown-time distribution functions are determined (Fig. 5.27), which are generally approximated by two-parameter Weibull distributions (see Section 2.3.2.4) or log-normal distributions (see Section 2.3.2.3). Experience in recent years has clearly shown that the Weibull distribution has mathematical advantages over the log-normal distribution when conversions have to be performed via the life relationship (see Section 5.5.2) or using the enlargement law

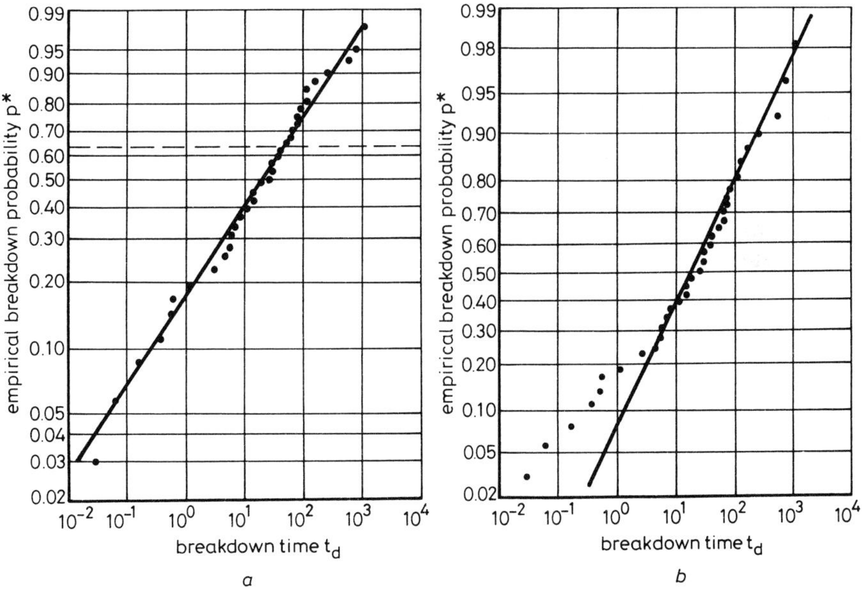

Fig. 5.27 Approximation of distribution function of breakdown of model polythene cable insulation

a Approximation by two-parameter Weibull distribution (Weibull paper)
b Approximation by log-normal distribution (Gauss log paper)

Alternating breakdown voltage: $u_d = 49$ kV; model cable: $r_i = 1.5$ mm; $r_a = 3.5$ mm; $l = 1$ m (see Chapter 6). The model of extreme-value distributions, moreover, corresponds more closely to physical concepts if electrical breakdown develops from a (or the extreme) point of disturbance in the structure of the solid body.

Cumulative-frequency and performance functions of the breakdown voltage can be advantageously described with Weibull distributions (Fig. 5.28), and approximation with a log-normal distribution is also possible. If the boundary conditions of the test (see Section 5.5.1) vary greatly, it may also be possible for the measured breakdown voltages to be normalised. This is then, however, less a consequence of the breakdown process itself than that of its boundary conditions.

As was the case with liquid insulation, we can only give details in the following concerning testing techniques and the basic statistical problems. Data for estimating distribution functions of breakdown voltage and breakdown time prior to experiments cannot be generalised sufficiently to be presented in the space available here. Generally speaking, one works on the basis of 'orientational' experiments (e.g. previous short-duration tests in the case of long-term investigations).

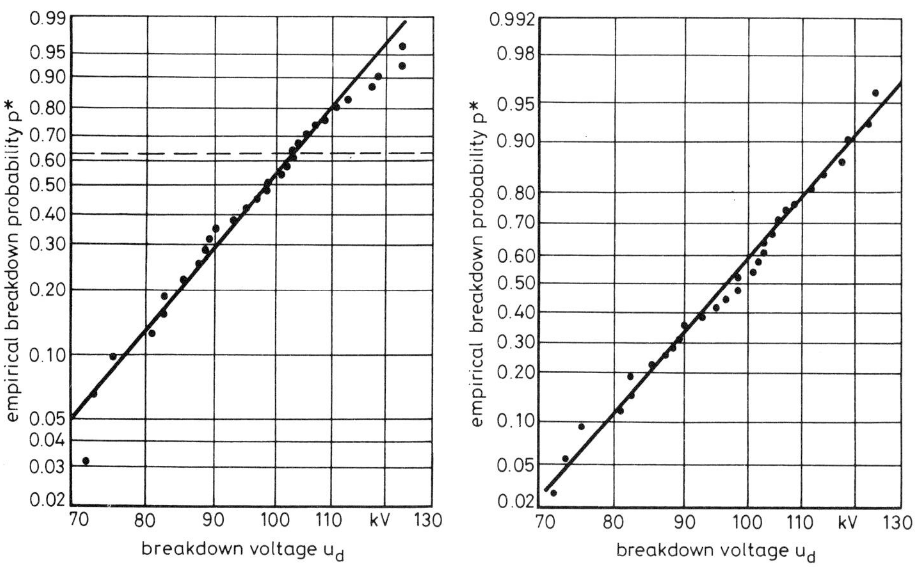

a　　　　　　　　　　*b*

Fig. 5.28 Approximation of distribution function of alternating breakdown voltage of model polythene cable insulation

a) Approximation by two-parameter Weibull distribution (Weibull paper)
b) Approximation by log-normal distribution (Gauss log paper)

Rate of rise of voltage $v_u = 1\,\text{kV}/10^{-2}\,\text{h}$; model cable: $r_i = 1.5\,\text{mm}$; $r_a = 3.5\,\text{mm}$; $l = 1\,\text{m}$

5.5.1 Testing problems

Every electrical discharge produces an irreversible destruction of the structure of the solid body, i.e. solid insulating materials are not self-restoring. A separate test piece is consequently needed for each individual test. Even when no discharge is observed, polarisation and conductivity processes can produce space charges that exist for a long time and that can affect the result of a subsequent test on the same test piece. To eliminate such effects, one test piece per stress (voltage application) should be used (where the 'stress' can also be a well defined impulse sequence (Fig. 3.3) or a rising voltage (Fig. 3.22)). Since solid test pieces are generally comparatively expensive, one is usually forced to make do with a small number of test pieces. Consideration of the choice of sample size (see Sections 3.2.1 and 3.3.1) should therefore be exercised with the greatest care.

The condition of the test piece is determined in the first instance by its technology. It is therefore essential to produce the test piece using the same technology as that by which the insulation is subsequently to be manufactured and to make it match the latter as closely as possible in its construction. For example, model polythene (polyethylene) coaxial cables, possibly with smaller dimensions, are as far as possible used as test pieces in the development of polythene cables. If one is only interested in data relating to the

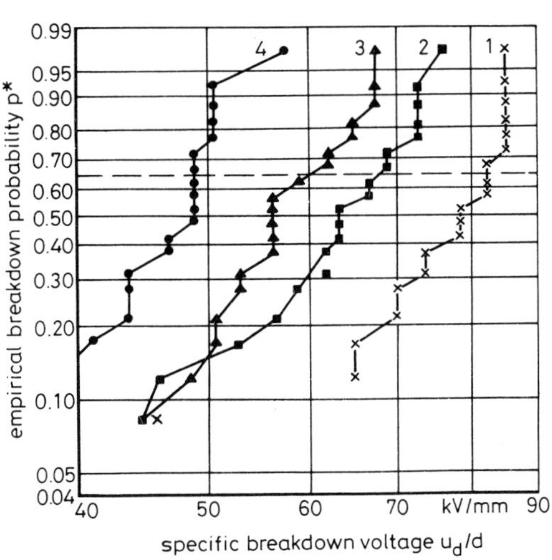

Fig. 5.29 Effect of copper admixtures in model polythene (polyethylene) insulation on empirical cumulative-frequency function of breakdown voltage

Model cable: $r_i = 1.2$ mm; $r_a = 2.5$ mm; $d = 1.3$ mm; $l = 300$ mm
Rate of rise of voltage $v_u = 0.5$ kV/s
Copper particle sizes:
1 No admixture (technically pure)
2 40 μm
3 70 μm
4 140 μm

basic behaviour of insulating materials, it is possible to work with basic arrangements, such as point/plane ('needle test'), sphere/plane or plane/plane.

The independence of tests is primarily influenced by the test pieces themselves. Between different batches there may possibly be certain differences in chemical composition, in the proportion of filler, in the proportion of impurities (see below), in the geometry of the electrodes and the electrode gap. There is a close relationship between insulating capacity and the mechanical stresses that may be produced in the solid body by the setting process and by the different coefficients of expansion of electrodes and insulating material, particularly in the case of casting resins. Careful planning is absolutely essential in this instance, for which valuable advice can be derived from the results of Georgi and Schirr.

Impurities, especially abraded metal, can greatly reduce insulating capacity (Fig. 5.29). The larger the particles added, the lower the 63% breakdown voltages. There is only a slight difference between the small quantiles of the breakdown voltage, however (Fig. 5.29), since there are impurities even in technically 'pure' insulating materials.

Further chemical processes and structural changes occur during storage after manufacture. In the case of polythene, for example, it has been established that the channel inception voltage is reduced as the storage

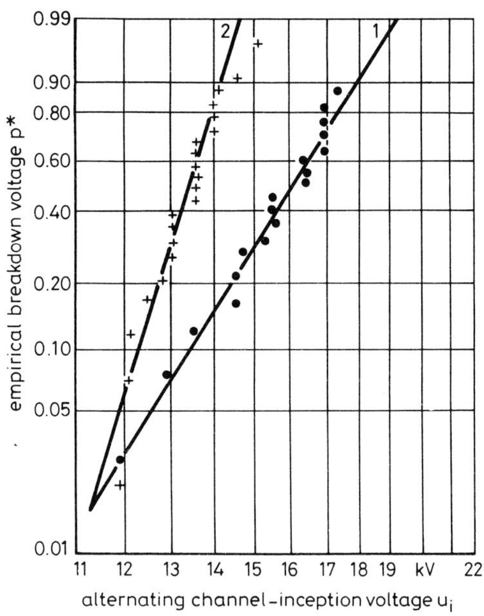

Fig. 5.30 Distribution function of channel inception voltage of a needle/plate arrangement ($d = 3$ mm) in polythene. Rising-voltage test; Weibull paper

1 Freshly vulcanised (cross-linked) polythene
2 Polythene stored for one year

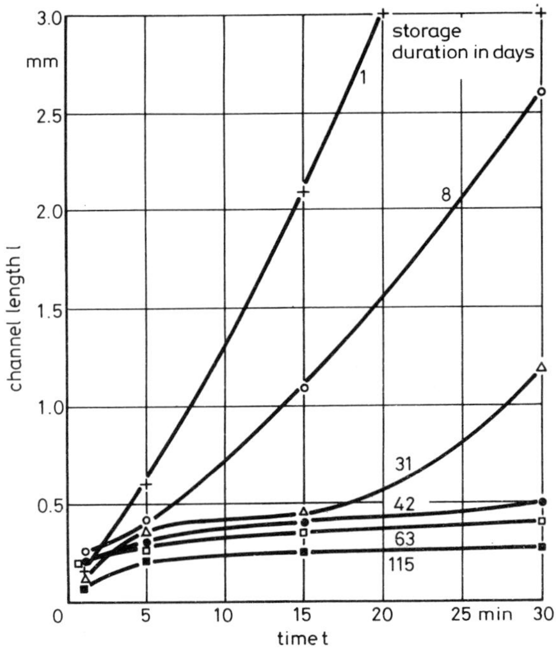

Fig. 5.31 Channel length, as a function of storage duration of a needle/plate arrangement ($d = 3$ mm) in epoxy resin

duration increases (Fig. 5.30). On the other hand, the growth rate of partial-discharge channel inception voltage is reduced as the storage duration increases (Fig. 5.30). On the other hand, the growth rate of partial-discharge channels slows down as the storage duration after manufacture increases (Fig. 5.31). These two trends can have opposite effects on the breakdown voltage, which demonstrates the importance of the time interval between manufacture of the test piece and performance of tests.

The test results are greatly affected by the ambient conditions, and especially the ambient medium, which can diffuse into the test piece or can occur with the plasma-chemical reactions. The life characteristics (Fig. 5.32) determined by Rüffer in 'needle tests' on polythene foil insulation, with air as the ambient medium, exhibit a downward inflexion after about 50 hours (curve 1); with silicon oil as the ambient medium, on the other hand, there is an upward inflexion after about 10 minutes, because of diffusion. Since the phenomena referred to are time-dependent, the distribution functions of breakdown voltage and breakdown time (Fig. 5.33) are also influenced by the storage duration in the ambient medium. If the condition of the test piece changes during measurement, mixed distributions can be caused by this dependence. The ambient medium should therefore be chosen as it occurs in subsequent service, and the diffusion processes should be completed by the beginning of the test. Otherwise one must always test the effect of the ambient medium on the test.

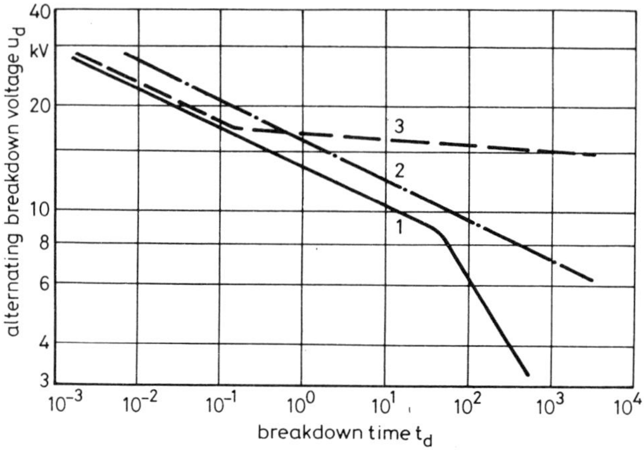

Fig. 5.32 Effect of impregnating medium on life characteristic of polythene foil insulation

Needle/plate; four polythene foils (each 50 μm thick)
Impregnating medium:
1 Air
2 SF_6
3 Silicon oil

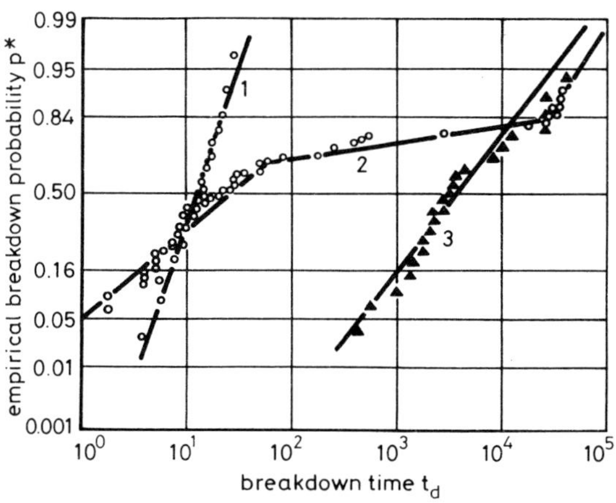

Fig. 5.33 Distribution functions of breakdown for a point/plane arrangement in polythene with various ambient media

1 Ambient medium silicon oil
2 Ambient medium transformer oil (short storage duration)
3 As *2*, but storage duration until conclusion of diffusion

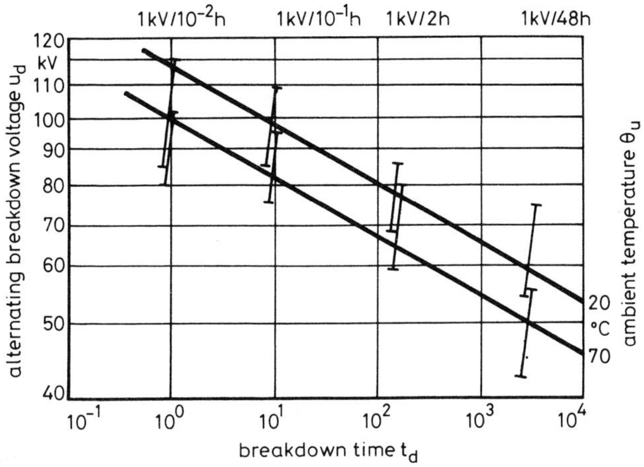

Fig. 5.34 Determination of life characteristics in rising-voltage test, showing effect of temperature

Epoxy resin; sphere ($r = 5$ mm)/plane ($d = 1$ mm)

The insulating capacity of solid materials is reduced as the ambient temperature rises (Fig. 5.34); a thermal breakdown mechanism occurs above a limiting temperature.

The choice of test procedure is determined by the aim of the test. Constant-voltage tests provide reliable comprehensive data for the distribution function of the breakdown time, but determination of a life characteristic is very time-consuming. Rising-voltage tests (Fig. 5.34) can be performed less expensively, but generalisation of the test result is appreciably more problematical. Questions relating to test procedure and evaluation are considered in the next two Sections (see also Section 5.1 and Fig. 5.1).

5.5.2 Distribution functions of breakdown time and breakdown voltage

Let us first consider a constant-voltage test at voltage u_{d1}, performed on n test pieces. In the test result, n realisations of the variate breakdown time are obtained.

The empirical distribution function of the breakdown time (Fig. 5.35*a*) is obtained from the test: as has already been mentioned, it is advantageously described by a Weibull distribution (cf. Fig. 5.27*a*):

$$F(t_d\,;\,u_{d1}) = 1 - \exp\left[-\left(\frac{t_d}{t_{d63}(u_{d1})}\right)^{\delta_t}\right]. \tag{5.10}$$

The breakdown-voltage/breakdown-time diagram, the so-called 'life characteristic', can be constructed from selected quantiles of this distribution; experience has shown that it forms a straight line on a double-logarithmic scale (Figs. 5.35*b* and 5.36). If the confidence regions are known for the quantiles concerned, they can be transferred to the life characteristic. For

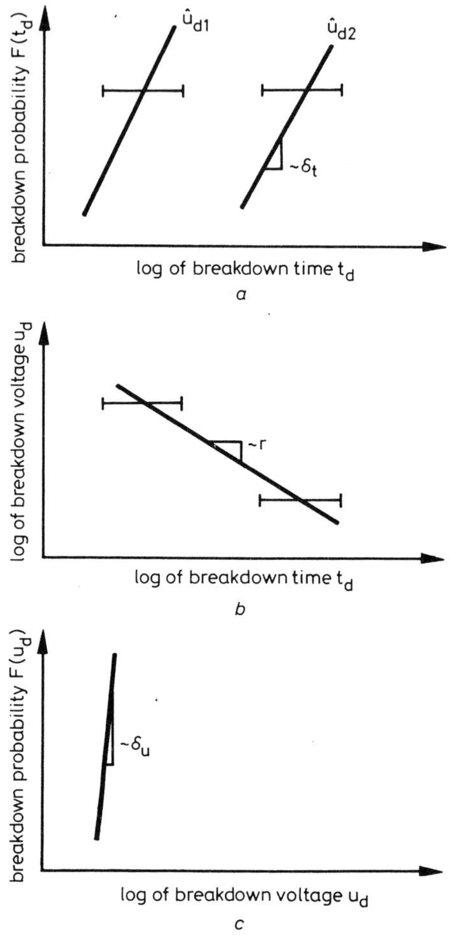

Fig. 5.35 Determination of distribution functions of breakdown time (a), life characteristic (b) and performance function of breakdown voltage (c)

(Diagrammatic after [5.25])

each p-order quantile of the breakdown time, the life characteristic is described by

$$u_{dp} = k_{dp} t_{dp}^{-1/r} \tag{5.11}$$

where k_{dp} is a constant characterising the geometry of the arrangement and r is the life exponent mainly dependent on the insulating material (e.g. $r \approx 9$ for polythene; $r \approx 12$ for filled epoxy resin). Inflexions in the life characteristic indicate changes in the ageing mechanism (breakdown process).

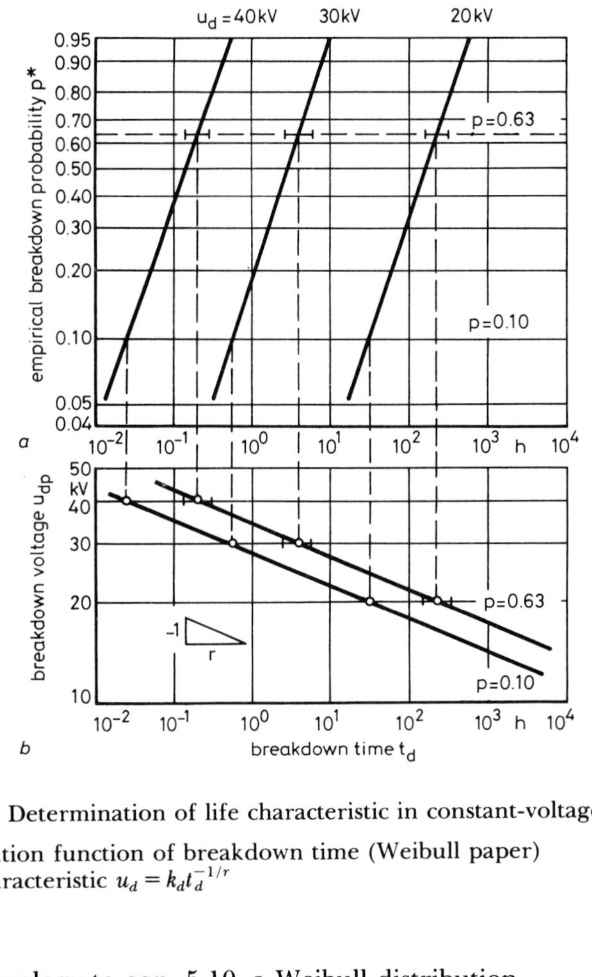

Fig. 5.36 Determination of life characteristic in constant-voltage test

a Distribution function of breakdown time (Weibull paper)
b Life characteristic $u_d = k_d t_d^{-1/r}$

If, by analogy to eqn. 5.10, a Weibull distribution

$$F(u_d\,;\,t_{d1}) = 1 - \exp\left[-\left(\frac{u_d}{u_{d63}(t_{d1})}\right)^{\delta_u}\right] \tag{5.12}$$

is also assumed for the breakdown voltage u_d with a fixed breakdown time t_{d1}, then, for the same breakdown probabilities $F(t_d\,;\,u_{d1}) = F(u_d\,;\,t_{d1})$:

$$u_{d63}(t_{d1})[t_{d1}]^{\delta_t/\delta_u} = u_{d1}[t_{d63}(u_{d1})]^{\delta_t/\delta_u}. \tag{5.13a}$$

With the life law (eqn. 5.11), assuming the exponent r is applicable to all quantiles, the following expression holds for the same relationship and a pair of values ($u_{d1} = u_{d3}$; t_{d1}):

$$u_{d63}(t_{d1})[t_{d1}]^{1/r} = k_{d63}. \tag{5.13b}$$

Comparing the coefficients on the left-hand sides of eqns. 5.13*a* and 5.13*b* provides a relationship between the Weibull exponents for breakdown time and breakdown voltage (δ_t; δ_u), and the life exponent r, in the form:

$$r = \frac{\delta_u}{\delta_t}. \tag{5.14}$$

This relationship is only correct if both variates T_d and U_d are Weibull-distributed (Figs. 5.28, 5.35*c*). It should be expressly pointed out that eqn. 5.14 and the model presented can only be used if r applies equally well to all quantiles. Köppe, for example, found that, with epoxy-resin insulation, r is reduced with the order of the quantile concerned. It is consequently a more reliable approach to determine the performance function of the breakdown voltage directly by experiment, as in Fig. 5.1 (see Section 5.1).

5.5.3 Relationship between constant-voltage and rising-voltage tests

In a rising-voltage test, the typical breakdown-voltage cumulative-frequency function is obtained for the rate of rise of the voltage v_u. It can also be used to determine life characteristics by plotting the v_u used in a u_d/t_d diagram and marking the relevant quantiles (or their confidence regions) on the straight-line graphs obtained (Fig. 5.34). The lines of best fit drawn through these points can be interpreted as life characteristics of the rising-voltage test. They are far more economical to determine than those in the constant-voltage test, but they provide less information. There is consequently a need to establish the relationship between the results of rising-voltage and constant-voltage tests.

The conversion is attributable to Starr and Endicott, and has been theoretically reinforced by Pilling and others with the model representation of damage accumulation. Damage accumulation characterises the development of the irreversible destruction of the solid-body structure with the quantity relative life consumption

$$l_v = \frac{t_b}{t_d}, \tag{5.15}$$

where t_d is the breakdown time and $t_b \leqq t_d$ the insulation stress time. Using eqn. 5.11, it follows for any quantile not specially characterised here, that:

$$t_b = k_{dp}^r l_v (u_b)^{-r},$$

and, consequently, for known constant values of the life exponent r, pairs of stress values (u_b, t_b) can be converted to equivalent stresses (u_b^*; t_b^*), i.e. stresses with the same life consumption:

$$t_b^* = t_b \left(\frac{u_b}{u_b^*} \right)^r. \tag{5.16}$$

Heavy electrical stress of short duration can therefore produce the same damage (the same life consumption) as lower stress of long duration. How this comes about will depend in the first instance on r: if, for example, solid insulation has been stressed for 25 years with a voltage u_b and for 100 hours with $u_b\sqrt{3}$ (earth fault), then the life consumption for $r < 8$ is determined

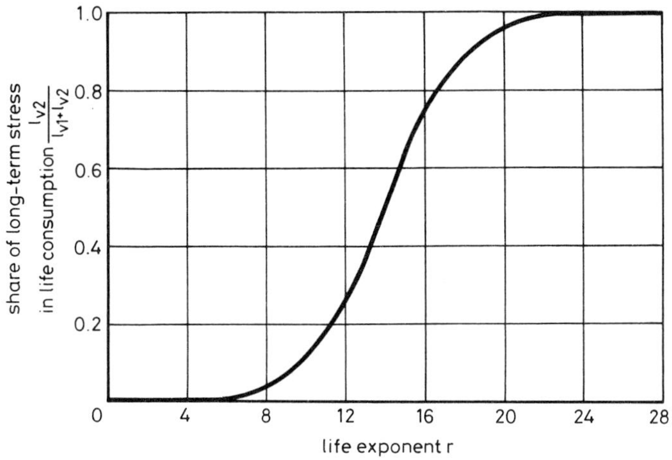

Fig. 5.37 Components of life consumption for a combined stress (u_{b1}; $t_{b1}=219{,}000$ $h=25$ years) and ($u_{b2}=u_{b1}\sqrt{3}$; $t_{b2}=100$ h)

l_{v1}: life consumption by (u_{b1}; t_{b1})
l_{v2}: life consumption by (u_{b2}, t_{b2})

by the long-term stress with u_b, and for $r>20$ by the earth-fault stress with $u_b\sqrt{3}$ (Fig. 5.37). Breakdown occurs when the relative life consumption (eqn. 5.15) reaches unity.

If the life-consumption model is applied to the entire rising-voltage test (u_{ds}; t_{ds}), then, by analogy to eqn. 5.16 for various constant-voltage tests, the breakdown time and breakdown voltage can be calculated for an equivalent constant-voltage test (u_{dk}; t_{dk}). If we make $u_{dk}=u_{ds}$ for a rise from zero, then for any quantile:

$$t_{dk}=\frac{t_{ds}}{r+1}; \tag{5.17}$$

if we make $t_{dk}=t_{ds}$, then

$$u_{dk}=\frac{u_{ds}}{\sqrt[r]{r+1}}. \tag{5.18}$$

If the voltage rise starts from a value $u_0>0$, one has to ensure that the life consumption l_{v0} calculated for the same voltage rise in the range $0<u\leqq u_0$ can be neglected. With a suitable value, e.g. $l_{v0}=0.01$, and an estimated value u_{ds}^* for u_{ds}, the initial voltage should be specified as:

$$u_0=u_{ds}^*\sqrt[r+1]{l_{v0}}. \tag{5.19}$$

Applicability of the model to conversion (eqns. 5.17, 5.18) rigidly presupposes that r has one known fixed value over the entire range of the life characteristic concerned and for every quantile of the distribution function. If there are kinks or discontinuities in the life curve, one must not work

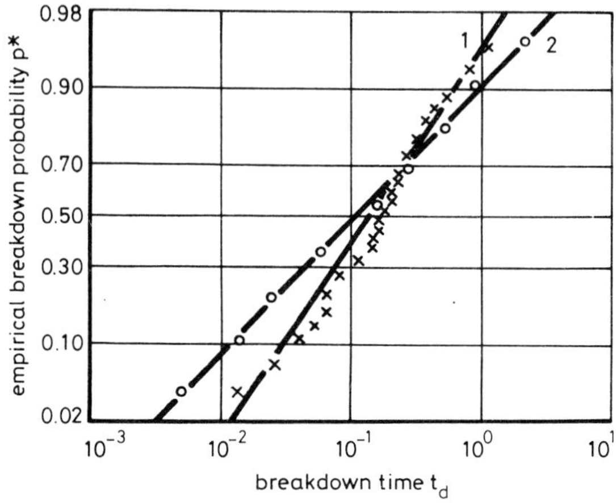

Fig. 5.38 Breakdown-time distribution function for a 200 μm-thick polythene foil (needle test); Gauss log. paper

1 Determined in constant-voltage test
2 Calculated from rising-voltage tests (○)

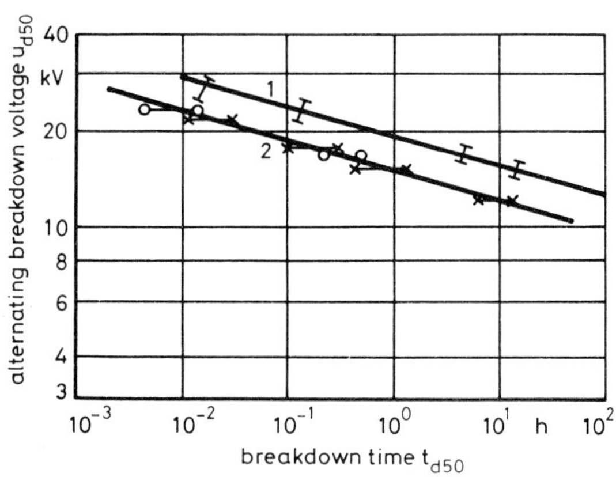

Fig. 5.39 Life characteristics from rising-voltage (*1*) and constant-voltage tests (*2*) for a polythene foil

Needle/plate; $d = 200$ μm; ambient medium air
Confidence regions:
|——| for u_{d50} in rising-voltage test (*1*)
×|—|× for t_{d50} in constant-voltage test (*2*)
○- - -○ calculated from (1) for (*2*)

with eqns. 5.17 and 5.18. If, however, the conditions are satisfied, then breakdown-time distribution functions can be advantageously determined (Fig. 5.38) for constant-voltage tests from the breakdown times determined for rising-voltage tests using eqn. 5.17. Agreement is particularly good in the region of the mean value, which is clearly demonstrated by the life curves corresponding to the mean value (Fig. 5.39).

5.6 Statistics of partial discharges

With the increasing utilisation of the electric strength of insulating materials, partial-discharge measurements are becoming increasingly important. With modern systems for partial-discharge measurement, various characteristic quantities of partial discharge can be measured (impulse charge, maximum impulse charge, apparent charge, cumulative impulse charge, mean partial-discharge current, partial-discharge energy, impulse frequency, etc.), all of which (including the 'mean' quantities) can be treated as variates. Within the scope of the present monograph, only a few fundamental remarks can be made about the statistics of partial discharges, with reference to the most important measured quantity, the impulse charge (apparent charge). In so doing, we shall pay particular attention to the interaction between the random process and the measuring circuit.

With most partial-discharge measuring equipment, the measuring interval ('gate time') can be adjusted to the desired phase angle in relation to the test voltage and duration (Fig. 5.40*a*, *b*). The random discharge-process produces in the measuring circuit a flow of current, generally consisting of individual current-pulses, which are of random amplitude, duration, charge and repetition rate (Fig. 5.40*c*). The impulse charge (current/time/area of impulse) is determined electronically in the partial-discharge measuring instrument. Within the instrument, each current impulse has a corresponding charge impulse, whose amplitude is proportional to the impulse charge (Fig. 5.40*d*). The shape of the charge impulse depends on the measuring principle involved: the narrower the bandwidth of the partial-discharge measuring instrument, the longer they will be compared with the original current pulses. With broadband instruments, too, its duration is greater than that of the current pulses.

The distribution function can now be determined for the individual measured pulse charges. Since the evaluation of general-view oscillograms is very troublesome and subject to error, devices for pulse-amplitude analysis are used. These simultaneously evaluate the pulse charge and pulse frequency (as a measure of the time interval between two charge-pulses). To a certain extent, the relationship between the pulse charge and the absolute pulse frequency is obtained as an absolute frequency distribution (Fig. 5.41*a*), the amplitudes of the pulse charges being prescribed as the level in the experiment, and the frequency of the charge pulses exceeding that level being determined. On the basis of this experimentally determined characteristic, the distribution function of the pulse charge $F(q_i)$ (Fig. 5.41*b*(1)) is more rarely reproduced in evaluations than its complementary

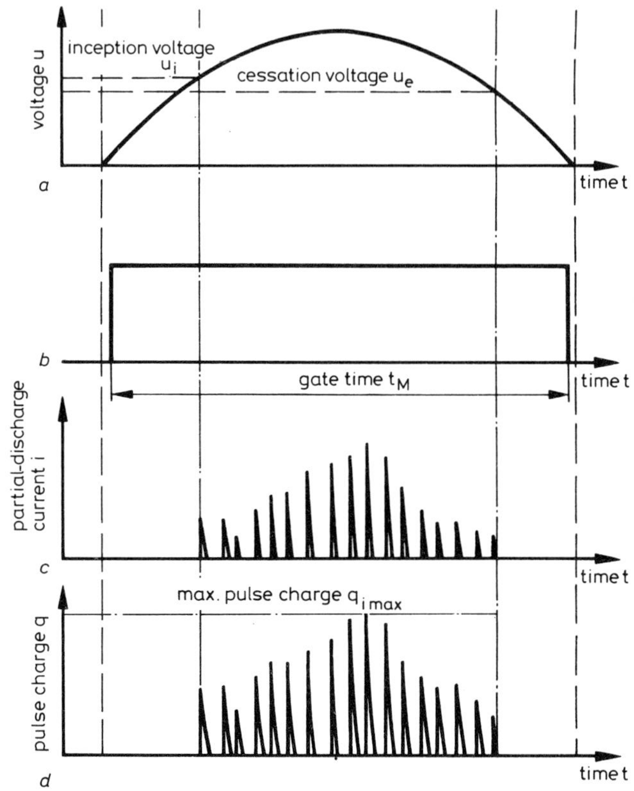

Fig. 5.40 Formation of partial-discharge measured quantity pulse charge (diagrammatic)

exceeding-frequency $H_{\ddot{u}}(q_i)$ (Fig. 5.41*b*(2)):

$$H_{\ddot{u}}(q_i) = 1 - F(q_i).$$

In many cases a normal distribution can be assumed for $F(q_i)$, especially when there is only a single discharge point. Particularly large pulse charges are naturally of special interest, which in the case of $H_{\ddot{u}}(q_i)$ are associated with small probabilities, and in the case of $F(q_i)$ occur as large quantiles. Since they are comparatively rare, $H_{\ddot{u}}(q_i)$ must be regarded as the technically clearer, and $F(q_i)$ on the other hand the mathematically more advantageous method of representation. With a degree of familiarisation, there will hardly be any objection to working here on every occasion with the distribution function $F(q_i)$.

The maximum pulse charge can be interpreted as a quantile, whereas it is a measured quantity in most partial-discharge measuring devices and is displayed on them (cf. Fig. 5.40*d*). The reading on the devices represents

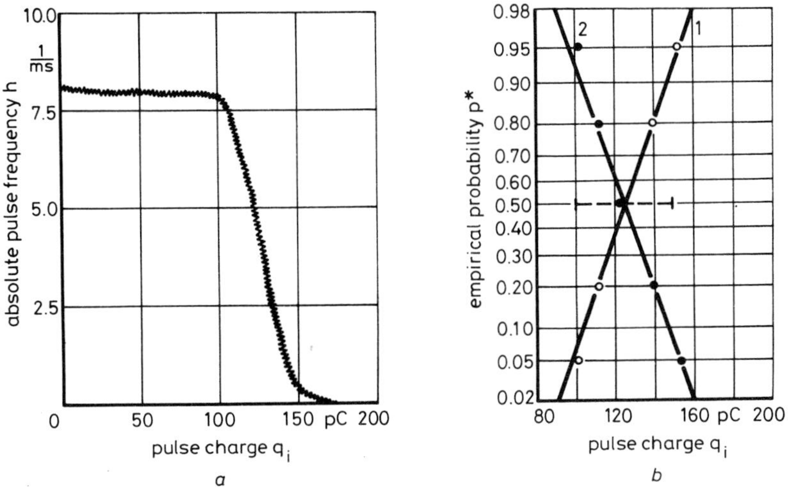

Fig. 5.41 Distribution functions of pulse charge

Point/plane in air

a Measured result of pulse-amplitude analysis
b Distribution function (*1*) and exceeding-frequency (*2*) (Gauss paper)

a mean value over a large number of gate times, but it nevertheless fluctuates considerably (Fig. 1.1). This indeed on the one hand characterises the pronounced randomness of the discharge processes, but on the other hand it can also be caused by the measuring system. If the time interval between two original partial-discharge pulses is very small, the charge pulses are superimposed in the measuring instrument, and the apparent charge will be far too great (Fig. 5.42(*a*)). In the case of intensive or widely branched discharges, and especially in the case of parallel discharge points, a large number of current pulses are often superimposed to form group pulses, whose evaluation in the measuring instrument can also be problematical (Fig. 5.42(*B*, *Ü*)). With partial-discharge measurements, it is largely a matter of the transmission performance and the device's evaluation principle being clear to the measurer and being appropriate to the measuring task. When assessing partial-discharge measurements, it is absolutely essential to take into consideration the evaluable pulse repetition frequency, the integration times (when forming the charge from current and time), the transmission behaviour and especially the effect of disturbing signals. When one can ensure that the partial-discharge measuring instrument does not affect the result and that the superimposition of measured quantities does not occur, statistical methods — e.g. the enlargement law — can be successfully applied to the results of partial-discharge measurements. It is an advantage to evaluate not just one measured quantity, but to use partial-discharge measuring systems and to measure as many characteristic quantities as possible. The suitability of the partial-discharge measuring instrument will depend

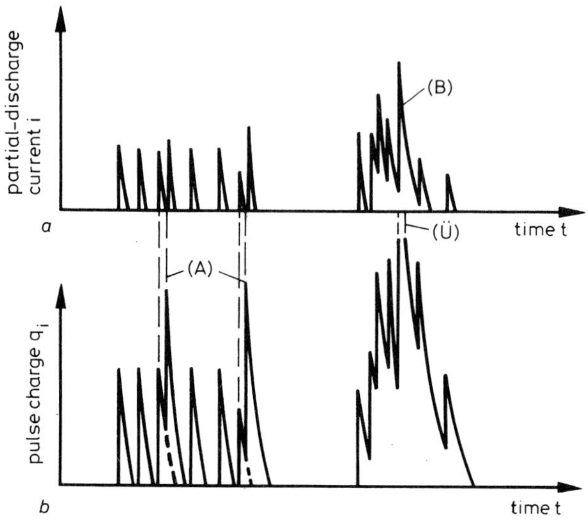

Fig. 5.42 Production of superimposition in case of pulse charge (*A*) and of group pulses in current (*B*), together with extreme values of charge (*Ü*) (diagrammatic).

on how true to the original the electronic processing of the partial-discharge signals is, and how well spurious signals can be suppressed. Since interference suppression is becoming more efficient, better information about the condition of insulation can certainly be anticipated in the future from broadband partial-discharge measuring instruments than from narrowband instruments, which greatly distort the original current pulses prior to amplification.

Chapter 6

Enlargement law

6.1 Problem

Up to now, the basis has always been that a single discharge in the insulation arrangement concerned develops into breakdown in a random manner, this random process being described by the distribution function of a suitably specified variate (U_d, E_d, T_d, etc.).

If, however, the insulation arrangements are large and the voltage sources are sufficiently powerful, the discharges which may possibly produce breakdown can develop in parallel, both in space and time. The discharge point at which discharge develops most quickly in a random manner will produce breakdown of the entire arrangement. Examples of large insulation arrangements consisting of discrete individual insulating components include the large numbers of insulators in parallel in transformer substations and on overhead lines, the winding insulation in parallel on the high-voltage side in power transformers, instrument transformers, reactors and rotating machinery, or the insulation of the large number of individual capacitors in high-voltage test plants and compensation equipment. Large insulation arrangements with a continuous structure include, for example, the insulation of cables, SE_6-insulated switchgear and tubular gas-filled cables, as well as on large-area screening electrodes in high-voltage test plants. In both cases, the basic problem is that, in the laboratory or test plant, it is always only the performance function of an individual insulating component or a part of a continuous insulation arrangement that is determined, whereas it is the performance function of the large-volume insulation itself that is relevant to practical applications. The application of laboratory measurements to models when dimensioning practical insulation arrangements also suggests that it is desirable to describe the effects of the increase in volume of an insulation arrangement (e.g. in the case of a cable, not just the extension, but also the increase in its diameter) for the performance function. The problems are similar when extending the duration of voltage applications to insulation arrangements, say when comparing test durations in the laboratory with the duration of stress in the system. Conversely, one may be interested in a reduction of the three-dimensional dimensions of an insulation arrangement or the shortening of the stress duration.

From a statistical standpoint, all these questions can be dealt with using the enlargement law, which represents a practical application of the multiplication law for non-dependent probabilities (eqn. 2.12) $P(A \cap B) = P(A)P(B)$. (The term enlargement law, 'Vergrösserungsgesetz', is attributable to Widmann. The expression 'growth law' (Wachstumsgesetz) is also used (by

Dokopoulos), as is 'area or volume laws' (Flächen- oder Volumengesetze).) The non-dependence of the discharge processes, which take place in parallel with respect to space and time, or consecutively when the time is extended, is of course assumed. In many practical cases the assumption is indeed justified, but discharges occurring physically close together produce mutual influences (dependences), which prevent the enlargement law from being used. In the case of time extensions, the assumption of complete independence of consecutive processes appears particularly questionable: the successive destruction (life consumption) of solid insulation is a process precisely presupposing dependences in a statistical sense. Handling insulation life problems purely with the enlargement law will consequently only be successful in particular cases.

When investigating any enlargement effects, the question of non-dependence must first of all be clarified without fail, the corresponding tests (see Section 2.5.3) only being able to supplement physical investigations. If independence is assured, the relationships subsequently represented can be fully applied to the problem. To produce a clearer presentation, the enlargement law for major theoretical distribution functions will be considered after the statistical fundamentals. This will be followed by separate presentation of the parallel connection of discrete insulation components, the enlargement of continuous installation arrangements with an electrode area determining discharge (area effect) and with volume-determining discharge (volume effect), as well as consideration of the extension of the duration of stress (time effect).

6.2 Statistical fundamentals

To handle statistically the enlargement or reduction of an arrangement with respect to space (volume V) and time (T), a four-dimensional enlargement factor n is introduced, representing the relationship between the enlarged or reduced state (V_n, T_n) and the initial state (V_1, T_1):

$$n = \frac{V_n T_n}{V_1 T_1}, \qquad 0 < n < +\infty. \tag{6.1}$$

In the following the starting point will only be this enlargement factor n, but naturally still including all special cases of the three-dimensional, two-dimensional or one-dimension enlargement factor, such as volume effect ($N = V_n/V_1$; $T_n = T_1$), area effect ($n = A_n/A_1 = V_n/V_1$; $T_n = T_1$), time effect ($n = T_n/T_1$; $V_n = V_1$) and the parallel connection of discrete insulation components (n is the number of parallel insulation components).

When insulation having a breakdown probability $p_1 = F_1(x_0)$ and a stress x_0 (e.g. $x_0 = u_0$) is enlarged by a factor n, the breakdown probability $p_n = F_n(x_0)$ is derived directly from the multiplication law (eqn. 2.12), consideration being given to the complementary non-breakdown events. Non-breakdown of the enlarged arrangement $(1-p_n)$ presupposes the non-breakdown of all the initial arrangements $(1-p_1)_i$; $i = 1, \ldots, n$:

$$(1-p_n) = (1-p_1)_1(1-p_1)_2 \ldots (1-p_1)_n, \tag{6.2}$$

i.e. because $(1-p_1)_1=(1-p_1)_2=\ldots=(1-p_1)_n$

$$p_n=1-(1-p_1)^n. \tag{6.3}$$

If one considers the complete distribution functions $F_1(x)$ and $F_n(x)$, rather than a discrete probability, then:

$$F_n(x)=1-(1-F_1(x))^n. \tag{6.4}$$

Enlargement has up to now assumed identically distributed variates in the n elements concerned. This is not always so in practice, however: the breakdown probabilities are different, for example (p_{1i} with $i=1,\ldots,n$) when there is a change in the field-strength distribution with an increase in volume, say when the radius is increased in a coaxial system. Eqn. 6.2 then assumes the form:

$$(1-p_n)=(1-p_1)(1-p_2)\ldots(1-p_i); \qquad i=1\ldots n, \tag{6.5}$$

from which the general form of the enlargement law for discrete insulation components can be derived:

$$p_n=1-\prod_{i=1}^{n}(1-p_{1i}) \tag{6.6}$$

or

$$F_n(x)=1-\prod_{i=1}^{n}(1-F_{1i}(x)). \tag{6.7}$$

The use of eqns. 6.3, 6.4, 6.6 and 6.7 presents no problems in the case of discrete individual insulation components (see Section 6.4) if only $F_1(x)$ or $F_{1i}(x)$ are known. With continuous insulation arrangements, however, the use of this expression calls for discretisation with a certain degree of randomness. This is easy, for example, when a coaxial cylinder arrangement is extended by the addition of n identical cylindrical sections (Fig. 6.1*a*) with the same breakdown probability, but presents great difficulties if enlargement is accompanied by a change in the field-strength distribution. If, say, the coaxial cylinder arrangement is enlarged in a radial direction, its length remaining constant, one must assume randomly selected cylindrical 'shells' of different dimensions (Fig. 6.1*b*), which will naturally also have a different breakdown probability. It is consequently an advantage to dispense with random discretisation in such cases and to stick to the continuous structure by using integral calculus. Naturally, only a random volume element V_1 of a uniform field can be used as a reference element for a freely selected reference time T_1 (reference quantity V_1T_1). No nonuniform field will be suitable as a reference-and-comparison element, because of its nonuniform field-strength distribution.

Example 6.2.1 The maximum breakdown field strengths are measured on model polythene cables with a coaxial field, with different internal radii, lengths and insulating material thicknesses. The methods described below can be used to calculate the distribution function for a reference element stressed in the uniform field, e.g. the quantity $V_1=1\text{ cm}^3$, from the empirical

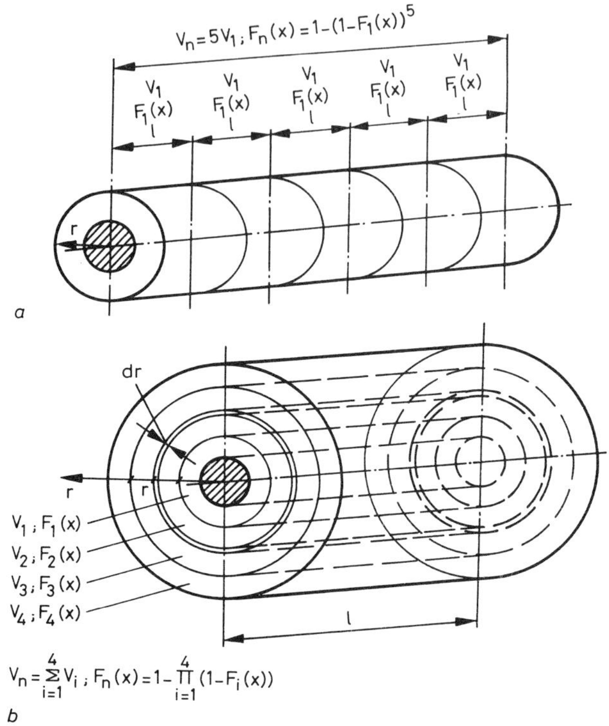

Fig. 6.1 Enlargement of a coaxial cylinder arrangement

a Enlargement in axial direction (extension)
b Enlargement in radial direction

distribution functions of the various model cables. The distribution-function parameters calculated for V_1 are compared with each other. On the basis of the distribution function derived in this way for V_1, the performance functions can be calculated for any insulation arangements, provided the statistical model used is valid.

The reference element $(V_1 T_1)$ is assigned a distribution function of the variate X used, with the parameters α and β, which are initially still unknown and are determined as indicated in Example 6.2.1 (see Section 6.6):

$$F_1(x) = F_1(x;\ \alpha;\ \beta).$$

For any differentially small elements of size $(dv\ dt)$ in the nonuniform field concerned, a variate X_e must be introduced, for which a physically expedient relationship to variate X of the reference element can be produced. The variate's maximum breakdown field strength $X = E_{dh}$ and the electric strength $X = E_d$ are particularly suitable. One should generally assume a relationship between X_e and X with a geometry and/or time function $f(v;\ t)$

in the form

$$X_e = Xf(v;\, t) = X_e(U_d;\, v;\, t). \tag{6.8}$$

The variate X_e will furthermore have a distribution identical to that of reference element X, i.e. $F(x_e;\, \alpha;\, \beta)$ applies to the differential element, with x_e according to eqn. 6.8. The arrangement as a whole consists of an infinite number of differential elements, all of which have to be taken into consideration.

Before one can integrate, eqn. 6.7 must be transformed for discrete elements:

$$\ln\,(1 - F_n(x)) = \ln \prod_{i=1}^{n} (1 - F_{1i}(x)) = \sum_{i=1}^{n} \ln\,(1 - F_{1i}(x)) \tag{6.9}$$

with

$$n = \frac{V_n T_n}{V_1 T_1}.$$

When changing over to differentially small elements $(dv\, dt)$, $F_{1i}(x) \to F(x_e;\, \alpha;\, \beta)$ and the integration for the generalised enlargement law is derived directly from the summation in eqn. 6.9:

$$\ln\,(1 - F_n(x)) = \frac{1}{V_1 T_1} \int^{T_n} \int^{V_n} \ln\,(1 - F(x_e;\, \alpha;\, \beta))\, dv\, dt, \tag{6.10}$$

$$F_n(x) = 1 - \exp\left[\frac{1}{V_1 T_1} \int^{T_n} \int^{V_n} \ln\,(1 - F(x_e;\, \alpha;\, \beta))\, dv\, dt\right]. \tag{6.11}$$

The representation of the enlargement law for insulation arrangements with a continuous structure, which is here necessarily rather abstract, is illustrated by application to the area effect (see Section 6.5), volume effect (see Section 6.6) and time effect (see Section 6.7).

6.3 Application to theoretical distribution functions

We shall now apply the general relationship described by eqns. 6.4 and 6.7 to selected theoretical distribution functions. One is first struck by the basic differences produced by the enlargement law when applied to different types of distribution: if the processes producing breakdown are not random and are therefore deterministically described, i.e. by a single-point distribution, then there are no enlargement effects (Fig. 6.2*a*). The deterministic value applies to any quantities of the arrangement concerned.

In a distribution function $F_1(x)$ with a lower limit (lower limit x_0, e.g. log-normal distribution, Weibull distribution), the distribution function for enlargement factors $n \to \infty$ converges to a single-point distribution at x_0 (Fig. 6.2*b*). The initial value x_0 thus assumes major significance, since it represents the point below which, on insulation of any size, no further breakdown is possible. If x_0 is incorrectly (too high, say) estimated from a sample, the technical consequences could be considerable.

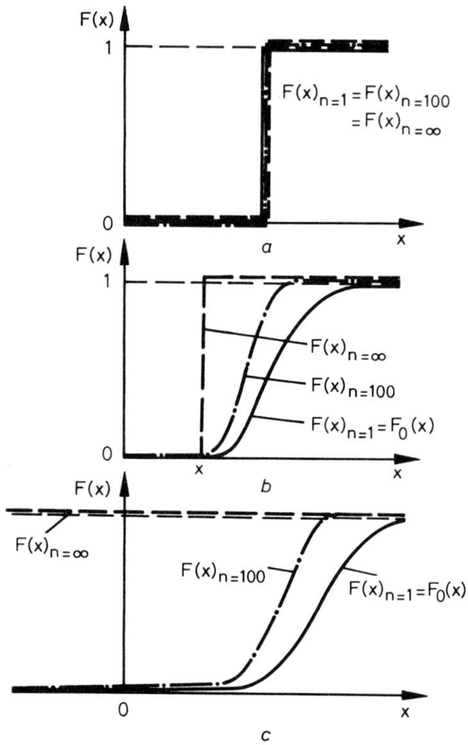

Fig. 6.2 Application of enlargement law to various types of distribution (diagrammatic)

a Single-point distributions
b Distribution functions with lower limit
c Unlimited distribution functions

In a distribution function $F_1(x)$ with no lower limit (e.g. normal distribution, double exponential distribution), the distribution function $F_n(x) \rightarrow 1$ for $n \rightarrow \infty$ (Fig. 6.2*c*). In technical problems, one can rarely reckon on enlargement factors $n > 10^4$, so that this borderline case is itself of minimal relevance. When applying the enlargement law with large enlargement factors to unlimited distributions, there is always a danger that distribution functions will be obtained which are excessively pessimistic in a technical sense or are even physically nonsensical (e.g. with a 'negative' mean value).

Extreme-value distributions (see Sections 2.3.2.4 and 2.3.2.5) have a mathematical model based on the enlargement law. If the enlargement law is applied to extreme-value distributions, with identical elements, the type of distribution is preserved, and only one parameter changes. If the enlargement law is applied to other distributions, the type of distribution will change with the enlargement.

If a double-exponential distribution $F_1(x) = 1 - \exp[-\exp((x - x_{163})/\gamma)]$ is substituted for identical initial distributions $F_1(x)$ in the enlargement law (eqn. 6.4), it follows that

$$F_n(x) = 1 - \exp\left[-\exp\left(\frac{x - (x_{163} - \gamma \ln n)}{\gamma}\right)\right], \tag{6.12}$$

where x_{163} and γ are the parameters of the initial distribution, and n is the enlargement factor. When there is an enlargement, only the mode changes to

$$x_{n63} = x_{163} - \gamma \ln n, \tag{6.13}$$

while the dispersion $\gamma_1 = \gamma_n = \gamma$ remains the same. In the probability grid, the distribution functions derived from the standardised double exponential distribution ($x_{163} = 0$; $\gamma = 1$, Table 2.17) are parallel to the initial distribution

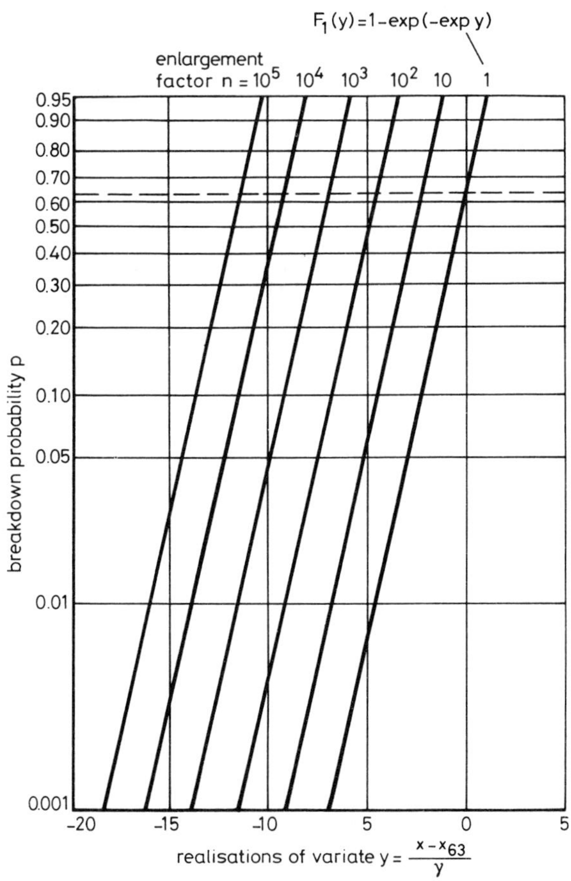

Fig. 6.3 Enlargement law for double exponential distribution, represented on double exponential paper

$F_1(x)$ (Fig. 6.3). The change in the mode accompanying enlargement (eqn. 6.13) is illustrated in relative form in Fig. 6.4.

If n elements are connected in parallel, for which the variates are distributed double-exponentially with different parameters (x_{1i63}; γ_{1i}), then application of the generalised enlargement law (eqn. 6.7) gives

$$F_n(x) = 1 - \exp\left[-\sum_{i=1}^{n} \exp\left(\frac{x - x_{1i63}}{\gamma_{1i}}\right)\right]. \tag{6.14}$$

Here the distribution function of the enlarged arrangement can be fully calculated, but the type of distribution of the double exponential distribution is lost.

If a Weibull distribution $F_1(x) = 1 - \exp[-((x - x_0)/\eta_1)^\delta]$ is substituted for identical initial distributions in the enlargement law (eqn. 6.4), it follows that:

$$F_n(x) = 1 - \exp\left[-\left(\frac{x - x_0}{\eta_1}\right)^\delta n\right], \tag{6.15}$$

where η_1, x_0 and δ are the parameters of the initial distribution. In the case of enlargement by a factor of n, the 63% quantile only changes to:

$$\eta_n = (x_{nm} - x_0)_{63} = \eta_1 n^{-1/\delta} = (x_{1m} - x_0)_{63} n^{-1/\delta}, \tag{6.16}$$

while the initial value x_0 and the Weibull exponent δ remain the same. In the probability grid, the distribution functions derived from the reduced Weibull distribution ($\eta_1 = 1$, $x_0 = 0$, Table 2.10) are parallel to the initial distribution $F_1(x)$ (Fig. 6.5). The change in the mode when there is an enlargement (eqn. 6.16) is illustrated in relative form in Fig. 6.6.

In the case of the parallel connection of n elements with a Weibull-distributed variate, but with different parameters η_{1i}, x_{0i} and δ_i, then, by

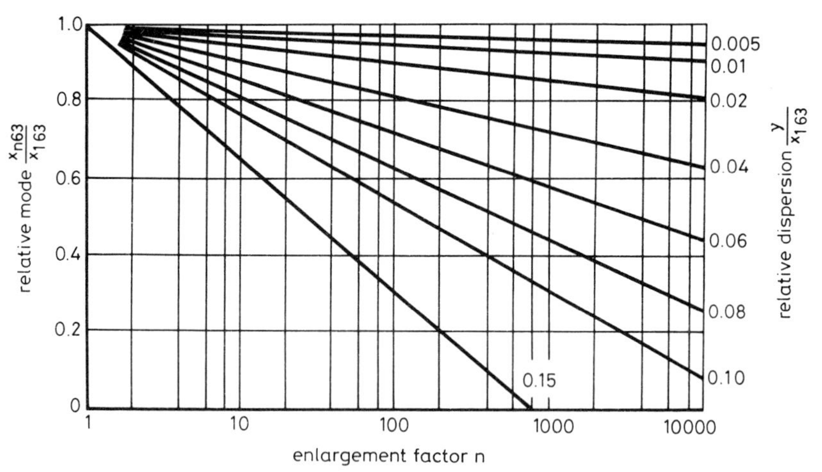

Fig. 6.4 Parameter x_{n63} of double exponential distribution, as a function of enlargement factor n

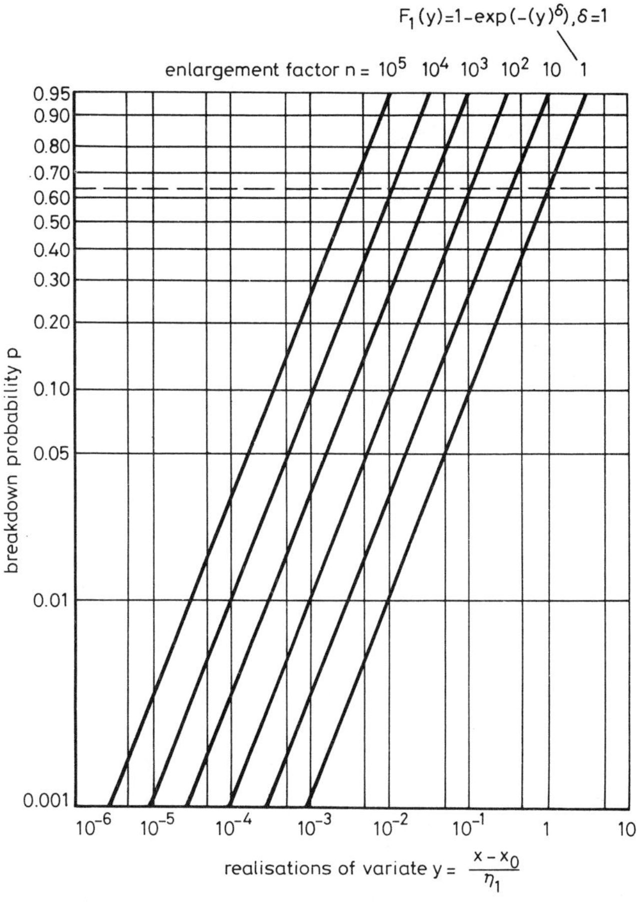

Fig. 6.5 Enlargement law for Weibull distribution, represented in probability grid (Weibull paper)

analogy to eqn. 6.14, the generalised enlargement law gives:

$$F_n(x) = 1 - \exp\left[-\sum_{i=1}^{n}\left(\frac{x - x_{0i}}{\eta_{1i}}\right)^{\delta_i}\right]. \tag{6.17}$$

In this case, as with the double-exponential distribution, the type of distribution is lost.

If a normal distribution (eqn. 2.71) is substituted in the enlargement law (eqn. 6.4), a full solution for the problem cannot be produced: numerical solutions to the problem are required. Corresponding results (from Schrader) illustrate the fact that, with a representation in a probability grid (Fig. 6.7), with a normally distributed initial distribution, the distribution functions of an enlarged arrangement are no longer normally distributed.

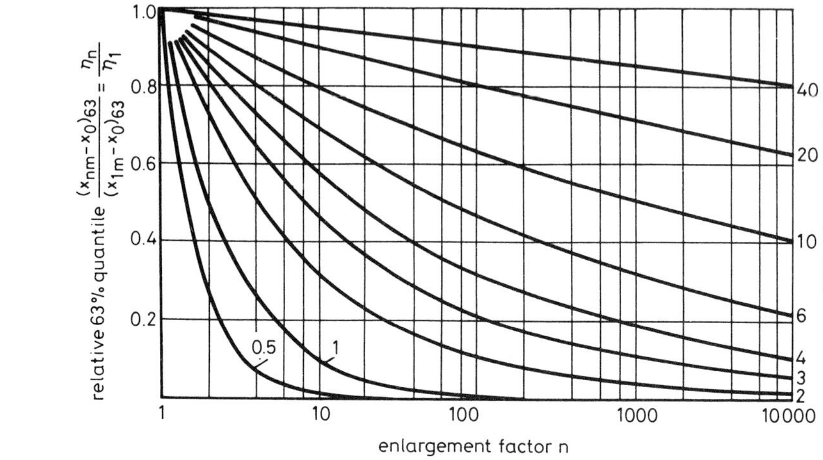

Fig. 6.6 Parameter $\eta_n = (x_{nm} - x_0)_{63}$ of Weibull distribution, as a function of enlargement factor n

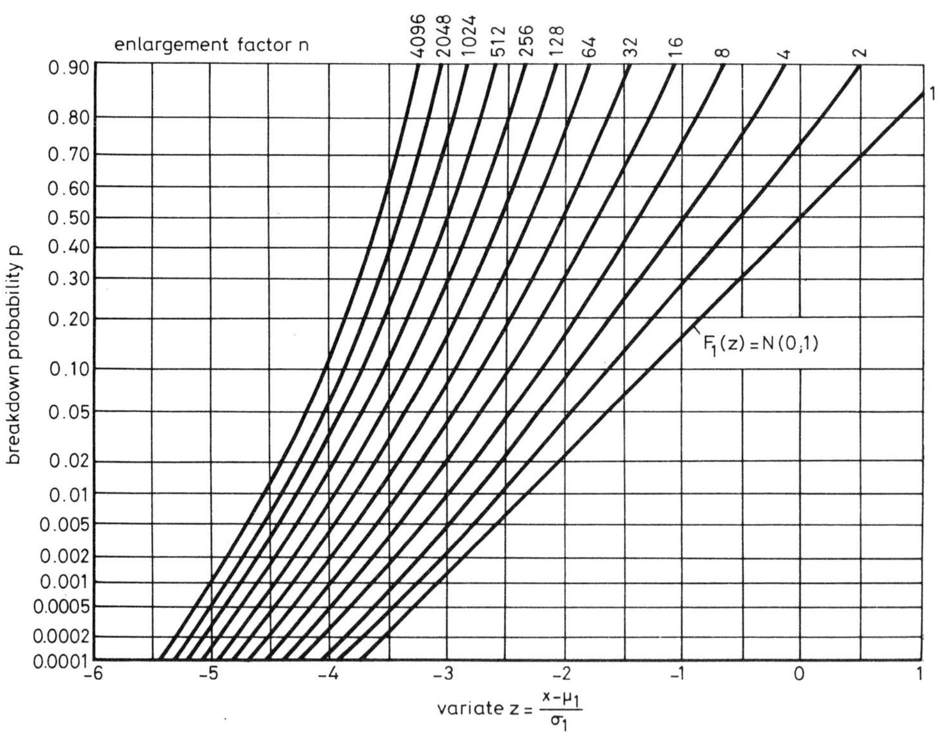

Fig. 6.7 Enlargement law for normal distribution, represented in a probability grid

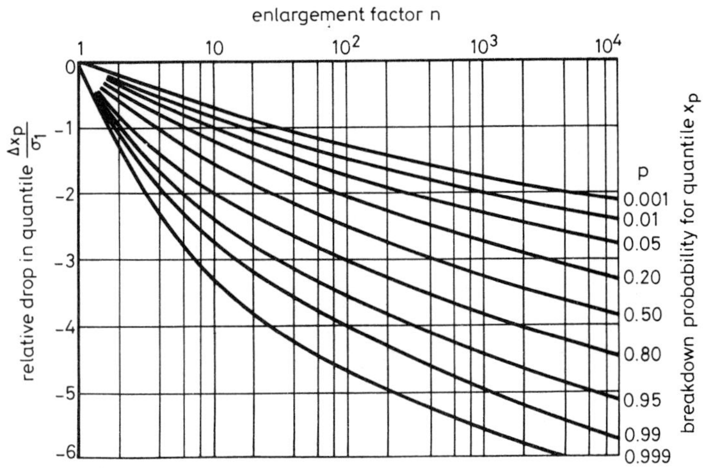

Fig. 6.8 Drop in characteristic quantiles of a normal distribution ($\Delta x_p/\sigma_1$), as a function of enlargement factor n

It can be shown that, with an increasing enlargement factor n, the distribution functions $F_n(x)$ converge towards a double exponential distribution (see Section 2.3.2.5). The curvature of the curves in the probability grid illustrates this convergence (Fig. 6.7).

Using a nomogram (Fig. 6.8) provided by Schrader, based on a standardised normal distribution, characteristic quantiles x_{np} can be determined for enlargement up to $n = 10^4$. The relative drop in the quantile ($\Delta x_p/\sigma_1$) indicated in the nomogram gives directly, for an initial distribution $N(\mu_1; \sigma_1)$ with the quantiles x_{1p}, the required quantile:

$$x_{np} = x_{1p} - \left(\frac{\Delta x_p}{\sigma_1}\right) \sigma_1 . \tag{6.18}$$

If, however, the distribution function $F_n(x) \neq N(\mu; \sigma^2)$ obtained point-by-point using eqn. 6.18 is replaced by a normal distribution, then with the factors a and b according to Table 6.1 and the parameters of the initial

Table 6.1 Factors a and b for approximating distribution function of enlarged arrangements by a normal distribution, using eqns. 6.19 and 6.20 from Pearson

Enlargement factor n	1	10	10^2	10^3	10^4
Factor a	0	1.54	2.50	3.25	3.85
Factor b	1	0.59	0.43	0.35	0.30

distribution $N(\mu_1; \sigma_1^2)$ we obtain:

$$\mu_n = \mu_1 - a\sigma_1, \tag{6.19}$$

$$\sigma_n = b\sigma_1. \tag{6.20}$$

In the case of parallel connection of elements whose variates are in fact

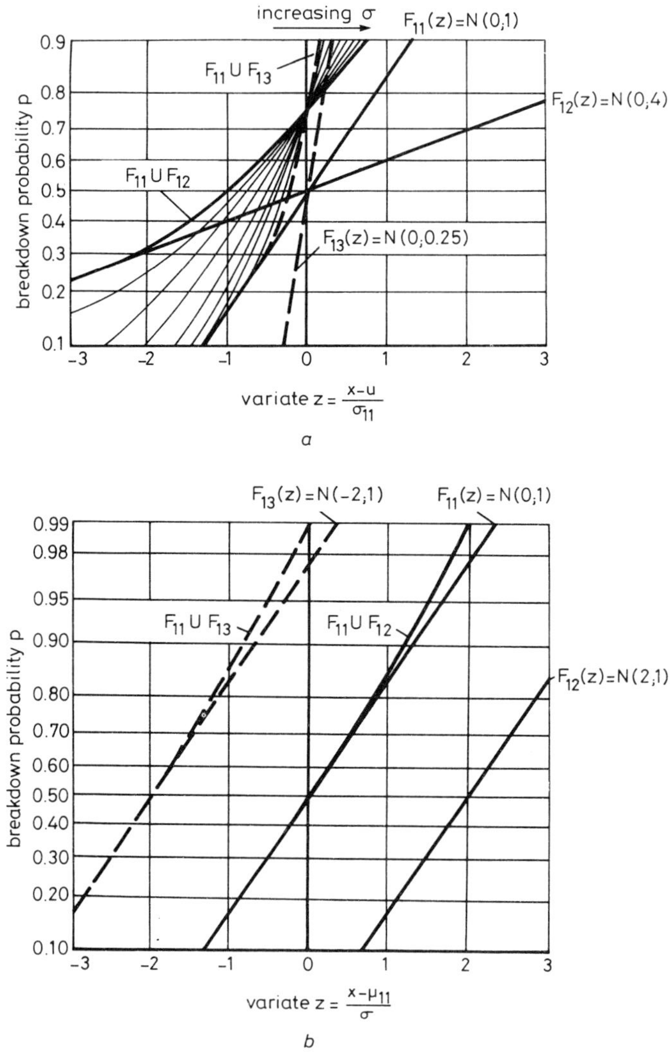

Fig. 6.9 Resultant distribution functions for parallel connection of two elements with normally-distributed variates

a) $\mu_{11} = \mu_{12} = \mu_{13} = \mu$; $\sigma_{11} \neq \sigma_{12} \neq \sigma_{13}$
b) $\mu_{11} \neq \mu_{12} \neq \mu_{13}$; $\sigma_{11} = \sigma_{12} = \sigma_{13} = \sigma$

normally distributed but have different parameters (μ_{1i}; σ_{1i}^2), conditions are quite complicated if the problem is to be handled, not point by point (see Section 6.4), but as far as possible in a complete manner. For the special case of only two parallel-connected insulation units, Fig. 6.9*a* shows the resultant distribution functions for two cases with $\mu_{11} = \mu_{12}$; $\sigma_{11} \neq \sigma_{12}$, and Fig. 6.9*b* the resultant distribution functions for $\mu_{11} \neq \mu_{12}$; $\sigma_{11} = \sigma_{12}$. In each case, the resultant distribution function is always determined by the initial distribution that has the greater breakdown probability in the section concerned.

The relationships described for a normal distribution can be directly applied to log-normally distributed variates *Y*, if one works with the transformed variate $X = \log Y$. Use of the nomogram in Fig. 6.8 and of eqn. 6.18 must be followed by transformation back to variate *Y* (see Section 2.3.2.3). As a limited distribution function (initial value $y_0 = 0$), the log-normal distribution converges towards a two-parameter Weibull distribution (initial value $x_0 = 0$) as the enlargement factor *n* is increased.

The enlargement law is applied by Dokopoulos to additive mixed distributions derived from Weibull distributions and by Schrader to additive mixed distributions consisting of normal distributions. Fig. 6.10 demonstrates that even a very small proportion (0.8%) with a small mean value in the mixed distribution is sufficient to produce exceptional reductions of $F_n(x)$ (cf. Fig.

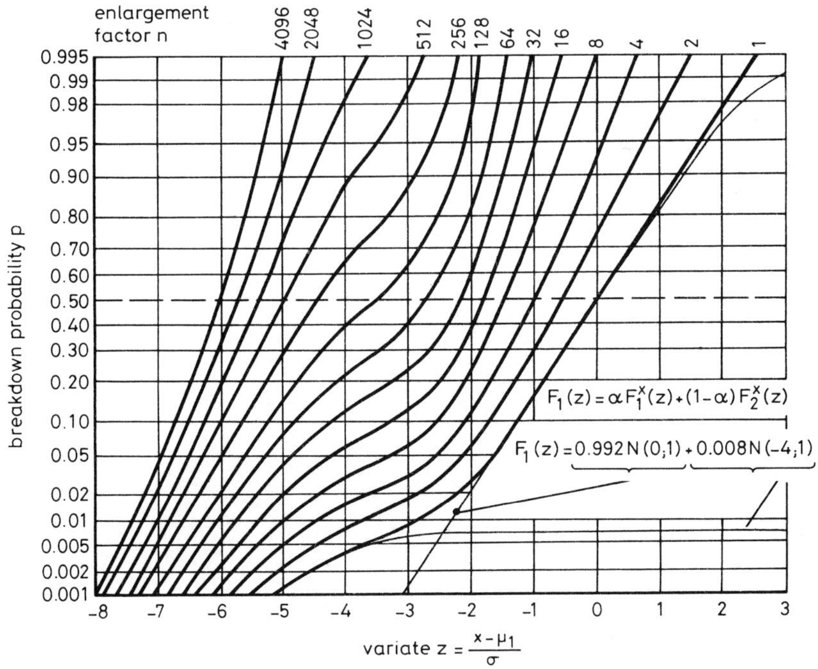

Fig. 6.10 Enlargement law for an additive mixed distribution consisting of two normal distributions in a probability grid

6.7). If the initial mixed distribution is broken down into its components (e.g. normal distributions), then the relationships dealt with for distribution functions can be applied to the components.

So-called multiplicative mixed distributions are obtained simply by using the enlargement law (e.g. eqns. 2.14 and 2.17). They do not, therefore, form basic initial distributions, but are themselves composed of such. For them, as for all other distribution functions, the distribution functions of the enlarged arrangements can either be fully calculated using eqn. 6.4 or can be determined point by point using eqn. 6.3 — preferably with the nomogram (Fig. 6.11) explained in Section 6.4.

6.4 Discrete parallel discharge points

In the case of parallel connection of discrete, independent discharge points, the enlargement factor is one-dimensional and is equal to the countable number n of discharge points. Discrete discharge points may be conceived, on the one hand, as all insulation arrangements composed of individual elements (parallel insulators, parallel turns, etc.), and, on the other, all clearly 'discretisable' continuous insulation arrangements (cables, coaxial metal-clad busbars etc., when extended), wherein any chosen technically

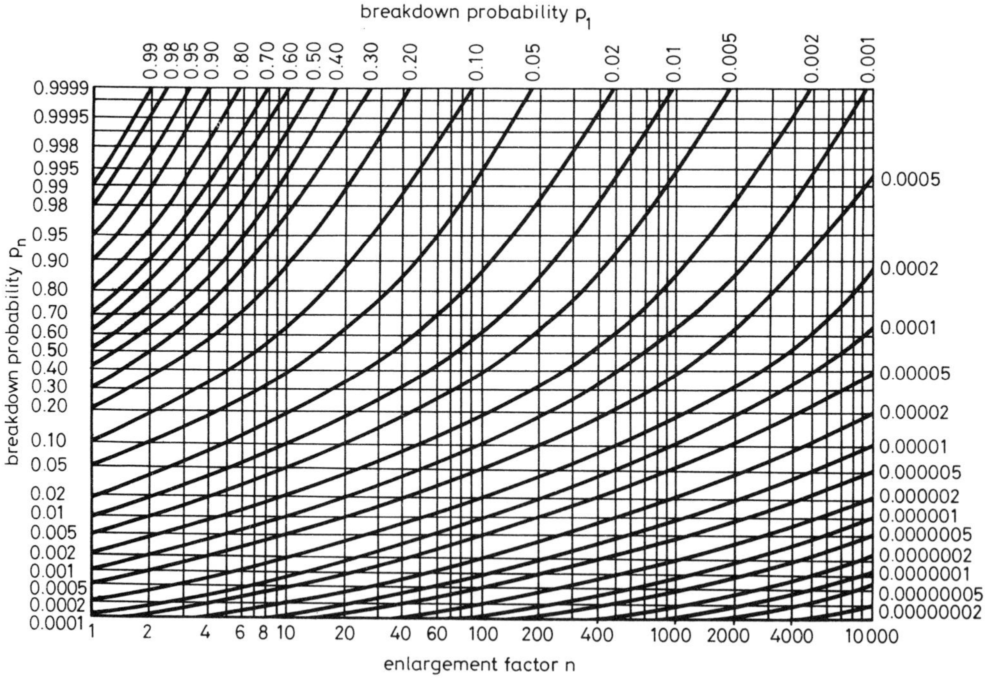

Fig. 6.11 Breakdown probability and enlargement factor for identical parallel elements

meaningful component (e.g. section of cable 2 m long, tested in the laboratory) can be used as a reference element for the latter. If such a discretisation of continuous insulation arrangements is possible, then treating the enlargement effect as a parallel connection is always preferable to continuous consideration as an area effect or volume effect (see Sections 6.5 and 6.6), since the calculation is far simpler.

Eqns. 6.3 and 6.4 are the basis in the case of parallel connection of identical elements. With extremely small breakdown probabilities ($p_1 \ll 1/n$), the simple expression

$$p_n = np_1 \tag{6.21}$$

or

$$F_n(x) = nF_1(x) \quad \text{in the range} \ll \frac{1}{n} \tag{6.22}$$

is obtained approximately from series developments, for the enlarged insulation.

Evaluation of point by point consideration (eqns. 6.3 and 6.21) is represented in Fig. 6.11, from which the breakdown probability for $n \leqq 10^4$ parallel-connected elements (p_n) can be directly obtained for a given breakdown probability of the individual element (p_1). The probability p_n applies to the entire arrangement and characterises the breakdown of any element, but two or more elements may break down simultaneously with a certain, very small probability.

In the first place, use of these relationships permits conversion from one element to the distribution function of n parallel elements. The enlargement law based on a normal distribution (Fig. 6.12) is applied to parallel point spark gaps, but parallel spacers in a vacuum must be based on a Weibull distribution. The 'discretisation' of a coaxial cylinder arrangement into elements of length l_1, which are connected in parallel in the case of extension ($n = l_n/l_1$), is a particularly important practical problem for cables and metal-clad busbars. In the case of model polythene cables, the applicability of the enlargement law was confirmed on the basis of a Weibull distribution for breakdown voltage and breakdown time (Fig. 6.13), whereas the basis for the electric strength of SF_6-insulated coaxial cylinders has to be a double exponential distribution (Fig. 6.14). The number of examples is endless.

Reversal of the relationships is also possible, moreover, the following being derived for one element $F_1(x)$ from the distribution function $F_n(x)$ for n parallel elements:

$$p_1 = 1 - \sqrt[n]{1 - p_n}, \tag{6.23}$$

$$F_1(x) = 1 - \sqrt[n]{1 - F_n(x)}. \tag{6.24}$$

If the breakdown probability is to be determined for an insulation arrangement down to very small values, it is advisable to investigate n parallel insulating components and to arrive at a conclusion for the required insulation using eqn. 6.23 or eqn. 6.24.

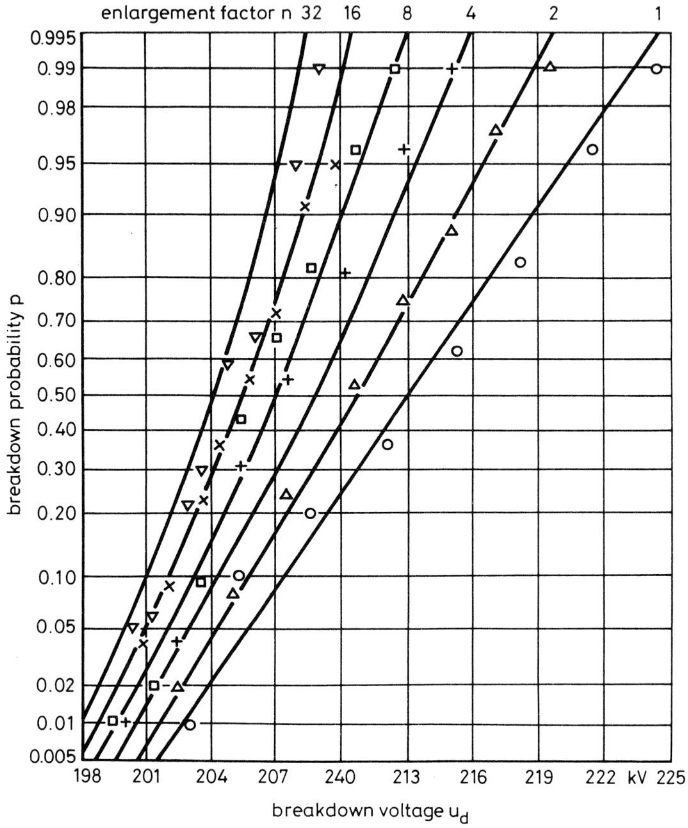

Fig. 6.12 Empirical enlargement law for parallel air spark-gaps

Point/plane, $d = 40$ cm
Lightning voltage 1.2/50

Example 6.4.1 The performance function of the lightning breakdown voltage is to be determined down to a value $p_1 = 0.0005$ for a point/plane spark gap ($d = 40$ cm). This would require at least $j = 2000$ voltage applications. Expense can be reduced considerably by calculating back point by point to one spark gap (Fig. 6.15), using eqn. 6.23, on the basis of the empirical breakdown probabilities ($j = 100$) reproduced in Fig. 6.12 for $n = 32$, 16, 8, 4 and 2 parallel spark gaps. It is, of course, basically possible to calculate back to the individual spark gap from only one parallel connection (e.g. $n = 32$). Fig. 6.11 can also be advantageously used. To check accuracy, eqn. 6.23 should be applied to the limits of the confidence regions for the breakdown probability determined for example for $n = 32$ spark gaps with $j = 100$ (see Section 2.3.1.3) and be compared with the limits determined directly, i.e. for $j = 2000$.

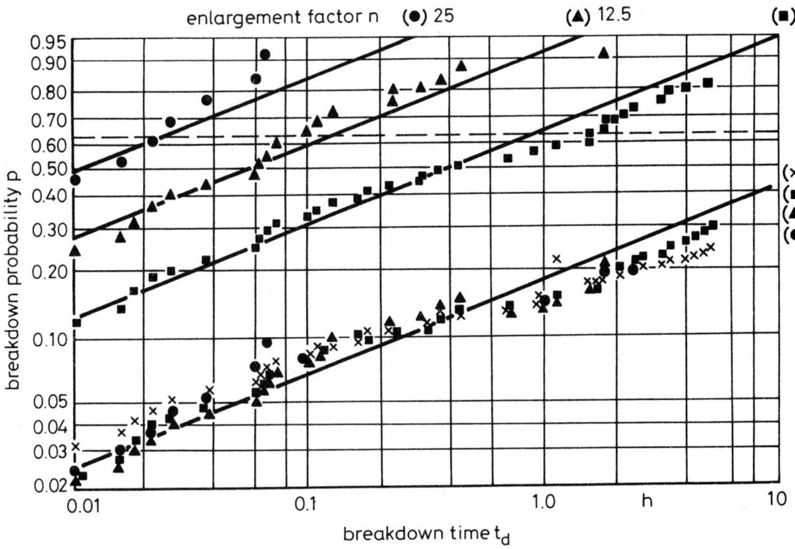

Fig. 6.13 Extension of a polythene cable: enlargement law for Weibull-distributed variate breakdown time
$r_i = 1.5$ mm; $r_a = 3.5$ mm; $l_1 = 1000$ mm

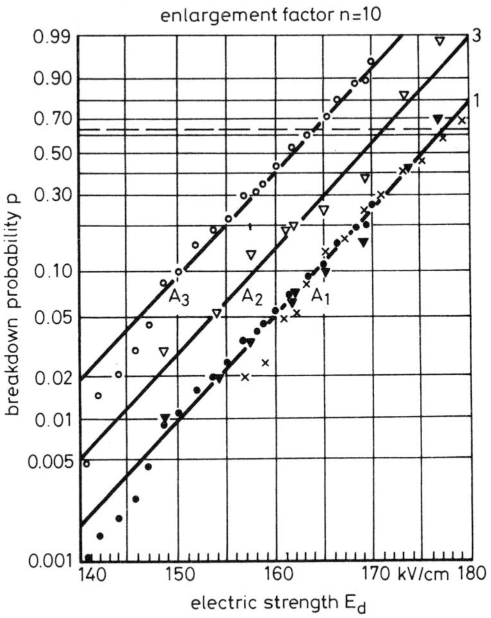

Fig. 6.14 Enlargement law for extension of a coaxial cylinder arrangement in SF_6
$r_i = 6$ mm; $r_a = 20$ mm; $l_1 = 10$ mm
● A_3 values corrected to A_1 $(n = 10)$
▼ A_2 values corrected to A_1 $(n = 3)$

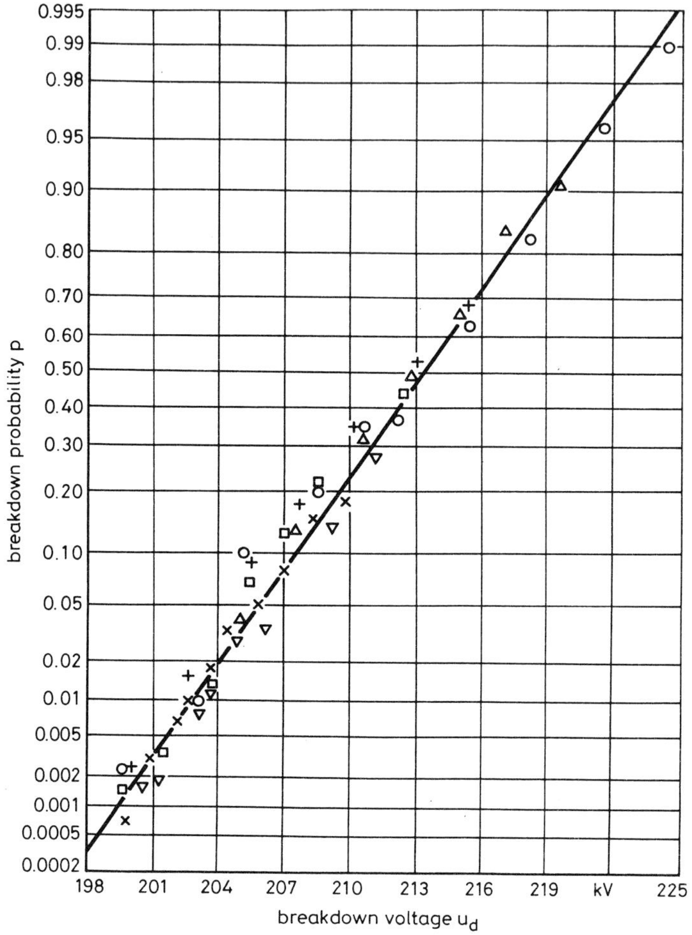

Fig. 6.15 Empirical determination of performance function of an air spark-gap consisting of n parallel gaps (Fig. 6.12)

Comparison of the variation coefficients determined empirically for one and n parallel insulating components can also be used to assess the type of distribution. The relationship between the variation coefficient for the parallel connection $v_n = s_n/\bar{x}_1$, obtained for various theoretical distribution functions (Table 6.2 and graphical representation in Fig. 6.16) is used and a check is made as to which theoretically anticipated characteristic $v_n = f(n)$ corresponds closest to the empirically determined values.

Example 6.4.2 In investigations of the breakdown time of parallel point spark gaps, the following variation coefficients of the breakdown time were

Table 6.2 Dependence of variation coefficient on enlargement factor

Distribution	Conversion from $\left(\frac{s}{\bar{x}}\right)_1$ to $\left(\frac{s}{\bar{x}}\right)_n$	Remarks
(1) Weibull distribution (2 parameters) $x_0 = 0$	$\left(\frac{s}{\bar{x}}\right)_n = \left(\frac{s}{\bar{x}}\right)_1$	Variation coefficient is not dependent on enlargement factor
(2) Weibull distribution (3 parameters)	$\left(\frac{s}{\bar{x}}\right)_n = \left(\frac{s}{\bar{x}}\right)_1 \dfrac{1}{1 + \frac{x_0}{\bar{x}}(\sqrt[\delta]{n} - 1)}$	Case (1) applies to $x_0 = 0$; x_0, δ parameters of Weibull distribution
(3) Double exponential distribution	$\left(\frac{s}{\bar{x}}\right)_n = \left(\frac{s}{\bar{x}}\right)_1 \dfrac{1}{1 - \left(\frac{s}{\bar{x}}\right)_1 \frac{\sqrt{6}}{\pi} \ln n}$	—
(4) Normal distribution	$\left(\frac{s}{\bar{x}}\right)_n = \left(\frac{s}{\bar{x}}\right)_1 \dfrac{k_{1n}}{1 - \left(\frac{s}{\bar{x}}\right)_1 k_{2n}}$	$k_{1n} = \frac{1}{2}({}^{\lambda}\sqrt[n]{0.84} - {}^{\lambda}\sqrt[n]{0.16})$ $k_{2n} = \frac{1}{2}({}^{\lambda}\sqrt[n]{0.84} + {}^{\lambda}\sqrt[n]{0.16})$ ${}^{\lambda}\sqrt[n]{0.84}$ and ${}^{\lambda}\sqrt{0.16}$ are the quantiles of order $\sqrt[n]{0.84}$ and $\sqrt[n]{0.16}$ of an $N(0; 1)$

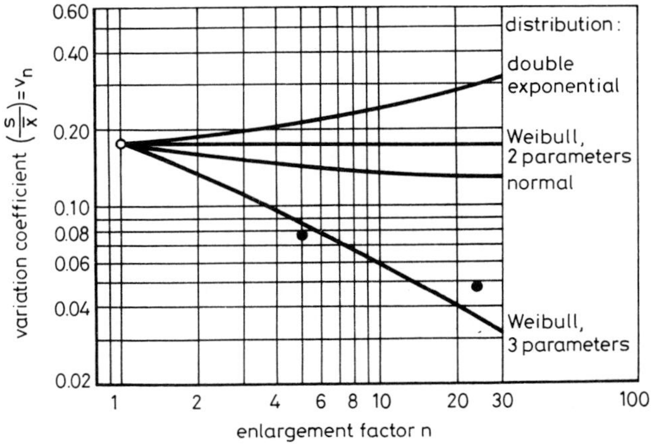

Fig. 6.16 Dependence of variation coefficient on enlargement factor for various theoretical distribution functions
Weibull 3 distribution very variable: here $\delta = 1.7$

determined as a function of the enlargement factor:

$$n = 1: \qquad v_n = 0.175$$

$$n = 5: \qquad v_n = 0.078$$

$$n = 25: \qquad v_n = 0.048.$$

Comparison of the values with the theoretically anticipated trends (Fig. 6.16 from Table 6.2), shows that the empirical characteristic $v_n = f(n)$ corresponds closest to that with a three-parameter Weibull distribution. The breakdown time for the point spark gap investigated is therefore Weibull-distributed.

Related to practical requirements (say different parallel insulators in switching stations or the parallel connection of the insulation of transformer turns having different insulations, the parallel connection of different insulating components is of considerable relevance, but the problem can only be handled on the basis of eqns. 6.6 and 6.7. For the special case of the parallel connection of two insulating components with breakdown probabilities p_{11} and p_{12}, the resultant breakdown probability p_2 can be obtained from the nomogram in Fig. 6.17 (example: $p_{11} = 0.05$ with $p_{12} = 0.15$ gives $p_2 = 0.20$).

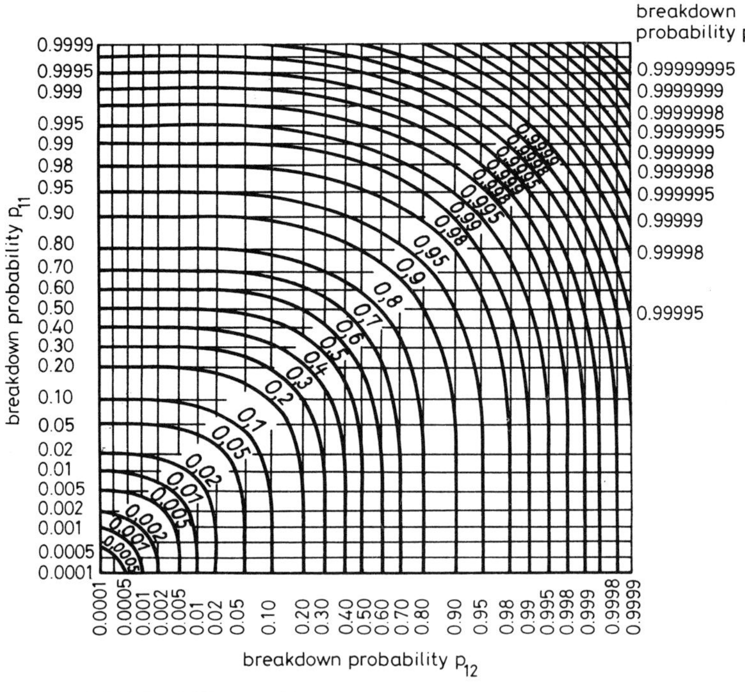

Fig. 6.17 Breakdown probabilities for parallel connection of two different elements

6.5 Area effect

In nonuniform fields, electrical discharges develop to an intensity critical for breakdown in a volume formed on the one hand by the electrode surface (or a part of the electrode surface with a sufficiently high field-strength) and on the other by a section of path Δx along a field line (Fig. 6.18). If Δx is very small in relation to d, or if the field strength hardly changes along the breakdown path, the 'breakdown-effective volume' can be reduced to a 'breakdown-effective area'. Instead of the mathematically cumbersome volume effect (see Section 6.6), it is sufficient to investigate the area effect.

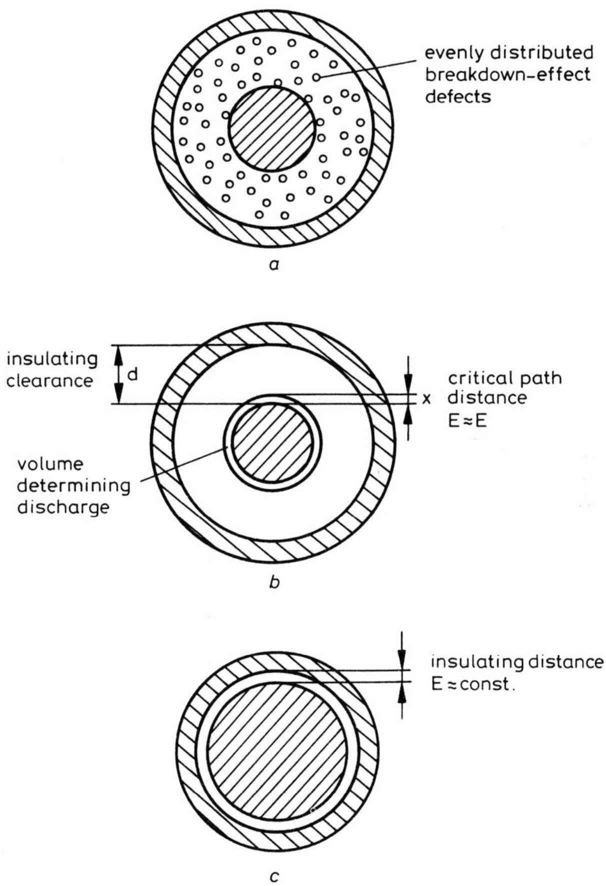

Fig. 6.18 Volume and area effect

a Volume effect (solid material)
b Area effect as a result of $\Delta_x \ll d$ (compressed gas)
c Area effect as a result of $E \approx \text{const.}$ (thin-layer insulation)

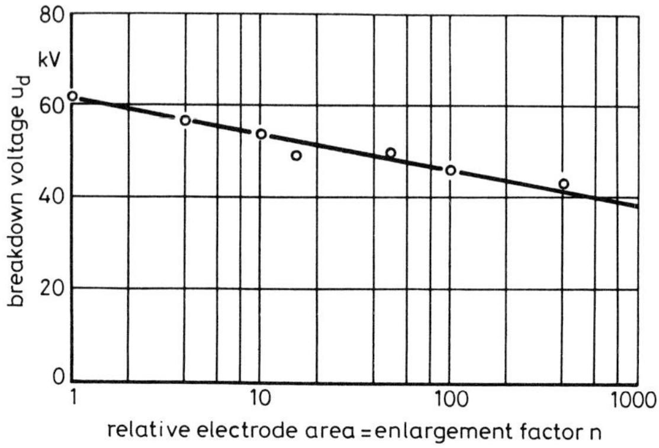

Fig. 6.19 Empirical area effect in insulating oil, described by a double exponential distribution

The ideas associated with the area effect were confirmed experimentally for quasi-uniformly stressed, in most cases thin-layer solid and liquid insulation arrangements. They apply, for example, to laminated insulation in capacitors, cables, bushings, and also to oil channels in transformer insulation (Fig. 6.19), where normally distributed, double-exponential distributed or Weibull-distributed breakdown voltages are mainly assumed. The model of the area effect has, moreover, proved particularly useful in the case of slightly nonuniform compressed-gas insulation, where a critical electron avalanche producing breakdown develops over a section of path which may be less than 1 mm. The use of a double-exponential distribution (Fig. 6.20) and a Weibull distribution has proved most satisfactory.

The mathematical treatment of the area effect presupposes the following assumption concerning the discharge mechanism: breakdown is governed by the field strength prevailing at the electrode surface. If this field strength

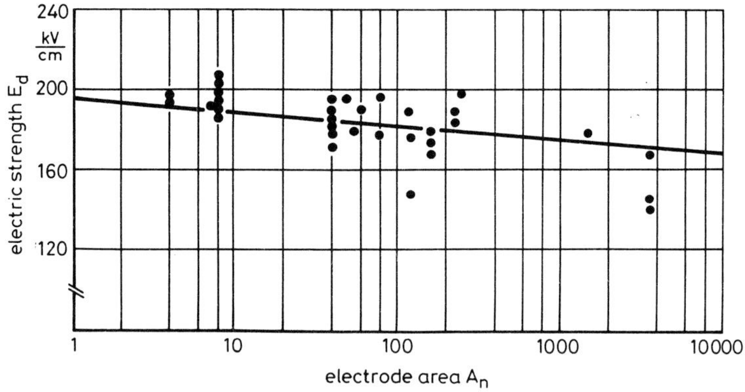

Fig. 6.20 Area effect in SF_6, described by a double exponential distribution

is universally constant for electrode enlargements (e.g. in a coaxial field, Fig. 6.1*a*), appropriate discretisation is advisable, treating the problem as a parallel connection (see Section 6.4).

If, on the other hand, the field strength varies at the electrode surface, eqn. 6.11 must be taken as the basis, now assuming the form:

$$F_n(x) = 1 - \left[\exp \frac{1}{A_1} \int^{A_n} \ln\left[1 - F(x_e;\, \alpha;\, \beta)\right] dA\right]. \tag{6.25}$$

To be in a position to provide solutions to this equation, it must be possible to represent the variate X_e of the differentially small area-element dA as the product of the variate X of the reference element and a geometrical function, i.e. it follows from eqn. 6.8 that:

$$x_e = xf(y, z), \tag{6.26}$$

where y and z are the co-ordinates of the electrode area. The electrode area can be advantageously represented by selecting appropriate co-ordinate systems. The procedure is illustrated by the following example:

Example 6.5.1 The electric strength of SF_6 can be described by a double-exponential distribution, where the parameter values $E_{d63} = 182$ kV/cm and $\gamma = 6$ kV/cm apply to a reference element with electrode area $A_1 = 10\ \text{cm}^2$ in a uniform field at an SF_6 pressure $p_{20} = 0.25$ MPa and a certain degree of electrode roughness. We are required to determine the performance function of the electric strength of a coaxial cylinder arrangement with a hemispherical end (Fig. 6.21*a*), with the dimensions $r_i = 10$ cm, $r_a = 20$ cm and length $l = 10\,000$ cm (e.g. tubular gas cable).

The variate used is the electric strength, i.e. $x_e = E_d(z)$ or $E_d(\phi)$. To be able to compare the various sections of the arrangement concerned, one needs to know the field strength at the surface of the inner electrode (Fig. 6.21*b*). The following characteristic sections can be distinguished:

- the coaxial-cylinder field [length $(l - l_1)$] (1)
- a transitional region (length l_1) (2)
- the concentric spherical field (3)

The maximum field strength $E_h = E_{A3}$ occurs in the spherical field (3). To reduce the volumetric extent of the governing discharge-process to an area problem, the electric strength in section (3) is determined from the maximum breakdown field strength E_{dh3} with a curvature factor $e_{h3} = 1.050$ and used as a reference quantity for the system as a whole (eqn. 6.26):

$$E_{d3} = \frac{E_{dh3}}{e_{h3}} = E_d. \tag{6.27}$$

A strength value satisfying eqn. 6.26 then prevails in the cylindrical field (1) (because $U =$ constant):

$$E_{d1} = \frac{e_{h3}(r_a - r_i)E_d}{e_{h1} r_a \ln \dfrac{r_a}{r_i}} = 0.732 E_d. \tag{6.28}$$

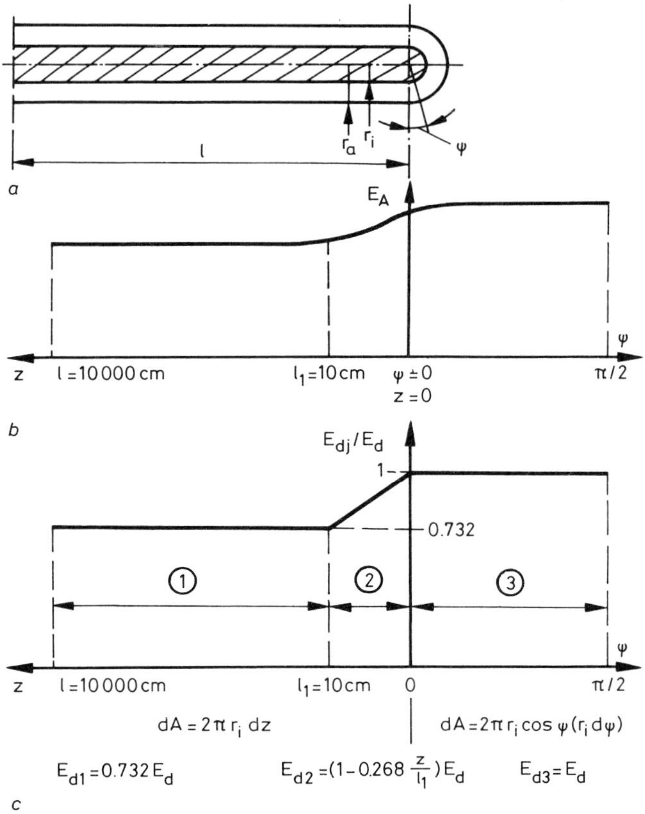

Fig. 6.21 Assumptions when dealing with the area effect in the case of an SF_6-insulated tubular gas-cable (Example 6.5.1)

a Arrangement
b Maximum field-strength at surface
c Strength E_{dj}, as a function of location of area element dA

(If the correction using e_h is dispensed with, there is little change in this numerical value, since $e_{h1} = 1.035$ is close to e_{h3}. The correction is absolutely essential, however, if there is an appreciable difference between the radii of curvature of sections in a system.) A linear transition of the strength from E_{d1} to E_{d3} is assumed for the transitional region (2) (Fig. 6.21*c*), i.e. eqn. 6.26 assumes the form:

$$E_{d2} = \left(1 - 0.268\frac{z}{l_1}\right) E_d. \tag{6.29}$$

To integrate using eqn. 6.25, appropriate co-ordinate systems must be chosen section by section. In sections (1) and (2), one must calculate using cylinder co-ordinates and the area element

$$dA_{1,2}=2\pi r_i\, dz \tag{6.30}$$

whereas in section (3) spherical co-ordinates with

$$dA_3=2\pi r_i \cos\varphi r_i\, d\varphi \tag{6.31}$$

are recommended.

If a double-exponential distribution strength is assumed in the surface elements dA_j (cf. Section 6.3),

$$F(E_{dj})=1-\exp\left[-\exp\left(\frac{E_{dj}-E_{d63}}{\gamma}\right)\right] \tag{6.32}$$

with E_{dj} according to eqns. 6.27, 6.28 and 6.29, eqn. 6.25 assumes the form

$$\begin{aligned}F_n(E_d)=1-\exp\Bigg[\frac{1}{A_1}\Bigg\{&\int_{l_1}^{l}\left[-\exp\left(\frac{0.732E_d-E_{d63}}{\gamma}\right)\right]2\pi r_i\, dz\\&+\int_0^{l_1}\left[-\exp\left(\frac{(1-0.268z/l_1)E_d-E_{d63}}{\gamma}\right)\right]2\pi r_i\, dz\\&+\int_0^{\pi/2}\left[-\exp\left(\frac{E_d-E_{d63}}{\gamma}\right)\right]2\pi r_i^2\cos\varphi\, d\varphi\Bigg\}\Bigg]\end{aligned} \tag{6.33}$$

and the solution

$$\begin{aligned}F_n(E_d)=1-\exp\Bigg[\frac{1}{A_1}\Bigg\{&2\pi r_i(l-l_1)\left[-\exp\left(\frac{0.732E_d-E_{d63}}{\gamma}\right)\right]\\&+2\pi r_i\left\{-\exp\left(\frac{E_d-E_{d63}}{\gamma}\right)\left(\frac{\gamma l_1}{0.268E_d}\right)\left[1-\exp\left(\frac{-0.268}{\gamma}E_d\right)\right]\right\}\\&+2\pi r_i^2\left[-\exp\left(\frac{E_d-E_{d63}}{\gamma}\right)\right]\Bigg\}\Bigg].\end{aligned} \tag{6.34}$$

It is apparent from the solution that, within sections (1) and (3), the enlargement law applies as the parallel connection of similar elements with enlargement factors $n_1=2\pi r_1(l-l_1)/A_1$ and $n_3=2\pi r_i^2/A_1$. In section (2), on the other hand, because $E_{d3}=E_d(z)$ (eqn. 6.29), the enlargement factor is a function of the value of the variate:

$$n_2=\frac{2\pi r_i}{A_1}\left(\frac{l_1\gamma}{0.268E_d}\right)\left[1-\exp\left(-\frac{0.268}{\gamma}E_d\right)\right]. \tag{6.35}$$

Numerical evaluation of eqn. 6.35 shows that, in the relevant range $140\ \text{kV/cm}\leqq E_d\leqq 180\ \text{kV/cm}$, the enlargement factor n_2 is between 10.0 and 7.8 for section (2) and is thereby appreciably less than $n_1=62\,770$ and $n_3=63$. Whereas in the normal case one should calculate using the variate-dependent enlargement factor (eqn. 6.35), in this situation it is advisable to use the fixed enlargement factor $n_2=8.8$ in an approximate calculation. (In purely geometrical terms this would give $n_2=63$.) Eqn. 6.34 thus assumes the form of eqn. 6.14, i.e. the parallel connection of elements whose strengths are double exponentially distributed in a different manner:

$$F_n(E_d)=1-\exp\left[-\exp\left(\frac{E_d-158}{8.2}\right)-\exp\left(\frac{E_d-169}{6}\right)-\exp\left(\frac{E_d-157}{6}\right)\right]. \tag{6.36}$$

Graphical representation of this relationship (Fig. 6.22) clearly shows that, in the region of the withstand voltage, the distribution function of the transition (2) has hardly any effect on $F_n(x)$. Eqn. 6.36 can therefore be simplified to

$$F_n(E_d)=1-\exp\left[-\exp\left(\frac{E_d-158}{8.2}\right)-\exp\left(\frac{E_d-157}{6}\right)\right]. \tag{6.37}$$

The double-exponential distribution is gone. By virtue of the greater scatter, the large-area section (1) governs the small quantiles, while the

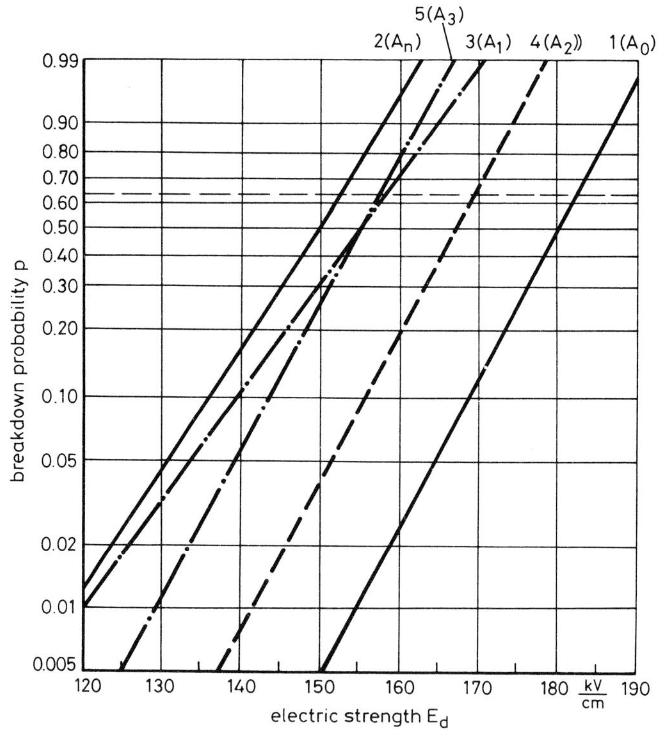

Fig. 6.22 Area effect on an SF_6-insulated tubular gas cable (Example 6.5.1)

1 Uniform reference element $A_0 = 10\ \text{cm}^2$
2 Complete arrangement $A_n = 628\,956\ \text{cm}^2$
3 Cylindrical field $A_1 = 627\,700\ \text{cm}^2$
4 Transition $A_2 = 628\ \text{cm}^2$
5 Spherical field $A_3 = 628\ \text{cm}^2$

average trend is also clearly influenced by the hemispherical end (3). The result permits conclusions relating to a more favourable design, particularly for section (1) ($r_a/r_i = e = 2.7\ldots$) and to the shaping of section (3), which we do not intend to consider in detail here.

The example vividly demonstrates that, with a location-dependent surface field strength, the area effect produces distribution functions that do not correspond to the theoretical distribution functions introduced, are difficult to handle and awkward to comprehend. Since the area effect is of great significance in the case of gas-insulated, slightly nonuniform systems, eqn. 6.25 should be numerically handled using a computer.

Nitta *et al.* use a method for handling the area effect which is basically similar, but is not quite consistent in its execution. The electrode area is weighted with the surface field strength, and the enlargement law in its customary form is applied to the weighted area. The average trend is ignored, so that the enlargement factor can be a function of the field strength (cf. eqn. 6.35).

In many cases the results agree with those obtained by the method described here, but in particular cases (e.g. control electrodes in a coaxial system), there is a risk of error with the method described by Nitta.

6.6 Volume effect

If the discharge development producing breakdown cannot be represented as area dependent, as described above, but takes place three-dimensionally, the enlargement must be treated as a volume effect. This has been investigated experimentally on oil insulation of industrial purity with large volumes (Fig. 6.23), and especially on high-polymer solid insulation, where defects may be distributed over the entire volume of the insulation.

If the volumetric enlargement of an insulation arrangement with a parallel plane field takes place in an axial direction (e.g. the extension of a cable, as in Fig. 6.1*a*), the field-strength distribution is not affected, and the arrangement can be 'discretised' into sections of a technically appropriate length. One must ensure, however, that the discharge processes in adjoining sections do not affect each other. In such volumetric enlargements, one can then work with discrete elements (see Section 6.4), using the relationships for the theoretical distribution functions (see Section 6.3).

The situation is considerably more complicated in the case of volumetric enlargements associated with changes in the field-strength distribution. The general enlargement law (eqn. 6.11) then assumes the form

$$F_n(x) = 1 - \exp\left\{\frac{1}{V_1}\int^{V_n} \ln\left[1 - F(x_e;\alpha;\beta)\right] dV\right\}. \qquad (6.38)$$

The variate X_e of the differentially small volume-element $dV = du\,.\,dy\,.\,dz$ is now represented as the product of the variate X of the reference volume V_1 and a geometrical function. For its realisations:

$$x_e = xf(u;\, y;\, z). \qquad (6.39)$$

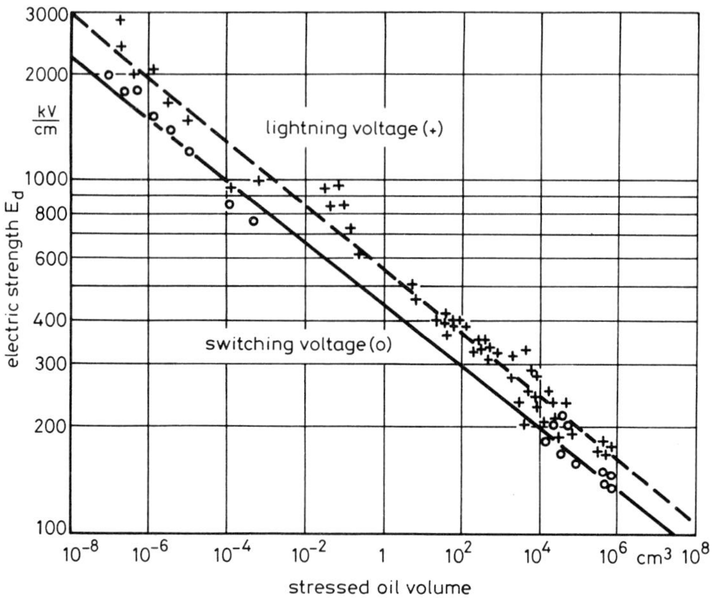

Fig. 6.23 Empirical volume effect of electric strength of insulating oil, described by a Weibull distribution

Advantageous co-ordinate systems should be selected for the geometrical function.

Example 6.6.1 The electric strength of polythene is to be described for a uniform reference volume $V_1 = 10\ \text{cm}^3$ by a two-parameter Weibull distribution

$$F(E_d) = 1 - \exp\left[-\left(\frac{E_d}{E_{d63}}\right)^{\delta}\right] \tag{6.40}$$

with the parameters E_{d63} and δ. We are required to determine the performance functions of the electric strength for cables with arbitrarily selected external and internal radii (r_a, r_i) and length (l). An essential assumption is that all the defects distributed over the volume of the cable are capable of producing breakdown by the same field-strength-dependent mechanism, and that breakdown of one element always implies complete breakdown. This assumption may be far more readily satisfied in arrangements with a high degree of uniformity (say $\eta > 0.7$) than in arrangements with a low degree of uniformity (say $\eta < 0.4$). Satisfaction of the prerequisite should be checked in each individual case.

The variate used is the electric strength, which ought here to be identical to the maximum breakdown field-strength. Since the stress only changes in a radial direction, rather than in an axial direction, the dependence of the strength variate from eqn. 6.39 in action in volume elements dV can be derived from the field-strength distribution applicable to coaxial cylinder systems:

$$E_{de} = \frac{E_d r_i}{r}. \tag{6.41}$$

The volume dependence is thereby reduced to a radius dependence. Accordingly, a volume element can be derived which is only dependent on the radius (cylinder shell *dr* thick: see Fig. 6.1*b*):

$$dV = 2\pi l r \, dr. \tag{6.42}$$

In accordance with the assumption made, breakdown of an element dV leads to breakdown of the cable. Using eqns. 6.40–6.42, eqn. 6.38 is then obtained in the form

$$F_n(E_d) = 1 - \exp\left\{\frac{1}{V_1}\int_{r_i}^{r_a}\left[-\left(\frac{E_d r_i}{E_{d63} r}\right)^{\delta}\right] 2\pi l r \, dr\right\} \tag{6.43}$$

and the solution ($\delta \neq 2$)

$$F_n(E_d) = 1 - \exp\left[-\left(\frac{E_d}{E_{d63}}\right)^{\delta} \frac{2\pi l r_i^2\left[\left(\frac{r_a}{r_i}\right)^{2-\delta} - 1\right]}{(2-\delta)V_1}\right]. \tag{6.44}$$

In this case the expression

$$n = \frac{2\pi l r_i^2\left[\left(\frac{r_a}{r_i}\right)^{2-\delta} - 1\right]}{V_1(2-\delta)}. \tag{6.45}$$

acts as the enlargement factor. As in the case of the extension of a cable (see Fig. 6.13), in a coaxial cylinder field, assuming a two-parameter Weibull distribution, the type of distribution is retained, and only the 63% quantile changes with n from eqn. 6.45, in accordance with eqn. 6.16. Thus:

$$E_{d63N} = E_{d63}\left\{\frac{V_1(2-\delta)}{2\pi l r_i^2\left[\left(\frac{r_a}{r_i}\right)^{2-\delta} - 1\right]}\right\}^{1/\delta}. \tag{6.46}$$

Note: It should be expressly pointed out that such a simple treatment of the volume effect is only possible under the assumptions made (cylindrical field, Weibull distribution). When different fields are handled and when different distribution functions are used, the type of distribution is lost (eqn. 6.34), as in the case of the area effect, for example. With the volume effect too, in the general case, a numerical treatment of the problem will have to be provided, on the basis of eqns. 6.38 and 6.39.

If the parameters of the Weibull distribution (E_{d63}, δ) are known for a uniform reference element of the quantity V_1, then the Weibull distribution can be calculated (Fig. 6.24) for any coaxial-cylinder cables (E_{d63n}, δ) in the validity range of the assumed breakdown mechanism.

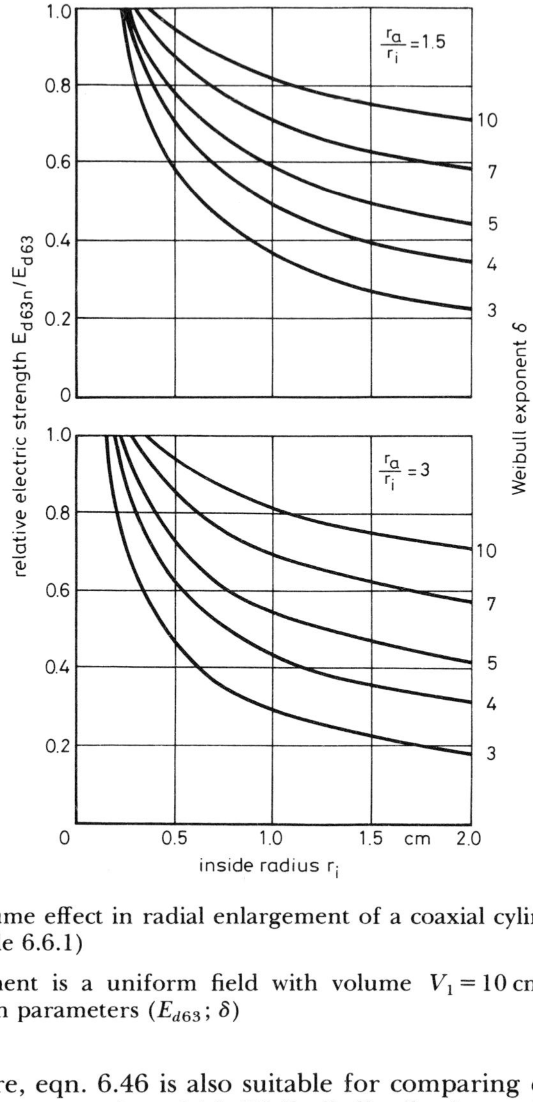

Fig. 6.24 Volume effect in radial enlargement of a coaxial cylinder arrangement $l = 1$ m (Example 6.6.1)

Reference element is a uniform field with volume $V_1 = 10\,\text{cm}^3$ and a Weibull distribution with parameters (E_{d63}; δ)

What is more, eqn. 6.46 is also suitable for comparing different coaxial cylinder arrangements for which Weibull distributions with the common component δ can be assumed. If two arrangements with (r_{a1}; r_{i1}; l_1) and (r_{a2}; r_{i2}; l_2) are considered, then the electric strength (which is identical to the maximum breakdown field strength) can be obtained for the ratio of the 63% quantiles:

$$\frac{E_{d63_1}}{E_{d63_2}} = \left(\frac{l_2}{l_1}\right)^{1/\delta} \left(\frac{r_{i2}}{r_{i1}}\right)^{2/\delta} \left[\frac{\left(\dfrac{r_{a2}}{r_{i2}}\right)^{2-\delta} - 1}{\left(\dfrac{r_{a1}}{r_{i1}}\right)^{2-\delta} - 1}\right]^{1/\delta} . \tag{6.47}$$

Since it is not maximum field strengths but breakdown voltages that are to be optimised, then, using the relationship between breakdown voltage and electric strength in a coaxial cylinder field ($E_{d63} = E_{dh63}$)

$$U_d = E_{d63} r_i \ln \frac{r_a}{r_i} \tag{6.48}$$

the ratio of the breakdown voltages can be obtained from eqn. 6.47:

$$\frac{U_{d63_1}}{U_{d63_2}} = \frac{\ln \dfrac{r_{a1}}{r_{i1}}}{\ln \dfrac{r_{a2}}{r_{i2}}} \left(\frac{l_2}{l_1}\right)^{1/\delta} \left(\frac{r_{i2}}{r_{i1}}\right)^{(2/\delta)-1} \left[\frac{\left(\dfrac{r_{a2}}{r_{i2}}\right)^{2-\delta} - 1}{\left(\dfrac{r_{a1}}{r_{i1}}\right)^{2-\delta} - 1}\right]^{1/\delta}. \tag{6.49}$$

Fig. 6.25 shows the evaluation of eqn. 6.49 for the special case $\delta = 5$; $l_1 = l_2 = l$; $r_{a1} = r_{a2} = r_a$, as a function of the conductor-radii ratio r_{i1}/r_{i2}. Under these conditions, the change in r_{i1}/r_{i2} naturally also corresponds to a change in the insulating distance

$$d = r_a - r_{i1} = r_a - r_{i2} \frac{r_{i1}}{r_{i2}}. \tag{6.50}$$

The maxima in Fig. 6.25 show that, for a given r_{i2}, there is an optimum value for r_{i1}/r_{i2}, i.e. an optimum insulating distance, which is determined not purely by the field-strength dependence (eqn. 6.48), but is governed by the statistical volume effect of the breakdown voltage. Further details are available in Küllig and Rüffer.

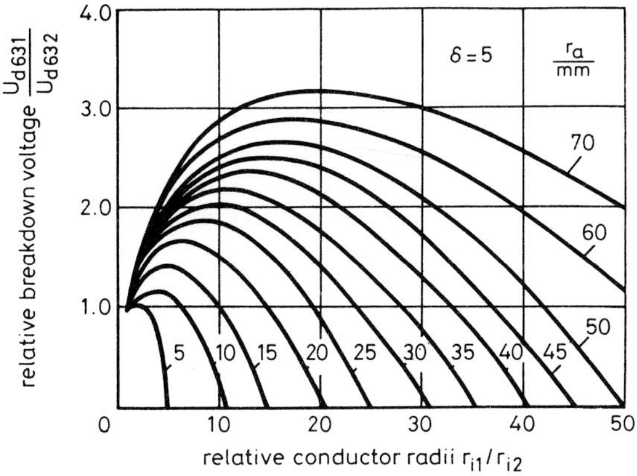

Fig. 6.25 Optimisation of breakdown voltages of coaxial cables, taking volume effect into account

Weibull distribution $\delta = 5$

$l_1 = l_2 = l$; $r_{a1} = r_{a2} = r_a$; $r_{i2} = 1$ mm,

v_{d632} then known

With regard to mathematical treatment, it should also be noted that the volume effect must be calculated using a physically appropriately chosen variate (in this instance $E_d = E_{dh}$). The result can at any time be transformed for the desired variate, using known relationships with other variates, with the breakdown voltage for example (eqn. 6.48). One must, of course, ensure that the time to breakdown is constant or physically negligible, otherwise the time effect (see Section 6.7) has to be taken into account. Such time influences are important, particularly in the case of high-polymer solid insulation. Optimising, say, the geometry of polythene cables purely on the basis of the volume effect (eqn. 6.49) and the experimental results of rising-voltage tests therefore calls for caution and must include conversion between constant-voltage tests and rising-voltage tests (see Section 5.5.3).

Handling the volume effect for any nonuniform fields assumes a knowledge of the areas of equal field strength. These can generally only be determined with numerical methods of field calculation, followed by numerical treatment of the volume effect. In the literature there are examples for sphere/plane, cylinder/plane, rounded-edge/plane and point/plane electrode arrangements.

6.7 Time effect

As with an enlargement of the geometrical dimensions, for statistical reasons an extension of the stress duration produces a reduction of insulating capacity. It is hardly to be expected that this time effect should statistically act in the same way as the volume effect, because spatially parallel processes are far more likely to be independent of each other, and may therefore generally be dealt with by the enlargement law as consecutive processes. The dependence of the processes in successive time periods is, in fact, often the object of model concepts, as, say the damage accumulation with solid material breakdown. Irreversible destruction of a component distance of the insulation occurs in a period being considered, and in the following period the discharge process develops from the component distance just destroyed. The range builds up in this way into complete breakdown.

Empirically, the time dependence of breakdown voltage or strength for high-polymer solid insulation is described by the so-called 'life law' (see Section 5.5.2)

$$\frac{U_{dn}}{U_{d1}} = \frac{E_{dn}}{E_{d1}} = \left(\frac{t_{dn}}{t_{d1}}\right)^{-1/r}, \tag{6.51}$$

where r is a value, the so-called 'life exponent', characteristic of each ageing mechanism. (r is sometimes denoted by n, but that would cause confusion with the enlargement factor here.)

If $t_{dn}/t_{d1} = n$ is regarded as an enlargement factor, then eqn. 6.51 describes the time influence to which breakdown in solid insulation is subject, without separating the damage-accumulation component and statistical time effect component (Fig. 6.26). If one tries to handle the problem statistically,

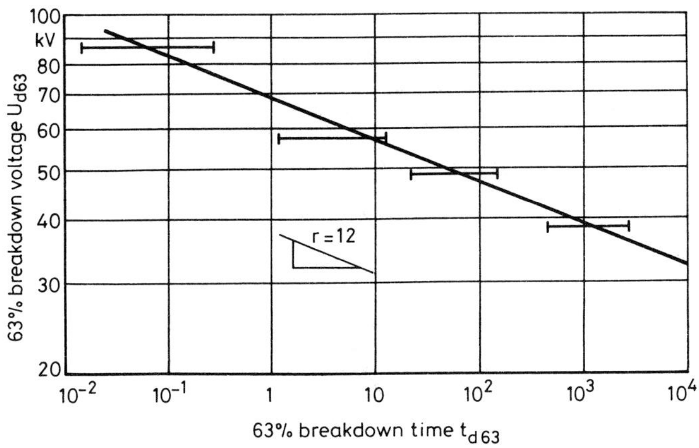

Fig. 6.26 Life law for model polythene cable-insulation
$r_i = 1.5$ mm; $r_a = 3.5$ mm; $l = 1$ m

interesting conclusions can be drawn from the difference between theory and experiment concerning the reasons for the time dependence.

Interesting test results relating to the time effect on insulating capacity are also available for liquid and gaseous insulating materials. Widmann observes that the enlargement law applies to insulating oil with repeated application of impulse voltages, provided the time interval between two applications is sufficiently long. The validity of the Weibull distribution and applicability of the life expression (eqn. 6.51) has been experimentally determined for long-term stress on oil insulation. In SF_6 insulation, long-term investigations have revealed a time influence on the electric strength, but it is far weaker than the enlargement law would lead us to expect (Fig. 6.27). This demonstrates the existence of dependences related to the movement of microscopic dust-particles in the insulating gas.

Although with long-term stresses, a purely statistical time effect hardly exists in the case of most insulating materials (because the dependences of the processes occur in consecutive time intervals), with impulse applications in the nanosecond and microsecond range the discharge processes are determined by random influences — particularly the provision of initiatory electrons. Statistical considerations were therefore applied at an early stage to describe the breakdown time in the case of impulse-voltage breakdown (Fig. 6.28), and more recently there have been attempts, especially in relation to SF_6 insulation, at stochastic modelling of the processes (see Section 5.3.2).

Despite its limited applicability, we intend to discuss the statistical time effect in relation to its mathematical treatment. For this purpose, the general enlargement law (eqn. 6.11) is given the form:

$$F_n(x) = 1 - \exp\left\{\frac{1}{t_1}\int_0^{t_n} \ln\left[1 - F(x_0; \alpha; \beta)\right] dt\right\}, \tag{6.52}$$

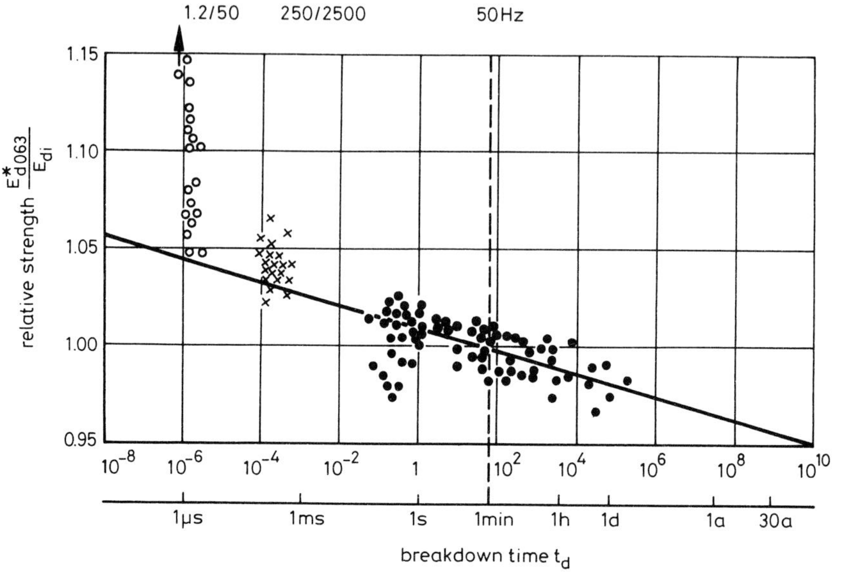

Fig. 6.27 Empirical time effect for electric strength in SF_6

$p_{20} = 0.4$ MPa

where the variate X_e of the differential time interval dt should be represented as the product of the variate X in the reference interval t_1 and a time function

$$X_e = Xf(t). \tag{6.53}$$

If the variate is not a time function ($f(t) = \text{const.}$), eqn. 6.52 assumes the form of the enlargement law for discrete elements with an enlargement factor $n = t_n/t_1$ (eqn. 6.4), (Section 6.4). In this case one should work with the known relationships for theoretical distribution functions (Section 6.3). For example, the following applies to the two-parameter Weibull distribution, which is very important in relation to solid insulation:

$$E_{d63n} = E_{d63_1}\left(\frac{t_n}{t_1}\right)^{-1/\delta}. \tag{6.54}$$

If the life law (eqn. 6.51) conforms to this relationship, i.e. $\delta = r$, the time dependence can on its own be reduced to a statistical time effect. The damage accumulation referred to above does not occur.

If the variate is a time function, eqns. 6.52 and 6.53 must be taken as the basis.

Example 6.7.1 The electric strength of a high-polymer solid insulation arrangement can be described for the reference period t_1 by a two-parameter Weibull distribution (eqn. 6.40). It is now postulated that the electric strength varies with the stress duration due to irreversible damage, as a function of the exponent m, in accordance with the life law. The following is accordingly

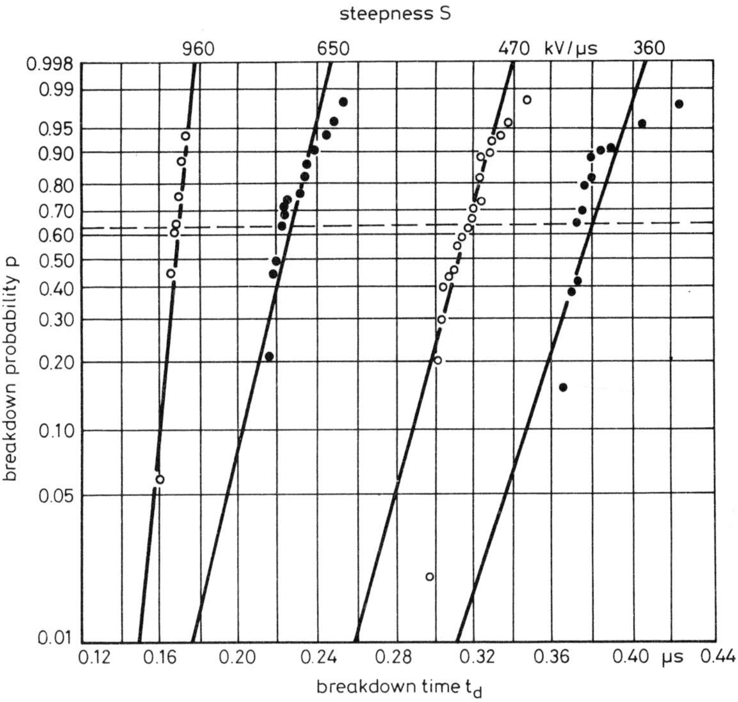

Fig. 6.28 Empirical distribution functions of breakdown time of a coaxial cylinder arrangement in SF_6; double exponential paper

Arrangement: $r_i = 7.5$ mm; $r_a = 20$ mm; $p_{20} = 0.1$ MPa
Parameter: rate of rise of impulse

derived from eqn. 6.51 for the variate effective during the periods *dt*, in accordance with eqn. 6.53:

$$\frac{E_{de}}{E_{d63e}} = \frac{E_d}{E_{d63}\left(\dfrac{t}{t_1}\right)^{-1/m}} = \left(\frac{E_d}{E_{d63}}\right)\left(\frac{t}{t_1}\right)^{1/m}. \tag{6.55}$$

Substituting eqn. 6.55 in eqn. 6.52 gives:

$$F_n(E_d) = 1 - \exp\left\{\frac{1}{t_1}\int_0^{t_n}\left[-\left(\frac{E_d t^{1/m}}{E_{d63} t_1^{1/m}}\right)^d\right] dt\right\} \tag{6.56}$$

with the solution

$$F_n(E_d) = 1 - \exp\left[-\left(\frac{E_d}{E_{d63}}\right)^\delta \left(\frac{t_n}{t_1}\right)^{(m+\delta)/m} \frac{m}{m+\delta}\right]. \tag{6.57}$$

In this case the expression

$$n = \frac{m}{m+\delta}\left(\frac{t_n}{t_1}\right)^{(m+\delta)/m}, \tag{6.58}$$

acts as an enlargement factor, and the 63% quantile is:

$$E_{d63n} = E_{d63}\left(\frac{m}{m+\delta}\right)^{-1/\delta}\left(\frac{t_n}{t_1}\right)^{-1/m-1/\delta} \quad . \tag{6.59}$$

Eqns. 6.57 to 6.59 express the simultaneous effect of the life consumption and the statistical time effect, where the expression in eqn. 6.55 chosen for the life assumption is naturally of a random nature for the elements dt, since it represents a macroscopic experience. In Fig. 6.29, the purely statistical time effect (eqn. 6.54), the empirical life law (eqn. 6.55) and the superimposition of statistical and dependent influences (eqn. 6.59) are jointly evaluated. For $\delta = \infty$ (curves 1 to 5), eqn. 6.59 supplies a time dependence corresponding to the pure life consumption (eqn. 6.55), while the pure statistical time effect (eqn. 6.54) occurs for $m = \infty$, $\delta = 5$ (curve 6). Curves 7 to 11 show that particularly pronounced time dependences occur when life consumption and statistical time effect are superimposed. The fact that these curves do not pass through the point (1 ; 1) is formally caused by eqn. 6.58, and appears to be physically due to the fact that the reference time t_1 is composed of differential times dt (see above).

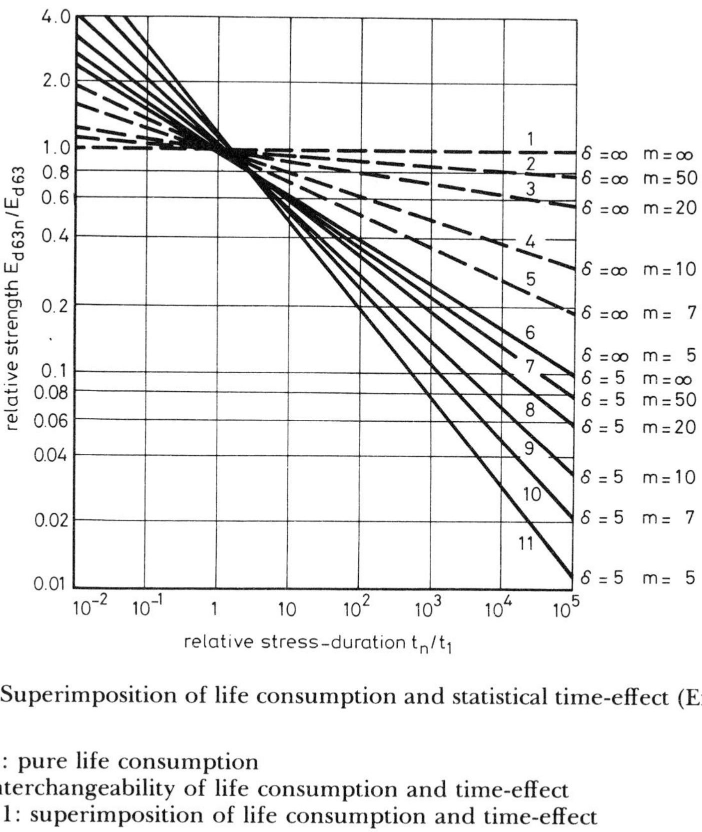

Fig. 6.29 Superimposition of life consumption and statistical time-effect (Example 6.7.1)

Curves 1–5: pure life consumption
Curve 6: interchangeability of life consumption and time-effect
Curves 7–11: superimposition of life consumption and time-effect

If the calculation procedure presented here is to be used to clarify whether a measured life characteristic is caused by statistical effects or by damage accumulation, it is advisable to carry out both experiments on volumetric enlargement (e.g. with cables of different lengths) and also long-term investigations (e.g. on cables of constant length).

To do so, the performance function of the electric strength is determined in constant-voltage tests (Fig. 6.30*a*), for example, for two stress durations

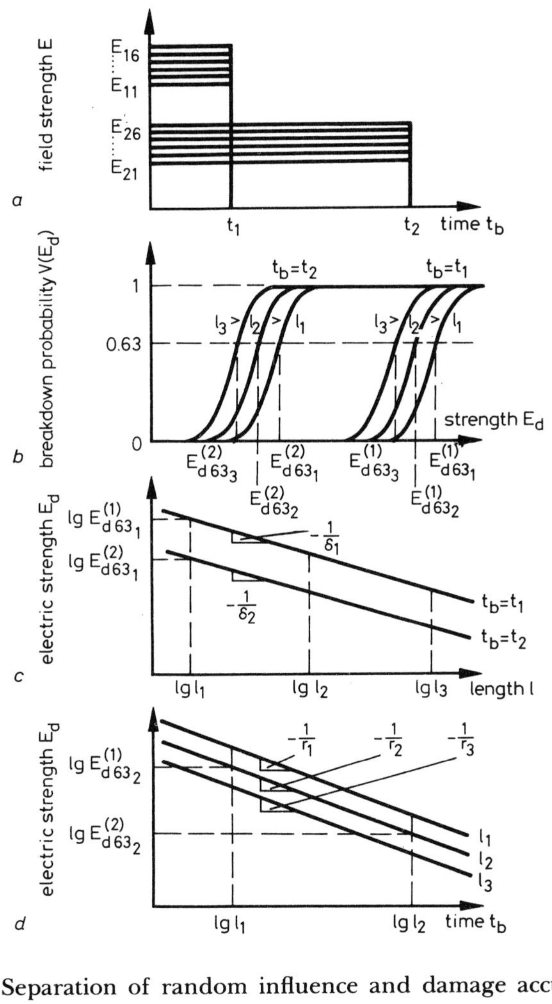

Fig. 6.30 Separation of random influence and damage accumulation (diagrammatic)

a Stresses (voltage applications)
b Empirical performance function
c Statistical volume-effect ($\delta_1 = \delta_2 = \delta$)
d Resultant time-effect ($r_1 = r_2 = r_3 = r$)

($t_b = t_1$; t_2) and three different volumes with the same field-strength distribution (e.g. cable lengths $l = l_1$; l_2; l_3). Both the statistical volume-effect and the life performance can be assessed from the test result (Fig. 6.30*b*). If the 63% quantile of the relative strength is plotted against the relative cable length (Fig. 6.30*c*), the Weibull exponent is obtained, which describes the volume effect:

$$\delta = \frac{\lg\left(\frac{l_n}{l_1}\right)}{\lg\left(\frac{E_{d63_1}}{E_{d63n}}\right)}. \tag{6.60}$$

To use the model described here, the quantity δ must be independent of the stress duration t_b (i.e. $\delta = \delta_1 = \delta_2$). If this is so, then δ is also responsible for the statistical time effect. If the 63% quantile of the strength from Fig. 6.30*b* is now plotted against the stress duration (Fig. 6.30*d*), then, from the empirical inclination r of the characteristic

$$E_{d63} = kt_b^{-1/r},$$

because

$$\frac{1}{r} = \frac{1}{m} + \frac{1}{\delta}, \tag{6.61}$$

the exponent characterising the life consumption is obtained as

$$m = \frac{\delta r}{\delta - r}. \tag{6.62}$$

Negative values for m when $\delta < r$ in eqn. 6.62 mean that the insulating material does not age with time, but improves (conditioning).

The use of the enlargement law in relation to geometrical and time differences between laboratory test and system service is becoming increasingly important in assessing the reliability of insulation. The more accurately the performance functions of the breakdown voltage of the individual elements have been determined (bearing in mind the statistical criteria we have outlined), the better the description that can be provided of the overall performance of the composite insulation of a large apparatus. Only in this way is there any prospect of being able to estimate the performance of the breakdown voltage for service conditions for structures that are extremely complex in insulation-engineering terms, such as metal-clad switchgear or power transformers. In the future, the high-voltage engineer will have to work even more intensively with statistical methods.

Bibliography

Chapter 1

[1.1] *Müller, P.H.*, u.a.: Wahrscheinlichkeitsrechnung und mathematische Statistik — Lexikon der Stochastik. Berlin: Akademie-Verlag 1975.

[1.2] *Gnedenko, B.W.; Beljajew, J.K.; Solowjew, A.D.*: Mathematische Methoden der Zuverlässigkeitstheorie I, II. Berlin: Akademie-Verlag 1968.

[1.3] *Dummer, G.W.A.; Griffin, N.B.*: Zuverlässigkeit in der Elektronik. Berlin: VEB Verlag Technik 1968.

[1.4] *Reinschke, K.*: Zuverlässigkeit von Systemen (Band 1 und 2). Berlin: VEB Verlag Technik 1976.

[1.5] *Mann, N.R.; Schafer, R.E.; Singpurwalla, N.D.*: Methods for statistical analysis of reliability and life data. New York: John Wiley & Sons 1974.

[1.6] *Preuß, H.*: Zuverlässigkeit elektronischer Systeme. Berlin: VEB Verlag Technik 1980.

[1.7] *Buslenko, N.P.; Schreider, J.A.*: Die Monte-Carlo-Methode und ihre Verwirklichung mit elektronischen Digitalrechnern, Leipzig: B.G. Teubner Verlagsgesellschaft 1964.

[1.8] *Buslenko, N.P.*: Simulation von Produktionsprozessen, Leipzig: B.G. Teubner Verlagsgesellschaft 1971.

[1.9] *Melnyk, M.: Schwitter, J.P.*: Weibullverteilung und Monte-Carlo-Methode zur Ermittlung der Lebensdauer von Maschinen. Die Unternehmung 19 (1965) S. 206–212.

[1.10] *Lahres, H.*: Einführung in die diskreten Markoff-Prozesse und ihre Anwendung. Braunschweig: Vieweg-Verlag 1964; Leipzig: BSB B.G. Teubner Verlagsgesellschaft 1964.

[1.11] *Karlin, S.*: A first course in stochastic processes. New York, London: Academic Press 1968.

[1.12] *Parzen, E.*: Stochastic processes. San Francisco 1962 (also Russian: Moscow: Mir 1971).

[1.13] *Gnedenko, B.W.*: Lehrbuch der Wahrscheinlichkeitsrechnung. Berlin: Akademie-Verlag 1970.

[1.14] *Bharucha-Reid, A.T.*: Elements of the theory of Markov processes and their applications. New York, Toronto, London: McGraw-Hill 1960.

[1.15] *Mosch, W.; Hauschild, W.*: Hochspannungsisolierungen mit Schwefelhexafluorid. Berlin: VEB Verlag Technik 1979; Heidelberg, Basel: Dr. Alfred Hüthig Verlag 1979.

[1.16] *Hauschild, W.*, u.a.: Zum Einfluß stochastischer Prozesse auf das zeitliche Durchschlagverhalten schwach inhomogener Felder im SF_6. Z. elektr. Informat.- und Energietechn. 12 (1982) 4, S. 289 bis 318; 5, S. 385–403.

Chapter 2

[2.1] *Müller, P.H.*: Wahrscheinlichkeitsrechnung und mathematische Statistik — Lexikon der Stochastik. Berlin: Akademie-Verlag 1975.

[2.2] *Storm, R.*: Wahrscheinlichkeitsrechnung — mathematische Statistik — statische Qualitätskontrolle. 6.Aufl. Leipzig: VEB Fachbuchverlag 1976.

[2.3] *Winkler, W.; Ebersberger, H.*: Mathematische Grundlagen des komplexen Produktionsprozesses. Baustein: Wahrscheinlichkeitstheorie. KDT-Fernkurs 1972.

[2.4] *Wunsch, G.*: Systemanalyse. Band 2: Statistische Systemanalyse. Berlin: VEB Verlag Technik 1970.

[2.5] *Fisz, M.:* Wahrscheinlichkeitsrechnung und mathematische Statistik. Berlin: VEB Deutscher Verlag der Wissenschaften 1965.

[2.6] *Smirnow, N.W.; Dunin-Barkowskij, I.W.:* Mathematische Statistik in der Technik. Berlin: VEB Deutscher Verlag der Wissenschaften 1963.

[2.7] *Bauer, H.:* Wahrscheinlichkeitstheorie und Grundzüge der Maßtheorie. Berlin, New York: Walter de Gruyter 1974.

[2.8] *Schmetterer, L.:* Mathematische Statistik. Wien, New York: Springer-Verlag 1966.

[2.9] See Ref. [1.15].

[2.10] *Pfanzagl, H.:* Allgemeine Methodenlehre der Statistik. Teil I und II. Berlin, New York: Walter de Gruyter 1974 (Sammlung Göschen Band 7047).

[2.11] *Müller, H.P.; Neumann, P.; Storm, R.:* Tafeln der mathematischen Statistik. Leipzig; VEB Fachbuchverlag 1973.

[2.12] *Boutteau, M.*, u.a.: Anleitung zur statistischen Behandlung von Versuchsergebnissen, die im Konstantspannungsversuch erhalten wurden (franz.). EdF Bull. Serie B (1971) 3, S. 57–72.

[2.13] *Dragan, G.*, u.a.: Betrachtungen zur Bestimmung der Stoßüberschlagspannung. E und M 89 (1972) 11, S. 445–540.

[2.14] *Hauschild, W.:* Valuation of high-voltage impulse tests from a statistical point of view (German and English). Symposium on EHV Test Problems, Roorkee, India, 1980. IPH-Mitt, Sonderheft (1981) Nr. 23.

[2.15] *Heller, B.: Veverka, A.:* Die Anwendung der Poisson-Statistik in der Hochspannungstechnik. Acta Technica ČSAV (1972) 6, S. 613–620.

[2.16] *Wohlfahrt, O.:* Statistik als Instrument des Hochspannungs-Isolationstechnikers. E und M 74 (1957) 10, S. 223–228; 12, S. 267–272.

[2.17] *Woboditsch, W.:* Charakteristik technischer Funkenstrecken mit stark inhomogenem Feld. Diss. Techn. Hochsch. Dresden 1957.

[2.18] *Nemeth, E.; Csaki, E.:* Methods of mathematical statistics for evaluating electric breakdown measuring series. Periodica Polytechnica Budapest 7 (1963) 1, S. 9–35.

[2.19] *Widmann, W.:* Das Vergrößerungsgesetz in der Hochspannungstechnik. ETZ-A 85 (1964) 4, S. 97 bis 102.

[2.20] *Paderta, B.; Lesny, V.:* Statistische Methoden in der Höchstspannungstechnik (tschech.). Elektrotechn. Obzor 59 (1970) 6, S. 285–290.

[2.21] *Brown, G. W.:* Determination of critical flashover voltage and standard deviation from flashover probability data. IEEE Trans. PAS 88 (1969) 3, S. 189–194.

[2.22] *Brown, G.W.:* Testing for the cumulative flashover distribution. IEEE Trans. PAS 89 (1970) 6, S. 1180–1191.

[2.23] *Ebersberger, H.; Hauschild, W.; Förster, K.-H.:* Statistische Verfahren für die Bestimmung der Durchschlagwahrscheinlichkeit von Isolierstrecken. Wiss. Z. Elektrotechn. 17 (1971) 2/3, S. 117 bis 132.

[2.24] *Schrader, W.:* Das Durchschlagverhalten paralleler Funkenstrecken. Diss. Techn. Univ. Dresden 1974.

[2.25] *Hauschild, W.; Köppe, R.:* Statistiche Auswertung hochspannungstechnischer Messungen am Digitalrechner. ELTRA-Fortschrittsbericht 1973 (IPH Berlin), H. 7.

[2.26] *Rasquin, W.:* Statistische Auswertung der Meßergebnisse von Durchschlag-Untersuchungen. Bull. SEV 63 (1972) 5, S. 231–239.

[2.27] *Hauschild, W.:* Über die Schwierigkeiten bei der Schätzung von Verteilungsfunktionen der Durchschlagspannung. Z. elektr. Inform.- und Energietechn. 5 (1975) 3, S. 198–216.

[2.28] *Pilling, J.:* Ein Beitrag zur Interpretation der Lebensdauerkennlinien und zur dielektrischen Bemessung und Prüfung von hochpolymeren Feststoffisolierungen. Diss. B. Techn. Univ. Dresden 1976.

[2.29] *Tschacher, B.:* Zum Einfluß elektrischer Vorbeanspruchungen auf den Durchschlag von Epoxidharzisolierungen bei Beanspruchung mit Weschsel- und Impulsspannung. Diss. Techn. Univ. Dresden 1975.

[2.30] *Golinski, S.:* Ionisationserscheinungen und Fragen der Lebensdauer von Gießharziolationen. Elektrie 20 (1966) 9, S. 341–344.

[2.31] *Schuppe, W.:* Über die Alterungsbeständigkeit von Kunststoff-Folien. Bull. SEV 62 (1971) 16, S. 764–769.

[2.32] *Weibull, W.:* A statistical distribution function of wide applicability. J. of Applied Mechanics 18 (1951) 9, S. 293–297.

[2.33] *Gumbel, E.J.:* Statistics of extremes. New York: Columbia University Press 1958, Moscow: Mir 1965.

[2.34] *Bronstein, I.N.; Semendjajew, K.A.:* Taschenbuch der Mathematik (Neubearbeitung). Moskau: Nauka; Leipzig: BSB B.G. Teubner Verlagsgesellschaft 1979.

[2.35] *Stone, G.C.; Heeswijk, R.G. van:* Parameter estimation for the Weibull distribution. IEEE Trans. EI 12 (1977) 4, S. 254–261.

[2.36] *Reichelt, C.:* Rechnerische Ermittlung der Kenngrößen der Weibull-Verteilung. Fortschrittsberichte der VDI-Z. Reihe 1 (1978) 56.

[2.37] Angewandte-Statistik. — Regeln zur Bestimmung der Schätzwerte und Vertrauensbereiche für die Parameter der Weibullverteilung (vollständige Information über die Stichprobe). RGW-Standard Thema 01.913.10–76 (überarbeiteter Entwurf).

[2.38] *Veverka, A.; Kvasnicka, V.:* Zur Weibullverteilung (tschech.). Elektrotechn. Obzor 66 (1977) 4, S. 206–208.

[2.39] *Veverka, A.; Kvasnicka, V.:* Die Weibullverteilung und die konventionelle Abhängigkeit der Durchschlagwahrscheinlichkeit (tschech.). Elektrotechn. Obzor 67 (1978) 3, S. 135–137.

[2.40] *Wohlmuth, F.:* Einige Anwendungen der Wahrscheinlichkeitsrechnung und der mathematischen Statistik in der Hochspannungstechnik (tschech.). Habilitation Techn. Hochsch. Prag 1978.

[2.41] *Tomčik, J.:* Zur Bestimmung der Parameter der Weibullverteilung (tschech.). Elektrotechn. Obzor 66 (1977) S. 208–209.

[2.42] *Mann, N.R.; Schafer, R.E.; Singpurwalla, N.D.:* Methods for statistical analysis of reliability and life data. New York: Wiley-Interscience 1974.

[2.43] *Billman, B.R.; Antle, C. E.; Bain, L. J.:* Statistical inference from censored Weibull samples. Technometrics 14 (1972) 11, S. 831–840 (also: Technometrics 11 (1969) 8, S. 445–460).

[2.44] *Wanser, G.; Wiznerowicz, F.:* CIGRE 1972 — Aktuelle Kabelfragen auf der internationalen Hochspannungskonferenz. Elektrizitätswirtschaft 71 (1972) 25, S. 771–777.

[2.45] *Kreuger, F.H.; Bentvelsen, P.A.C.:* Durchschlagerscheinungen in PE-isolierten Kabeln (engl.). CIGRE 21–05 1972.

[2.46] *Brookes, A.S.:* The Weibull-distribution: Effect of length and conductor size of test cables. Electra Nr. 33, S. 49–61.

[2.47] *Lapschin, W.A.:* Statistische Gesetzmäßigkeiten des Durchschlages von Polyäthylenisolierungen bie bergrenzter Verteilungsfunktion (russ.) Isv. WUSOW Energetika (1973) 12, S. 21–26.

[2.48] *Brakelmann, H.:* Zündvolumina inhomogen beanspruchter Festoffisolierungen. Bull. SEV 68 (1977) 12, S. 595–599.

[2.49] *Rüffer, K.:* Zur elektrischen Alterung hochpolymerer Schichtisolierungen. Diss. Techn. Univ. Dresden 1976.

[2.50] *Köppe, R.:* Ein Beitrag zur elektrischen Alterung von Modellisolierungen aus Epoxidharz und zur Übertragbarkeit der Ergebnisse auf technische Isolierungen. Diss. Techn. Univ. Dresden 1978.

[2.51] *Occhini, E.:* A statistical approach to the discussion of the dielectric strength in electric cables. IEEE Trans. PAS 90 (1971), S. 2671–2682.

[2.52] *Wilson, W.R.:* A fundamental factor controlling the unit dielectric strength of oil. AIEE Trans. Part. III 72 (1953) 2, S. 68–74; AIEE Trans. Part. III 74 (1955) 8, S. 677–688.

[2.53] *Dokopoulos, P.:* Wachstumsgesetze der Durchschlagwahrscheinlichkeit von Hochspannungsisolierungen. Diss. Techn. Hochsch. Braunschweig 1967 (s. auch ETZ-A 89 (1968) S. 1945).

[2.54] *Nitta, T.,* u.a.: Statistical approach to the breakdown characteristics of large scale gas insulated systems. ISGD Knoxville (1978) S. 338–354.

[2.55] *Nitta, T.; Shibuya, Y.; Fujiwara, Y.:* Voltage-time characteristics of electrical breakdown in SF_6. IEEE Trans. PAS 93 (1974) 1, S. 108–115.

[2.56] *Nitta, T.; Yamada, N.; Fujiwara, Y.:* Area effect of electrical breakdown in compressed SF_6. Trans. IEEE PAS 93 (1974) 2, S. 623–629.

[2.57] *Ikeda, M.; Inoue, T.:* Statistical approach to breakdown stress of transformer insulation. 3. ISH Mailand (1979) 23.18.

[2.58] *Tsumoto, M.; Okiai, R.:* A new application of Weibull distribution to impulse breakdown of oil-filled cable. IEEE Trans. PAS 93 (1974) 1, S. 360–366.

[2.59] *Weber, K.H.; Endicott, H.S.; Kaufmann, R.B.:* Area effect and its extremal basis for the electric breakdown of transformer oil. AIEE Trans. Pt. III 75 (1956) 6, S. 371–381; AIEE Trans. Pt. III 76 (1957) 8, S. 393–398; AIEE Trans. Pt. III 76 (1957) 12, S. 1091–1098.

[2.60] *Sisojew, M.I.:* Eine statistische Methode zur Bestimmung des Isoliervermögens druckgasgefüllter Apparate (russ.). Električestvo 83 (1962) 8, 52–55.

[2.61] *Bortnik, I.M.:* Zur Ermittlung von Betriebs- und Prüffeldstärken von Hochspannungsanordnungen mit SF_6-Isolierungen (russ.). Električestvo 91 (1970) 5, S. 83–85.

[2.62] *Vibholm, S.; Pedersen, A.; Christensen, J. M.; Thyregod, P.:* The effect of surface roughness on low probability first breakdown in compressed SF_6. 3. ISH Mailand (1979) 32.06.

[2.63] *Veverka, A.; Wohlmuth, F.:* Approximation der Normalverteilung mit Grenzen (tschech.). Elektrotechn. Obzor 68 (1978) 6, S. 355–356.

[2.64] *Wohlmuth, F.:* Zweigrenzenfunktion der Überschlagwahrscheinlichkeitsverteilung. Prace Naukowe Politechniki Warszawskiej, Elektryka 53 (1979), S. 5–19.

[2.65] *Daeves, K.; Beckel, A.:* Großzahl-Methodik und Häufigkeitsanalyse. Weinheim: Verlag Chemie GmbH 1958.

[2.66] *Graf, U.; Hennig, H.-J.:* Statistische Methoden bei textilen Untersuchungen. Berlin, Göttingen, Heidelberg: Springer-Verlag 1960.

[2.67] *Sachs, L.:* Statistische Auswertungsmethoden. Berlin, Heidelberg, New York: Springer-Verlag 1968.

[2.68] *Martin, H.:* Rationelle statistische Auswertung von Meßergebnissen bei 5 bis 20 Einzelwerten. Monatsberichte der DAW zu Berlin 1 (1959) 7–10, S. 389–393.

[2.69] *Hoppadietz, F.:* Vereinfachtes Auswerteverfahren von Häufigkeitskurven. Wiss. Z. der Techn. Hochsch. Dresden 9 (1959/60) 3, S. 729–733.

[2.70] *Schwarz, J.:* Ein Verfahren zur Gewährleistung der Unabhängigkeit beir der Messung der Impulsdurchschlagspannung und -zeit im SF_6. Elektrie 35 (1981) 1, S. 33–36.

[2.71] *Paderta, B.:* Über die Beglaubigung der Unabhängigkeit von einzelnen Versuchen. Acta Technica ČSAV (1968) 1, S. 100–109.

[2.72] *Paderta, B.:* Über die Beglaubigung der Unabhängigkeit von einzelnen Versuchen in der Hochspannungstechnik. Acta Technica ČSAV (1969) 1, S. 18–24.

Chapter 3

[3.1] TGL 20618/07 Hochspannungsprüftechnik; Statistische Ermittlung von Durchschlagspannungen; Ausg. 6.78 (s. auch IEC-Publ. 60–2 (1973) Anhang A; VDE 0432/2 bzw. DIN 57432/2).

[3.2] *Hochrainer, A.:* Verhaltensfunktionen und Summenhäufigkeitsfunktionen. ETZ-A 90 (1969) 2, S. 25–33.

[3.3] *Weicker, W.:* Zur Kenntnis der Funkenspannung bei technischem Wechselstrom. Diss. Techn. Hochsch. Dresden 1905; ETZ 32 (1911) 18, S. 436.

[3.4] See Ref. [2.17].

[3.5] See Ref. [2.26].

[3.6] *Tetzner, V.:* Bestimmung einer für Isolieranordnungen unter Öl zweckmäßigen 50-%-Durchschlagstoßspannung. Arch. f. Elektrotechn. 44 (1958) 1, S. 52–56.

[3.7] *Thyregod, P.:* Statistik des ersten Durchschlages. Private Mitteilung vom 11.10.1979—Techn. Univ. of Denmark, Department of Physics, Section II.

[3.8] *Vibholm, S.; Pedersen, A.; Thyregod, P.:* Determination of low probability first breakdown voltages in compressed SF_6. ICPIG XIV (Grenoble) C7–289, Journal de Physics 40 (1978) 7.

[3.9] *Bakken, J.A.:* Determination of characteristic voltages in impulse and switching surge testing. IEEE Trans. PAS 86 (1967) 8, S. 962–968.

[3.10] *Fisher, R.A.:* The design of experiments. London, Edinburgh: Oliver and Boyd, 7. Aufl. 1960.

[3.11] *Scheffler, E.:* Einführung in die Praxis der statistischen Versuchsplanung. Leipzig: VEB Deutscher Verlag für Grundstoffindustrie 1974.

[3.12] *Box, G.E.P.:* Die 2^{k-p}-Faktor-Versuchsplanung (Teil I und II, engl.). Technometrics 3 (1961) 3, S. 311–351 und 3 (1961) 4, S. 449–458.

[3.13] *Bengalia, A.; Leva, U.; Vitali, A.; CESI:* Neue Technik der Erfassung und Verarbeitung von Meßdaten (ital.). CESI-Publ. N 79/22.

[3.14] *Wiesendanger, P.:* Automatische, digitale Aufzeichnung und Auswertung von transienten Signalen in der Hochspannungstechnik. Diss. Elektrotechn. Hochsch. Zürich, 1977.

[3.15] *Praxl, G.:* Automated evaluation of high voltage tests. 3.ISH Mailand (1979) 42.09.

[3.16] Meß- und Steuersystem für Gleich- und Wechselspannungsprüfanlagen des VEB TuR Dresden. Firmenschrift VEB Transformatoren- und Röntgenwerk "Hermann Matern" Dresden (Nr. 8090) 1980.

[3.17] Meß- und Steuersystem für Impulsspannungs-Prüfanlagen des VEB TuR Dresden. Firmenschrift des VEB Transformatoren- und Röntgenwerk "Hermann Matern" Dresden (Nr. 8091 d) 1979.

[3.18] Trigatron—automatisches Triggergerät für Impulsgeneratoren. Werbeschrift der Firma Haefely (HVTS E 134.1 und E 134.2).

[3.19] Steuergeräte für Impulsspannungs-Prüfanlagen in "Prüfeinrichtungen/Meßeinrichtungen". Firmenschrift des VEB Transformatoren- und Röntgenwerk "Herman Matern". Dresden, Nr. 805/5 d.

[3.20] *Thione, L.:* Anwendung von Lichtleitkabeln für die Übertragung von Meß- und Steuersignalen in Laboratorien und Hochspannungsanlagen (ital.). ANIPLA/CESI-Publ. N 79/07.

[3.21] TGL 20 620 Hochspannungsprüftechnik; Prüfung mit Wechselspannung; Ausg. 8.77 (s. auch IEC-Publ. 60–2 (1973); VDE 0432/2 bzw. DIN 57 432/2; ST RGW 1072 (1978)).

[3.22] IEC 60-2 High-voltage test techniques, test procedures, Section two: Tests with direct voltage. IEC-Publ. 60–2, 1. Ausgabe 1973 (s. auch VDE 0432/2 bzw. DIN 57 432/2; ST RGW 1072 (1978)).

[3.23] TGL 20 621 Hochspannungsprüftechnik; Prüfung mit Blitzspannung; Ausg. 12.78 (s. auch IEC-Publ. 60–2 (1973); VDE 0432/2 bzw. DIN 57 432/2; ST RGW 1072 (1978)).

[3.24] TGL 20 622 Hochspannungsprüftechnik; Prüfung mit Schaltspannung; Ausg. 12.78 (s. auch IEC-Publ. 60–2 (1978); VDE 0432/2 bzw. DIN 57 432/2; ST RGW 1072 (1978)).

[3.25] TGL 20 618/02 Hochspannungsprüftechnik; Bezugsatmosphäre und Korrekturen; Ausg. 12.76 (s. auch IEC-Publ. 60–1 (1973); VDE 0432/1; ST RGW 1071 (5.80)).

[3.26] TGL 20 618/03 Hochspannungsprüftechnik; Allgemeine prüftechnische Forderungen. Ausg. 12.76 (s. auch IEC-Publ. 60–1 (1973); VDE 0432/1; ST RGW 1071 (5.80)).

[3.27] TGL 20 623 Hochspannungsprüftechnik; Messung der Hochspannung; Bl. 1 u. 2, Ausg. 12.76; B1.4 Ausg. 6.77; Bl. 5 Ausg. 12.78 (s. auch IEC-Publ. 60–3 (1976); VDE 0432/2 bzw. DIN 57 432/3; ST RGW (Entwurf) zum Thema 01.507.01–78).

[3.28] *Kouno, T.; Oikawa, T.:* Standard deviation of flashover probability. J.I.E.E. (Japan) 87 (1967) 8, S. 84–89.

[3.29] *Hoppadietz, F.:* Rechnerische Ermittlung von statistischen Durchschlagspannungen und Stichprobenumfang. Elektrie 32 (1978) 9, S. 474–476.

[3.30] *Hoppadietz, F.:* Statistische Toleranzgrenzen, Stichprobenumfang und wahrscheinliche Überdimensionierung. Elektrie 32 (1978) 5, S. 265–269.

[3.31] *Vibholm, S.; Pedersen, A.:* On the reproducibility of negative lightning impulse breakdown voltage distributions for a rod–rod gap. ICPIG XI Prag (1973) paper 190.

[3.32] *Vibholm, S.; Pedersen, A.:* Some factors affecting the lightning impulse breakdown voltage distribution for a rod–rod gap in air at atmospheric pressure. 3. Int. Conf. Gas Discharges. London 1974.

[3.33] *Lennertz, H.:* Reproduzierbare Blitzstoßspannungsversuche bei einer inhomogenen Elektrodenanordnung und ihr statistischer Nachweis. ETZ-A 92 (1971) 9, S. 558–559.

[3.34] *Böhme, K.:* Vorentladungen im schrägen Luftspalt bei Wechselspannung. Diss. Techn. Univ. Dresden 1968.

[3.35] *Buchholz, K.-H.; Krey, B.:* Ein Beitrag zur Bemessung von schwach inhomogenen SF_6-Isolierungen mit Grenzflächen. Diss. Techn. Univ. Dresden 1974 (s. auch Eletrrie 28 (1974) 9).

[3.36] *Hauschild, W.; Tschacher, B.; Günther, H.:* Über die Ermittlung repräsentativer Verteilungsfunktionen der Durchschlagspannung. Elektrie 27 (1973) 6, S. 297–300.

[3.37] *Kahle, M.:* Ein Beitrag zur Aufklärung des Festkörperdurchschlagmechanismus. 25.IWK Ilmenau (1980) H.3 (A3), S. 3–7.

[3.38] *Ouyang, M.:* New method for the assessment of switching-impulse insulation strength. Proc. IEE 113 (1966) 11, S. 1835.

[3.39] *Ouyang, M.:* Interpretation of impulse tests. Electr. Times (1967) 2.

[3.40] *Oakeshott, D.F.*, u.a.: New method for the assessment of switching-impulse insulation strength. Proc. IEE 114 (1967) 11, S. 1734–1742.

[3.41] *Ouyang, M.; Carrara, G.:* Evaluation and application of impulse test results. Electra (1970) H. 13.

[3.42] *Hauschild, W.; Bürger, W.:* Statistische Modelle für den elektrischen Durchschlag im Isoliergas SF_6. Z. elektr. Inform.- und Energietechn. 5 (1975) 4, S. 283–295.

[3.43] Siehe [1.16].

[3.44] *Kouno, T.; Endo, M.:* Simulation of probability breakdown curve and presumption of the reliability of 50%-flashover voltage by means of electronic computer. J.I.E.E. (Japan) 84 (1964), S. 25–33.

[3.45] *Kučera, J.:* Statistical methods for evaluation of measurements with impulse voltage. 3.ISH Mailand (1979) 42.01.

[3.46] *Wohlfahrt, K.; Sautner, J.*, u.a.: Programmsysteme zur statistischen Auswertung von Hochspannungsversuchen. Jahresbericht 1979 des Hochspannungslaboratoriums der Techn. Hochsch. Graz, S. 14–16.

[3.47] *Carrara, G.; Marzio, L.:* Discharge probability under dielectric stress. CIGRE-Ber. 33–01 (1968), Appendix V.

[3.48] *Nollau, F.:* Statistische Analysen. Leipzig: VEB Fachbuchverlag 1977.

[3.49] *Ritzk, F.A.M.; Vincent, C.:* Testing for low breakdown probability with special reference to liquid insulation. IEEE Trans. PAS 96 (1977) 6, S. 1892–1899.

[3.50] *Hauschild, W.:* Zum Öldurchschlag im inhomogenen Feld bei Schaltspannung. Diss. Techn. Univ. Dresden 1970.

[3.51] *Sie, T.H.; Wohlfahrt, O.:* Contribution to the measurement of the impulse withstand voltage of oil-paper-insulation systems. IEEE Trans. PAS 88 (1969) 6, S. 862–868.

[3.52] *Bakken, J.A.:* Determination of the highest impulse withstand voltage for air gaps. The Norwegian Research Institute of Electricity Supply, TR No. 1292 E, 1965.

[3.53] *Dixon, W.J.; Mood, A.M.:* A method for obtaining and analysing sensitivity data. J. of American Statistical Assoc. (1948) S. 109–126.

[3.54] *Kučera, J.:* Confidence limits of 50-%-flashover voltage in the up-and-down method. Acta Technica ČSAV 17 (1972) 1, S. 61–72.

[3.55] *Kučera J.:* Confidence limits of the 50-% flashover voltage in the up-and-down method. Power Research Institute Praha, ČSSR, für CIGRE 33-71 (WG 03).

[3.56] *Brownlee, K.A.; Hodges, J.C.; Rosenblatt, M.:* The up-and-down method with small samples. J. American Statistical Assoc. (1953) 6, S. 262–276.

[3.57] *Anis, H.; Abo-El-Saad, M.:* Optimal up-and-down testing of external insulation. 3.ISH Mailand (1979) 42.08.

[3.58] *Carrara, G.; Dellera, L.:* Accuracy of an extended up-and-down method in statistical testing of insulation. Electra 23, S. 159–175.

[3.59] *Hancox, R.; Tropper, H.:* The breakdown strength of transformer oil under impulse voltage. Proc. IEE (1957) S. 250–262.

[3.60] *Hancox, R.:* The interpretation of the results of impulse voltage tests. Proc. IEE (1958) S. 404–406.

[3.61] *Klaus, R.; Plenio, C.:* Beurteilung vereinfachter Verfahren zur Ermittlung der Stehspannung und der 50-%-Durchschlagspannung von Hochspannungsisolatoren. Dipl.-Arbeit Sektion Mathematik, Wissenschaftsbereich Mathematische Statistik der Tech. Univ. Dresden (1971), unveröffentlicht.

[3.62] *Paderta, B.:* Bestimmung der Haltespannung mit Hilfe einer modifizierten Auf-und-ab-Methode (tschech.). Elektrotechn. Obzor 62 (1973) 5, S. 296–300.

[3.63] *Powell, C.W.; Ryan, H.M.:* Switching impulse strength of a 765 kV simulated tower window with V-string insulators under artificial rain. 3.ISH Mailand (1979) 52.11.

[3.64] *Wetherhill, G.B.:* Sequential methods in statistics. London: Chapman and Hall Ltd. 1975.

[3.65] *Fryxell, J.:* Determination of critical withstand voltages. CIGRE-Bericht 423 (1966).

[3.66] *Paderta, B.:* Beitrag zur Problematik der Haltespannung. Acta Technica ČSAV (1968) 6, S. 754 bis 769.

[3.67] *Paderta, B.:* A new method for the determination of impulse strength. Acta Technica ČSAV (1970) 4, S. 377–388.

[3.68] *Hylten-Cavallius, N.; Fonseca, R.J.:* Extreme value statistics and high voltage test techniques. 3.ISH Mailand (1979) 42.06.

[3.69] *Bandemer, H.; Bellmann, A.:* Statistische Versuchsplanung, Reihe Mathematik für Ingenieure, Naturwissenschaftler, Ökonomen und Landwirte. Leipzig: BSB B.G. Teubner Verlagsgesellschaft 1979.

[3.70] *Ahlers, H.; Schwartz, B.; Waldmann, J.:* Optimierung technischer Produkte und Prozesse. Berlin: VEB Verlag Technik 1981.

[3.71] *Böhme, H.:* Zur Nutzung der Methode "Statistische Versuchsplanung" in der Hochspannungs-Isoliertechnik. Elektrie 36 (1982) 8, S. 395.

[3.72] *Burgmann, M.:* Zur Anwendung der statistischen Versuchsplanung für die Bemessung von Hochspannungsisolierungen. Elektrie 36 (1982) 8, S. 396–398.

[3.73] *Digmayer, M.:* Optimierung schwach inhomogener Mehr-Elektroden-Anordnungen. Elektrie 36 (1982) 8, S. 397–403.

Chapter 4

[4.1] IEC 71 Insulation co-ordination, Part I: Terms, definitions, principles and rules (1976). Part II: Application guide (1976). IEC-Publ. 71–1 und 71–2.

[4.2] TGL 20445 Elektrotechnik; Isolationskoordination. Blatt 01 Begriffe; Ausg. 11.75, Blatt 02 Betriebsmittel und Anlagen mit Wechselspannung über 1 kV, Technische Forderungen; Ausg. 11.75; Blatt 03 Betriebsmittel und Anlagen mit Wechselspannung bis 1000 V; Ausg. 9.76.

[4.3] *Koettnitz, H.; Pundt, H.*, u.a.: Berechnung elektrischer Energieversorgungsnetze, Band I: Mathematische Grundlagen und Netzparameter. Leipzig: VEB Deutscher Verlag für Grundstoffindustrie 1973.

[4.4] *Gert, R.*: Aktuelle Probleme der Isolationskoordination. Elektrie 27 (1973) 11, S. 594–597.

[4.5] *Koettnitz, H.*: Beanspruchung elektrotechnischer Betriebsmittel; Lehrbriefe 4 und 5, Beanspruchungen durch Überspannungen und Koordination der Isolation. Berlin: VEB Verlag Technik 1976.

[4.6] *Hoy, C.*: Netzanalysator — Untersuchungen zur Ermittlung der Verteilungsfunktion von Einschaltüberspannungen auf Hochspannungsleitungen. Elektrie 27 (1973) 1, S. 31–34.

[4.7] *Bauer, H.; Drechsler, E.*: Nutzfähige Programme zur Berechnung von Blitzüberspannungen. Elektrie 32 (1978) 6, S. 312–313 (s. auch Elektrie 33 (1979) 6, S. 292–295).

[4.8] *Daus, W.*: Beitrag zur Registrierung der Häufigkeitsdichte von Überspannungen in elektrischen Energieversorgungsnetzen. Diss. Techn. Univ. Dresden 1975.

[4.9] IEC 60–1 High-voltage test techniques. Part I: General definitions and test requirements. IEC-Publ. 60–1, 1. Ausgabe 1973.

[4.10] TGL 20618/01 Hochspannungsprüftechnik; Allgemeine Begriffe; Ausg. 12.76 (s. auch IEC-Publ. 60–1 (1973); VDE 0432/1; ST RGW 1071 (5.80)).

[4.11] IEC 60–2 High-voltage test techniques. Part 2: Test procedures. IEC-Publ. 60–2, 1.Ausgabe 1973.

[4.12] Siehe [2.28].

[4.13] *Sieber, K.*, u.a.: Zur Bemessung und Zuverlässigkeit der Isolierung der feststoffisolierten Schaltzellen Typ ASIF 36. Elektrie 34 (1980) 2, S. 84–88.

[4.14] *Mosch, W.; Hauschild, W.*: Hochspannungs-Isoliertechnik, 1.Lehrbrief — Isoliervermögen als Zufallsgröße. Berlin: VEB Verlag Technik 1978.

[4.15] *Das, M.K.*: Typenprüfung mit Nenn-Stehstoßspannung und Nenn-Stehschaltspannung. ETZ-A 89 (1968) 2, S. 31–33.

[4.16] *Schwaiger, A.*: Elektrische Festigkeitslehre. Berlin: Springer-Verlag 1925.

[4.17] *Diesendorf, W.*: Insulation co-ordination of electrical networks. London: Butterworths 1974.

[4.18] *Koettnitz, H.*: Neue Grundsätze für die Isolationskoordination von Betriebsmitteln für Wechselstrom-Energieanlagen über 1 kV. Elektrie 26 (1972) 3, S. 58–62.

[4.19] *Koettnitz, H.*: Wesentliche Merkmale der internationalen Empfehlungen zur Isolationskoordination. Elektrie 28 (1974) 4, S. 197–200.

[4.20] *Hylten-Cavallius, N.*: Modern high voltage test techniques, standardization. WELC Moscow (1977), Beitrag 2–15.

[4.21] *Carrara, G.*, u.a.: Contribution to the study of insulation coordination from the probabilistic point of view. CIGRE-Ber. 421 (1966).

[4.22] IEC-TC 28: Supplement to IEC publication 71: Insulation co-ordination. Part III: Principles and rules for phase-to-phase insulation co-ordination. (Entwurf).

[4.23] *Küttner, H.*: Untersuchungen zur Spannungsbeanspruchung und Dimensionierung von Niederspannungsisolierungen. Diss. Techn. Univ. Dresden 1975.

[4.24] *Hutzler, B.; Gallet, G.*: Contribution of the physics of discharges to insulation coordination. CIGRE-Ber. 33–04 (1976).

[4.25] *Kostenko, M. W.; Michailow, J.A.*, u.a.: Ergebnisse von Messungen innerer Überspannungen in Hochspannungsnetzen. Energietechnik 29 (1979) 1, S. 8–11.

[4.26] *Rasewig, D.W.; Dmochowskaja, L.P.*: Statistik der Schaltüberspannungen (russ.). Theor. Probl. Elektroenergetika (1973), S. 403–414.

[4.27] *Weck, K.-H.*, u.a.: Phase-to-phase and longitudinal insulation testing technique. CIGRE-Ber. 33–09 (1976).

[4.28] *Takagi, T.*, u.a.: Dielectric strength of SF_6 gas and 3-core type CGI cables under inter-phase switching impulse voltage. IEEE Trans. PAS 93 (1974), S. 354–359.

[4.29] *Weck, K.-H.; Carrara, G.:* Design and testing of phase-to-phase insulation. Electra (1979) 64, S. 182–210.

[4.30] *Carrara, G.; Pigini, A.; Polo-Dimel, M.:* UHV disconnectors: Switching surge design and testing of external insulation. IEEE Trans. PAS 97 (1978) 6, S. 2094–2103.

Chapter 5

[5.1] *Schumann, W.O.:* Elektrische Durchbruchfeldstärke von Gasen. Berlin: Springer-Verlag 1923.

[5.2] *Hauschild, W.:* Beitrag zum Verständnis der elektrischen Festigkeit von SF_6 als Zufallsgröße. Elektrie 33 (1979) 6, S. 296–300.

[5.3] *Engelmann, E.:* Beitrag zum Entladungsverhalten großflächiger, störstellenbehafteter Elektroden in Luft bei positiver Schaltspannung. Diss. Techn. Univ. Dresden 1981.

[5.4] *Meek, J.M.; Craggs, J.D.:* Electrical breakdown of gases. Chichester, New York, Brisbane, Toronto: John Wiley & Sons 1978.

[5.5] *Schmiedl, G.:* Zum Einfluß der Luftfeuchte auf den Wechselspannungsdurchschlag meterlanger Luftisolierstrecken. Diss. Techn. Univ. Dresden 1969.

[5.6] *Kučera, J.; Fiklik, V.:* Correction of switching impulse flashover voltages for air humidity. IEEE Trans. PAS 89 (1970) 3, S. 441–447.

[5.7] *Feser, K.:* Einfluß der Feuchtigkeit auf das Durchschlagverhalten bei Wechselspannung. ETZ-A 91 (1970) 10, S. 584–586.

[5.8] *Aihara, Y.; Harada, T.; Aoshima, Y.; Jto, Y.:* Impulse flashover characteristics of long air gaps and atmospheric correction. IEEE Trans. PAS 97 (1978) 2, S. 342 bis 348.

[5.9] *Busch, W.:* Air humidity: An important factor for UHV design. IEEE Trans. PAS. 97 (1978) 6, S. 2086–2093.

[5.10] *Praxl, G.; Gradischnig, W.; Egger, H.:* Untersuchungen des Einflusses der Wechselwirkungen zwischen Temperatur und Feuchte auf die positive Durchschlagschaltspannung (engl.). 6. Gasentladungskonferenz (1980) Teil 2, S. 166–169.

[5.11] *Schreiber, G.; Bancos, J.:* Untersuchungen der Ionenkonzentration in atmosphärischer Luft (franz.). Rev. Gen. Electr. 86 (1977) 10, S. 745–748.

[5.12] *Newi, G.:* Streuung der Durchschlagspannung von Kugelfunkenstrecken in Luft bei negativen Stoßspannungen. 1.ISH München (1972) S. 327–333.

[5.13] *Müller, H.; Weisinger, J.:* Einfluß der kosmischen Strahlung auf das Durchschlagverhalten einer homogenen Luftfunkenstrecke. Bull. SEV 62 (1971) 9, S. 445–447.

[5.14] *Laub, S.:* Untersuchungen zum Einfluß der Wirkdauer und des zeitlichen Verlaufes der Spannung auf das Entladungsverhalten großflächiger, störstellenbehafteter Electrodenanordnungen in Luft. Diss. Techn. Univ. Dresden 1982.

[5.15] *Frank, S.:* Der Staubeinfluß bei Funkenstrecken. Archiv f. Elektrotechnik 28 (1934) S. 485–486.

[5.16] *Schulz, W.:* Fremdteilchen als Ursache für Tiefendurchschläge in Luft bei Gleich- und Wechselspannung. Diss. Techn. Univ. Braunschweig 1977.

[5.17] *Peier, K.; Dohnal, D.:* Der systematische Fehler von Kugelfunkenstrecken bei Wechselspannungsmessungen. ETZ-A 98 (1977) 4, S. 303.

[5.18] *Feser, K.:* Einfluß des Innenwiderstandes der Spannungsquelle auf das Durchschlagverhalten von Luftfunkenstrecken. ETZ-A 92 (1971) 8, S. 495–500.

[5.19] *Newi, G.:* Breakdown phenomena of switching impulse: Parameters of test circuit. IEEE Power Eng. Soc. (1974) 74 CH 0910–0–PWR, S. 81–85.

[5.20] *Böhm, A.:* Funkenstrecken-Vorwiderstände bei positiver Schaltstoßspannung. 2.ISH Zürich (1975), S. 483–487.

[5.21] *Voß, W.:* Untersuchungen zur Entwicklung paralleler Entladungen beim Stoßdurchschlag in Gasen. Diss. TU Braunschweig 1979.

[5.22] *Lalot, J.; Gallet, G.:* Breakdown phenomena at power frequency. 3.ISH Mailand (1979) 52.09.

[5.23] *Brasca, E.; Tellarini, M.; Zaffanella, L.:* The confidence limit of high-voltage dielectric test results. IEEE Trans. PAS 86 (1967) 8, S. 968–975.

[5.24] *Ritzk, F.A.M.:* Influence of rain on switching impulse sparkover voltage of large-electrode air gaps. IEEE Trans. PAS 95 (1976) 4, S. 1394–1401.

[5.25] *Mosch, W.*, u.a.: Hochspannungs-Isoliertechnik in: Taschenbuch Elektrotechnik, Bd. 6; Herausgeber E. Philippow. Berlin: VEB Verlag Technik 1982.

[5.26] TGL 20 624 Spannungsmessungen mit Kugelfunkenstrecken; Ausg. 12.66 (s. auch VDE 0430 und 0433).

[5.27] *Pingini, A.; Rizzi, G.; Thiome, L.:* Einfluß der Elektrodengeometrie und der Impulsform auf die Leadercharakteristik (franz., engl.). Electra (1977) 53, S. 51–59.

[5.28] Standard methods of high-voltage test techniques. IEEE-Studie 4 — 1978.

[5.29] *Lemke, E.:* Beitrag zur Abschätzung der Durchschlagspannung langer Luftfunkenstrecken. Z. elektr. Inform.- und Energietechn. 3 (1973) 4, S. 186–192.

[5.30] *Lemke, E.:* Ein Beitrag zur Abschätzung der räumlich-zeitlichen Entwicklung des Durchschlages bei Schaltspannung. Wiss. Z. Techn. Univ. Dresden 26 (1977) 1, S. 133–136.

[5.31] *Sprang, H.D.:* Analysis of breakdown mechanisms of long-air gaps under switching-impulse voltages with the help of a surface discharge analogy model. 3.ISH Mailand (1979) 51.07.

[5.32] *Paus, L.; Cortina, R.:* Switching and lightning impulse discharge characteristics of air gaps and station insulators. IEEE Trans. PAS 87 (1968) 4, S. 947–957.

[5.33] *Bürger, W.:* Beitrag zum Entladungsverhalten großflächiger, schwach gekrümmter Elektroden in Luft bei großen Schlagweiten. Diss. Techn. Univ. Dresden 1976.

[5.34] *Schneider, H.M.; Turner, F.J.:* Switching-surge flashover characteristics of long sphere-plane gaps for UHV stations design. IEEE Trans. PAS 94 (1975) 2, S. 551 bis 560.

[5.35] *Pingini, A.; Thione, L.; Brambilla, R.:* Corona phenomena on high voltage electrodes in air. Wiss. Beitr. WELC Moskau 1977, Sekt. 2, Pap. 21.

[5.36] *Ritzk, F.A.M.:* Effect of large electrodes on sparkover characteristics of air gaps and station insulators. IEEE Trans. PAS 97 (1978) 4, S. 1224–1231.

[5.37] *Carrara, G.*, u.a.: Switching impulse strength of phase-to-phase external insulation. (Arbeit der CIGRE-Working Group 33.03). Electra (1979) 64, S. 158–181.

[5.38] *Thione, L.*, u.a.: The influence of non standard conditions on the switching impulse strength of phase-to-phase insulation. (Arbeit des CIGRE-Task Force 33–33.03). Electra (1979) 64, S. 211–230.

[5.39] *Bondarenko, V.E.; Böhme, H.:* Durchschlag von Mehr-Elektroden-Anordnungen, Elektrie 34 (1980) 11, S. 581–584.

[5.40] *Streubel, H.:* Zum Überschlag beregneter Isolatoren bei Wechsel- und Impulsspannung. Diss. Techn. Univ. Dresden 1967.

[5.41] Guide for application of insulators to withstand switching surges. IEEE Trans. PAS 94 (1975) 1, S. 58–67.

[5.42] *Carrara, G.; Marzio, L.:* Flashover probability under dielectric stress. Anhang zum Bericht des CIGRE-Studienkomitees 8 (1968).

[5.43] TGL 20618/06 Hochspannungsprüftechnik; Prüfung fremdschichtbehafteter Isolierungen; Ausg. 3.75.

[5.44] *Obenaus, F.; Böhme, H.:* Pollution flashover tests on insulators in the laboratory and in systems and the model concept of creepage-path flashover. CIGRE-Ber. 407 (1966) (s. auch Elektrie 20 (1966) 11, S. 147 bis 422).

[5.45] *Issel, G.:* Neuere Erkenntnisse über das Wesen des Kriechüberschlages (Überschlagzeit, Löschlänge der Vorlichtbögen, abhängig vom Verschmutzungsgrad). Wiss. Z. Elektrot. 7 (1966), S. 105 bis 127 (s. auch *G. Issel*, Dissertation Techn. Univ. Dresden 1966).

[5.46] *Böhme, H.:* Kriechüberschlag von zylindrischen Isolatoren mit Schirmen, Elektrie 19 (1965) 6, S. 249–252.

[5.47] *Erler, F.*: Prüfung von verschmutzten Isolatoren bei Impulsspannungen. Elektrie 25 (1971) 7, S. 252 bis 255. (s. auch *F. Erler*, Dissertation Techn. Univ. Dresden 1970).

[5.48] *Reichel, R.*: Der Einfluß einer Parallelkapazität auf den Kriechüberschlag und seine Bedeutung für die Ermittlung der Kriechüberschlaggleichspannung. Diss. Techn. Univ. Dresden 1977.

[5.49] *Merchalev, S.D.; Solomonik, E.A.*: Isolierungen von Freileitungen und Stationen in Gebieten mit verschmutzter Atmosphäre (russ.). Leningrad: Energija 1973.

[5.50] *Böhme, H.; Pilling, J.; Streubel, H.*: Zur Interpretation der Kriechüberschlagkennlinie von Freiluftisolatoren. Elektrie 33 (1979) 12, S. 630–632.

[5.51] *Bärsch, R.; Böhme, H.; Schwarzer, K.*: Zum Langzeitüberschlag organischer Isolierungen mit Taubelag. 22.IWK Ilmenau 1977, S. 131–134 (s. auch *R. Bärsch*, Diss. Techn. Univ. Dresden 1979).

[5.52] *Jeske, J.*: Der Einfluß vorangegangener Stoßspannungsbeanspruchungen im Durchschlagstreubereich auf die elektrische Stoßbeanspruchung einer abgeschlossenen Gasmenge technisch reinen SF_6. Diss. Techn. Univ. Berlin (West) 1967.

[5.53] *Hauschild, W.; Bürger, W.*: An estimation of the distribution type for the electrical strength of sulphur hexafluoride. ICPIG XII Eindhoven (1975) S. 157.

[5.54] See [2.54].

[5.55] *Bortnik, I.M.; Gorjunow, B.A.; Panow, A.A.*: Reliability of gas insulation at high pressures. 2. Conf. Gas Discharges London (1972) S. 326–328.

[5.56] *Mosch, W.; Schwarz, J.; Hauschild, W.*: Interpretation of breakdown voltage-time-characteristics for a coaxial system in SF_6. ICPIG XV Minsk (1981) P-11-39.

[5.57] *Hauschild, W.; Speck, J.; Schierig, S.*: Experimentelle Modellierung des SF_6-Durchschlages mit Hilfe freibeweglicher Partikeln. Elektrie 30 (1976) 7, S. 354–356.

[5.58] *Blässig, H.*: Über das Durchschlagverhalten von kondensiertem SF_6 im Temperaturbereich von −20°C bis −40°C bei unterschiedlichen relativen Feuchten. ETG-Fachtagung Baden-Baden (1977) S. 96–100 und ETZ-A 98 (1977) 11, S. 763.

[5.59] *Peier, D.*: Untersuchung von Durchschlagvorgängen in flüssigem Stickstoff bei hohen Spannungen. Diss. Techn. Univ. Braunschweig 1975.

[5.60] *Gharabeiglu, B.*: Das Durchschlagverhalten des flüssigen Schwefelhexafluorids. Diss. Techn. Univ. Berlin (West) 1975.

[5.61] *Užakov, V.Ja.*: Impulsdurchschlag von Flüssigkeiten (russ.). Tomsk: Verlag Univ. 1975.

[5.62] *Schöne, G.*: Zum Verhalten von Gasblasen in Isolieröl im elektrischen Feld. Diss. Techn. Univ. Dresden 1966.

[5.63] *Günther, H.*: Zum Durch- und Überschlag von inhomogenen Isolieranordnungen unter verrußtem Isolieröl. Diss. Techn. Univ. Dresden 1973.

[5.64] *Günther, H.; Hauschild, W.*: Durch- und Überschlag von Isolieranordnungen mit feuchtem, gealtertem oder verrußtem Isolieröl bei Impuls- und Wechselspannung. Elektrie 25 (1971) 12, S. 470 bis 472.

[5.65] *Kratzenstein, M.G.*: Der Stoßdurchschlag in Isolieröl. Diss. Techn. Hochsch. München 1969.

[5.66] *Fiebig, R.*: Erscheinungsformen und Mechanismus der Teilentladungen bei Wechselspannung in Isolierflüssigkeiten im inhomogenen Feld. Diss. Techn. Univ. Dresden 1967.

[5.67] *Palmer, S.; Sharpley, W.A.*: Electric strength of transformer insulation. Proc. IEE 116 (1969) 12, S. 2029–2037.

[5.68] *Kawaguchi, Y.; Murata, H.; Ikeda, M.*: Breakdown of transformer oil. IEEE Trans. PAS 91 (1972) 1, S. 9–23.

[5.69] *Bossi, A.*, u.a.: Voltage-time relationship for PD inception in oil paper insulation for UHV transformers. WELC Moscow (1977), Beitrag 2–42.

[5.70] *Mosinski, F.*: Über die Anwendung der Weibullverteilung bei der Untersuchung von Isolier-systemen Papier–Öl (poln.). Przeglad Elektrotechniczny (1978) 2, S. 49–50.

[5.71] *Asenjo, E.; Eidelstein, G.:* Paper-oil insulation — New definition of damage. IEEE Trans. EI 13 (1978) 6, S. 179–183.

[5.72] *Küllig, P.:* Zum Einfluß von Fehlstellen und Isolierstoffvolumen auf die elektrische Lebensdauer von extrudierten Polyäthylen-Isolierungen bei Wechselspannungsbeanspruchung. Diss. Techn. Univ. Dresden 1982.

[5.73] *Eberhardt, M.:* Schlagweite-Spannungs-Kurven und physikalische Struktur von festen Isolierstoffen. Diss. Techn. Univ. Dresden 1964.

[5.74] *Mosch, W.:* Elektrische Isolierungen unter dem Einfluß von Konstruktion und Technologie. Elektrie 33 (1979) 12, S. 623–625.

[5.75] *Georgi, A.:* Ein Beitrag zum Einfluß der Verarbeitungstechnik auf den elektrischen Durchschlag von Epoxidharzisolierungen. Diss. Techn. Univ. Dresden (1977).

[5.76] Siehe [2.50]

[5.77] *Schirr, J.:* Beeinflussung der Durchschlagfestigkeit von Epoxidharzformstoff durch das Herstellungsverfahren und durch mechanische Spannungen. Diss. Techn. Univ. Braunschweig 1974.

[5.78] *Ulrich, J.:* Über den Einfluß der Temperatur auf das Durchschlagverhalten von Polyäthylen im inhomogenen elektrischen Wechselfeld. Diss. Techn. Univ. Hannover 1978.

[5.79] *Löffelmacher, G.:* Über die physikalisch-chemischen Vorgänge bei der Ausbildung von Entladungskanälen in Polyäthylen und Epoxidharz im inhomogenen Wechselfeld. Diss. Techn. Univ. Hannover 1976.

[5.80] *Behder, G.*, u.a.: Development of technology for the manufacture of crosslinked polyethylene insulated cables rated 138 through 345 kV. IEEE Trans. PAS 96 (1977) 6, S. 1741–1753.

[5.81] *Küllig, P.:* Einfluß von Verunreinigungen und Isolierstoffvolumen auf die elektrische Lebensdauer von extrudierten PE-Isolierungen. 25.IWK Ilmenau (1980) Vortragsreihe Elektr. Isoliertechn., S. 17–20.

[5.82] Siehe [2.49].

[5.83] *Feichtmayr, F.-J.; Würstlin, F.:* Die elektrische Zeitstandfestigkeit von Polyäthylen. ETZ-B 21 (1969) 6, S. 128–130.

[5.84] *Olshausen, R.; Westerholt, W.:* Dieletric strength of SF_6-impregnated polyethylene test samples under homogeneous a.c. stress. 3.ISH Mailand (1979) 23.03.

[5.85] Siehe [2.29].

[5.86] *Peschke, E.F.*, u.a.: Elektrische Langzeitfestigkeit von Kabeln mit PE-Isolierungen und die Problematik der Extrapolation. ETG-Fachtagung (1977) Baden-Baden, S. 151–155.

[5.87] *Artbauer, J.:* Elektrische Dauerfestigkeit und Kurzzeitfestigkeit. ETZ-A 91 (1970) 6, S. 326–331.

[5.88] *Brakelmann, H.:* Zündvolumina inhomogen beanspruchter Feststoffisolierungen. Bull. SEV 68 (1977) 12, S. 595–599.

[5.89] *Starr, W.; Endicott, H.:* Progressive stress — A new accelerated approach to voltage endurance. IEEE Trans. PAS 55 (1961) 8, S. 515–525.

[5.90] *Meyer, H.:* Zur Zeitabhängigkeit des elektrischen Durchschlages technischer Isolierungen. Diss. Techn. Hochsch. Hannover 1966.

[5.91] *Kojkov, S.N.:* Die elektrische Alterung von Dielektrika und die Zuverlässigkeit der elektrischen Isolierung (russ.). Električestvo (1978) 9, S. 33–37 (s. auch *Kojkov, S.N.* in: "Physik der Dielektrika" (russ.). Leningrad 1967).

[5.92] IEC 270 Partial discharge measurements. IEC-Publ. 270, Ausgabe 1981.

[5.93] *Lemke, E.:* Ein neues Verfahren zur Messung von Teilentladungen an langen Hochspannungskabeln. Elektrie 35 (1981) 7 (s. auch 3.ISH Mailand (1979) 43.13).

[5.94] *König, D.;* u.a.: Partial discharge measurements of SF_6 insulated high-voltage metal-enclosed switchgear on site a study based on fundamentals and experiences available up to now. CIGRE (1980) 23–09.

[5.95] Geräte zur breitbandigen Messung von Teilentladungen. Firmenschrift des VEB Transformatoren- und Röntgenwerk "Hermann Matern". Dresden 1981.

[5.96] *Lemke, E.*: Beitrag zur elektrischen Messung von Teilentladungen unter besonderer Beachtung eines Verfahrens zur breitbandigen Erfassung der Summenladung. Diss. B Techn. Univ. Dresden 1975.

[5.97] *Haller, R.*: Beitrag zur breitbandigen TE-Messung unter besonderer Berücksichtigung paralleler Störstellen. Diss. Techn. Univ. Dresden 1976.

Chapter 6

[6.1] *Voß, W.*: Untersuchungen zur Entwicklung paralleler Entladungen beim Stoßspannungsdurchschlag in Gasen. Diss. TU Braunschweig 1979.

[6.2] Siehe [5.53].

[6.3] *Kappeler, H.*: Recent forms of execution of 380 kV transformer bushings. CIGRE-Ber 126 (1958).

[6.4] *Paderta, B.*: Über die Koordination der Isolation paralleler Funkenstrecken bei Berücksichtigung der Statistik. Acta Technica ČSAV (1971) 3, S. 406–419.

[6.5] *Juchnewicz, J.; Mazurek, B.; Tyman, A.*: On some size effects of high voltage vacuum insulation. 3.ISH Mailand (1979) 31.02.

[6.6] *Paderta, B.; Vavrina, K.*: Über die Analyse der Überschlagzeiten in Abhängigkeit von Zahl und Ausmaß der Prüflinge bei Anwendung von Spannungen gleicher Art und Größe. Acta Technica ČSAV 17 (1972) 1, S. 11–30.

[6.7] *Heller, B.; Paderta, B.*: Die Durchschlagwahrscheinlichkeit von Transformatorwicklungen. Acta Technica ČSAV (1971) 3, S. 325–333.

[6.8] *Del Mar, W.A.*: Electric stresses in cables. AIEE Trans. 81 (1962) 6, S. 121–134.

[6.9] *Girling, D.S.*: Direct voltage instantaneous breakdown of oil-impregnated paper capacitors as a function of area. Electr. Communication 35 (1958) 2, S. 83–92.

[6.10] *Sie, T.H.; Wohlfahrt, O.*: Transference of test results from experiments on small models to larger test objects with insulation under oil. AIEE Trans. Part. III 81 (1962) 12, S. 601–608.

[6.11] *Mosch, W.; Hauschild, W.*: Die elektrische Festigkeit als Grundlage für die Berechnung von Durchschlagvorgängen im SF_6. 2.ISH Zürich (1975) S. 355–360.

[6.12] *Küllig, P.; Rüffer, K.*: Der Einfluß des Isolierstoffvolumens auf die Bemessung von festoffisolierten koaxialen Zylinderanordnungen. Der VEM Elektro-Anlagenbau 15 (1979) 4, S. 169–172.

[6.13] *von Laue, M.*: Bemerkungen zu K. Zubers Messungen der Verzögerungszeiten bei der Funkenentladung. Ann. Phys. 76 (1925) S. 261.

[6.14] *Strigel, R.*: Elektrische Stoßfestigkeit. Berlin. Göttingen, Heidelberg: Springer-Verlag 1955.

[6.15] *Boeck, W.*: Volumen-Zeit-Gesetz beim Stoßspannungsdurchschlag von SF_6. ETZ-A 96 (1975) 7, S. 300–305.

[6.16] *Boeck, W.; Taschner, W.*: Isolationsverhalten SF_6-isolierter koaxialer Zylinderanordnungen bei Stoßspannungsbeanspruchung. ETZ-A 97 (1976) 6, S. 335–340.

[6.17] *Knorr, W.*: A model to describe the ignition time-lag of slightly nonuniform arrangements in SF_6. 3.ISH Mailand (1979) 31.11.

Index